Thermodynamics of TECHNICAL GAS-REACTIONS

MAXWELL PRESS

Thermodynamics of TECHNICAL GAS-REACTIONS

F. HABER

Winner of Nobel Prize for Chemistry

Professor at the Technische Hochschule, Carlsruhe

Translated by

ARTHUR B. LAMB, Ph.D

Director of the Havemeyer Chemical Laboratory,
New York University, New York City

MAXWELL PRESS

Chennai Trichy NewDelhi

MAXWELL PRESS

This edition has been published in india by arrangement with Carson Books, UK

ISBN 978-93-90063-67-3 Maxwell Press

No. 44, Nallathambi Street,
Triplicane, Chennai 600 005

MJP 912

Publisher's Note

The legacy of a country is in its varied cultural heritage, historical literature, developments in the field of economy and science. The top nations in the world are competing in the field of science, economy and literature. This vast legacy has to be conserved and documented so that it can be bestowed to the future generation. The knowledge of this legacy is slowly getting perished in the present generation due to lack of documentation.

Keeping this in mind, the concern with retrospective acquiring of rare books has been accented recently by the burgeoning reprint industry. Maxwell Press is gratified to retrieve the rare collections with a view to bring back those books that were landmarks in their time.

In this effort, a series of rare books would be republished under the banner, "Maxwell Press". The books in the reprint series have been carefully selected for their contemporary usefulness as well as their historical importance within the intellectual. We reconstruct the book with slight enhancements made for better presentation, without affecting the contents of the original edition.

Most of the works selected for republishing covers a huge range of subjects, from history to anthropology. We believe this reprint edition will be a service to the numerous researchers and practitioners active in this fascinating field. We allow readers to experience the wonder of peering into a scholarly work of the highest order and seminal significance.

MAXWELL PRESS

TO

MY DEAR WIFE

CLARA HABER, PH.D.,

IN GRATITUDE FOR HER SILENT CO-OPERATION

TRANSLATOR'S PREFACE

THE fascinating possibility of predicting the course of a chemical reaction from a few characteristic constants of the reacting substances seemed very far from realization after the ill-starred attempt of Berthelot. Recent attacks upon this problem have been more successful, and the future is promising.

Prof. Haber's book entitled "Thermodynamik Technischer Gasreactionen" is a most important contribution to this subject. It is a pleasure for me to assist in making this book better known to the English-speaking world.

Professor Haber has thoroughly revised the German edition purposely for this translation. Many parts have been re-written, and the changes necessitated by the progress of the subject during the years 1905 and 1906 have been made. In those lectures where this involved too extensive alterations the original text has been adhered to, and the new results added in appendices. The appendix to the Third Lecture was translated by Dr. Maitland, that to the Fifth by Mr. R. Le Rossignol.

The translation of the first four Lectures was completed in its final form in the spring of 1906.

It gives me pleasure to acknowledge the valuable assistance rendered me by Dr. M. H. Hunter in the reading of the proofs.

ARTHUR B. LAMB.

NEW YORK CITY.
October 1st, 1907.

PREFACE

DURING February of this year I delivered a series of seven evening lectures on the thermodynamics of technical gas reactions, before several of my colleagues and a number of my younger research students. I could assume that my audience was familiar with the chemical and technical side of the subject, but was obliged to develop the mechanical theory of heat from its very foundations. These lectures, intended simply to introduce the subject for discussion by my colleagues and students, are here reproduced.

Questions asked me after the lectures, and certain difficulties which were encountered in making the subject clear, have lead me to make my explanations here somewhat fuller, and to adopt the style of an essay rather than of a lecture.

I have not made use of the atomic hypothesis in these lectures. This is not due to any antagonism on my part to this hypothesis. I am simply convinced that the application of the mechanical theory of heat to chemistry becomes easier and more comprehensible the closer we restrict ourselves to the heat and work effects of masses directly perceptible to our senses.

In presenting the fundamentals of this theory, I have chosen Hemholtz's point of view. From it a chemical reaction is considered to have a latent heat just as does any simple change in the state of aggregation. Consequently, in my presentation, the two parts into which total energy can be divided are not spoken of as *free* and *bound energy*, but as *reaction energy* and *latent heat*. Various reasons have lead me to adopt this less usual nomenclature. In the first place, latent heat is a concrete entity appealing directly to our senses. The idea of "bound energy" is an abstraction. Then, too, the intimate theoretical and practical connection between gas reactions and the dissociation and vaporization of solids can be much more readily appreciated from this standpoint. Finally, starting from it, there is no difficulty in grasping both the idea of a temperature coefficient of maximum work, upon which van't Hoff, Ostwald, and Nernst based certain special considerations, and the idea of entropy which underlies Planck's method of treatment.

In connection with the work of Helmholtz, and particularly because of an article by van't Hoff in the "Festschrift" published in celebration of Boltzmann's sixtieth birthday, I have discussed at length the influence of specific heats on the energy of gaseous reactions. The *special* importance of specific heats in this kind of reactions is fully discussed at the end of the Fourth Lecture. The methods of measuring specific heats, and the data we thus far possess in this field, have also been treated at length; indeed, very much at length in some places, in order to permit the reader to criticize the choice finally made. Our knowledge of specific heats is at present so scanty that we must often trust to a sort of expert instinct in selecting proper values, without being able in some instances to prove definitely that the chosen results are really better than those of some other investigator.

The book has been written for the sake of technical rather than theoretical chemistry. I hope that it may facilitate both teaching and experimental investigation of the subject of technical gas reactions. It is not a handbook, nor does it attempt a complete presentation of all the material, but rather a clear and exhaustive treatment of the more important cases. How well I have succeeded in this clarification must be left to the decision of my colleagues.

Messrs. Gottlob and Moser have assisted me greatly in the many numerical computations contained in the book. I shall be grateful for notification of any inaccuracies of any kind which may be found.

F. HABER.

CARLSRUHE,
May 20th, 1905.

TABLE OF CONTENTS

FIRST LECTURE

THE LATENT HEAT OF CHEMICAL REACTION AND ITS RELATION TO REACTION ENERGY

SECOND LECTURE

ENTROPY AND ITS SIGNIFICANCE IN GAS REACTIONS

THIRD LECTURE

ANOTHER DERIVATION OF THE FORMULA PREVIOUSLY OBTAINED, AND ITS BEARING ON REACTIONS BETWEEN SOLIDS

FIFTH LECTURE

SOME EXAMPLES OF REACTIONS INVOLVING A CHANGE IN THE NUMBER OF MOLECULES

SIXTH LECTURE

THE DETERMINATION OF THE SPECIFIC HEATS OF GASES

SEVENTH LECTURE

THE DETERMINATION OF GASEOUS EQUILIBRIA, WITH A THEORETICAL AND TECHNICAL DISCUSSION OF RELATED QUESTIONS

TABLE OF CONTENTS

THERMODYNAMICS

FIRST LECTURE

THE LATENT HEAT OF CHEMICAL REACTION AND ITS RELATION TO REACTION ENERGY

GENTLEMEN,—In these lectures I shall endeavour to make clear the significance of heat factors in gas reactions, with especial reference to the specific heats of the interacting substances and to the heat evolved during the reaction.

Thermochemistry concerns itself in general with the total amount of heat evolved (or absorbed) during the whole course of a reaction, but leaves quite untouched the question as to whether the reaction goes to completion in one direction or halts in an intermediate (equilibrium) condition.

All simple gas reactions do halt at such equilibrium conditions, and are consequently incomplete. Often, indeed as a rule, this incompleteness only becomes perceptible at high temperatures.[1] But phenomena at high temperatures are precisely what interest us here. In the region of low temperatures, gases often react very sluggishly with one another. The reaction velocities are then the governing factors of the process, and these are not to be predicted in advance from the standpoint of the theory of heat. The higher the temperature rises the less important do these factors become, while the equilibrium phenomena which are subject to theoretical treatment attain prominence. These equilibrium phenomena[2] are all that we shall concern ourselves with here.

[1] Theoretically every reaction belongs to this class, and may therefore be termed reversible. But it must be possible to demonstrate experimentally this reversibility under any conditions if its theoretical consideration is to be scientifically useful.

[2] Here we mean only "real" equilibria, which can be reached from both sides. The "false" equilibria, which can be reached from one side only, and their relation to passive resistance and reaction velocity, will be discussed in the last lecture.

They have often been discussed in relation to the mass action of the reacting substances, but the influence exerted by the specific heat of the substances has received but little attention. Le Chatelier,[1] to be sure, showed long ago, in a comprehensive work on dissociation, the importance of specific heats, but this work was not accorded the general notice which it deserved. The result is that our knowledge of the subject is limited and the number of cases at our disposal is scanty. The subject has, however, so broad a significance that it is well worth our while to become conversant with it. Only when we consider the joint effect of mass action and the specific heats of the reacting substances are we able, from observations made at any one temperature, to draw inferences regarding the phenomena which will take place at a temperature, say a thousand degrees higher. The theory even affords us the expectation of being able to find out all about a gas equilibrium at high temperatures, simply from a knowledge of the heat factors concerned, without any experimental determinations of mass actions.[2] Satisfactory confirmation, however, of this last step is lacking, because our knowledge of the specific heats of gases is as yet too incomplete.

Technical interest of gas reactions at high temperatures.

Gas reactions at high temperatures frequently possess a very great practical interest. We should mention, in the first place, those heating processes where carbon dioxide and water vapour are formed. These substances dissociate at very high temperatures into free oxygen, carbon monoxide, and hydrogen. The effect of temperature and the composition of the gas mixture on the dissociation of the carbon dioxide and the water-vapour is a first question of importance, which must be solved by means of the mechanical theory of heat. The two reactions—

$$2CO + O_2 \rightleftarrows 2CO_2 \quad . \quad . \quad . \quad . \quad . \quad (1)$$

and

$$2H_2 + O_2 \rightleftarrows 2H_2O \quad . \quad . \quad . \quad . \quad . \quad (2)$$

are connected with one another by the relation—

$$CO + H_2O \rightleftarrows CO_2 + H_2 \quad . \quad . \quad . \quad . \quad (3)$$

[1] *Annales des Mines*, viii. 13 (1888), 157.

[2] Since the publication of the German edition of this book, Nernst has given his views on this problem. See his "Thermodynamics and Chemistry" (Chas. Scribner's Sons, New York, 1907). An account of them will be found at the end of the Third Lecture.

We can decide in which direction this so-called "water-gas reaction" will proceed at a given concentration of the four substances in the gas mixture, and at a known high temperature, when the dissociation conditions represented in (1) and (2) are known. Dissociation of this kind also possesses a great significance in technical inorganic chemistry. The reaction—

$$2SO_2 + O_2 \rightleftarrows 2SO_3 \quad . \quad . \quad . \quad . \quad . \quad (4)$$

forms, as is well known, the foundation of the modern method of sulphuric acid manufacture. The two reactions—

$$2HCl \leftrightarrows Cl_2 + H_2 \quad . \quad . \quad . \quad . \quad . \quad (5)$$

and

$$2H_2O \leftrightarrows 2H_2 + O_2 \quad . \quad . \quad . \quad . \quad . \quad (6)$$

taken together, give us Deacon's chlorine process—

$$2H_2O + 2Cl_2 \leftrightarrows 4HCl + O_2 \quad . \quad . \quad . \quad . \quad (7)$$

The possibility of making ammonia from its elements, and nitric oxide, and consequently nitric acid and the nitrates, from the air, is governed by the relationships given by the mechanical theory of heat for the reactions—

$$N_2 + 3H_2 \rightleftarrows 2NH_3 \quad . \quad . \quad . \quad . \quad . \quad (8)$$

and

$$N_2 + O_2 \rightleftarrows 2NO \quad . \quad . \quad . \quad . \quad . \quad (9)$$

The fundamental thermodynamic relations which we will consider can be applied to the reactions of solids, liquids, and gases. They are employed under the most varied forms, especially for molten and dissolved substances. But they assume a particularly simple form for gaseous substances, because in these the pressure, temperature, and volume are connected by a very simple relation, while in the case of molten or dissolved substances no similar general relation connecting these three quantities is known. This relation is— **The fundamental gas law.**

$$pv = RT \quad . \quad . \quad . \quad . \quad . \quad . \quad . \quad (10)$$

where p = pressure and v = volume. If we consider a mol as the unit of mass, experiment shows us that at a pressure of one atmosphere (*i.e.* at 760 mm. Hg at sea-level and latitude 45°) this unit of mass occupies 22·412 litres at a temperature of 0°. Equation (10) requires that the temperature be reckoned in degrees on the absolute scale. The zero point of this scale is generally put at $-273°$ C. It is equal to minus the reciprocal value of the coefficient of expansion of an ideal gas ($\frac{1}{273}$). The coefficients of ordinary gases are very slightly different from this. By careful consideration of the deviation from the behaviour of ideal gases which ordinary gases show on compression and expansion it is possible to find the coefficient of expansion of ideal gases.[1] It amounts to $\frac{1}{273 \cdot 09}$. The absolute zero therefore is, strictly speaking, not at $-273°$ C., but 0·09° lower.[2] This difference will be too small to be considered in our later calculations, but we must take account of it here, in order to determine R with all possible precision. We then get—

The gas constant in litre-atmospheres.

$$\frac{22 \cdot 412 \times 1}{273 \cdot 09} = 0 \cdot 08207$$

This value of R is expressed in litre × atmosphere units (litre-atmospheres). If we compute the volume in cubic centimetres and the pressure in dynes per square centimetres, it becomes—

The gas constant in absolute units.

$$R = \frac{22{,}412 \times 76 \times 13 \cdot 596 \times 980 \cdot 6}{273 \cdot 09} = 0 \cdot 83155 \times 10^8 \text{ abs. units } \left(\frac{\text{erg}}{\text{T}}\right)$$

Here 76 is the height in centimetres of a column of mercury which exerts a pressure of one atmosphere; 13·596 the weight in grams of a cubic centimetre Hg; and therefore 76 × 13·596 the weight of the mercury column which exerts a pressure of one atmosphere per square centimetre. Since the gram weight at sea-level and lat. 45° corresponds to 980·6 dynes, then 76 × 13·596 × 980·6 is the pressure in absolute measure of an atmosphere on a unit of surface (1 sq. cm.); that is, the pressure on a piston 1 sq. cm. in area working in a cylinder containing 1 mol (22,412 c.c.) of gas.

For our purposes it is most convenient to measure R in heat units (gram calories). We therefore need to know the value of a gram calorie in absolute units.

[1] For the theory see Mach, "Prinzipien der Wäermelehre" (1900), 2nd edit., p. 309 (or translation).

[2] Daniel Berthelot, *Z. f. Electrochemie*, x. (1904), 621; and Nernst, *idem.*, 629.

This value depends upon the temperature at which it is determined. At present it is usual to define a calorie as that quantity of heat which will warm 1 gram water from 15° to 16°.[1] The quantity of heat required to raise 1 gram of water from 0° to 1° is greater than this by 0·06 per cent. This somewhat larger zero-point calorie used to be (till about 1880) preferred. In its stead Schuller and Wartha[2] substituted the quantity of heat necessary to heat 0·01 gram of water from 0° to 100°. (Ostwald[3] took a hundred times this value as a unit and called it K.) This "mean calorie" is very nearly equal to the 15° calorie.[4] The mechanical equivalent of heat has been set by the Committee on Units of the Deutschen Bunsengesellschaft as equal to $41 \cdot 89 \times 10^6$ erg.[5]

We get from this—

The gas constant in heat units.

$$R = \frac{0 \cdot 83155 \times 10^8}{41 \cdot 89 \times 10^6} = 1 \cdot 98507 \left(\frac{\text{gram-cal.}}{T}\right)$$

We shall use later, in our numerical calculations, the abbreviated value 1·98, which is very near the exact value of R computed on the basis of the 15° calorie, the zero-point calorie, and the mean calorie.

Equilibrium and driving force of a reaction.

The reactions (1) to (9) all proceed till they reach an equilibrium. Indeed, the characteristic of the equilibrium condition is that there the reaction will not of itself spontaneously progress in either the one or the other direction. That is, the driving force of a chemical reaction is zero at the equilibrium point. If we know the composition of the gas mixture at equilibrium for any given temperature, then for any other composition of the gas mixture at the same temperature the driving force of the reaction is given by the *distance from the equilibrium point.*

We should at first be inclined to call the reaction energy

[1] Or more accurately from $15\frac{1}{2}°$ to $16\frac{1}{2}°$. Compare Warburg's report on the unit of heat read before the Naturforscherversammlung in Munich (1899, "Bericht über die Verhandlungen," p. 62).

[2] *Wied. Ann.*, 2 (1877), 365.

[3] *Lehrb. d. allg. Chemie*, II_2 (Leipzig, 1893), p. 72.

[4] According to Behn (Drude's *Ann.*, 16 (1905), 653), the mean calorie is greater by 0·03 per cent., according to Dieterici (*idem.*, p. 593), 0·2 per cent. greater, than the 15° calorie. Other figures are given by Planck ("Thermodynamik" (Leipzig, 1905), 2nd edit., p. 31).

[5] *Zeitschr. f. Elektrochemie*, ix. (1903) 686.

the total work which a chemical reaction can do during its whole course. This definition, however, is unsuitable for theoretical treatment.

Reaction energy and composition. The driving force of a chemical reaction changes continuously with the changing composition of the reacting mixture. The nearer we approach the equilibrium point where the reaction stops, the smaller does this driving force become. We take as units of transformation, the transformation of such amounts as appear in equations (1) to (9). Thus, in reaction (2), the change of one mol O_2, two mols H_2, into two mols H_2O is the unit of transformation. The reaction energy of every infinitely small fraction dx of this unit quantity is equal to the driving force A multiplied by the quantity transformed, that is $A \times dx$. To get the total energy of transformation, we must find the sum of all the quantities $A \times dx$ with the aid of the integral calculus, having first expressed A as a function of x. Such an expression would be awkward and involved. It is therefore better to consider the quantity of gas so great that the transformation of a unit quantity would have no appreciable effect on the composition of the mixture, and consequently cause no appreciable change in its driving force. We then get for the product of the driving force into the unit quantity transformed (that is, one mol) the value $A \times 1 = A$, and consider this as the reaction energy. The value A of the reaction energy, therefore, characterizes only a single tiny phase in the progress of the reaction, during which the composition remains constant. It shows us, however, how far we are from the end of the reaction or the equilibrium point where A = zero, and this is the important thing.

Measure of reaction energy. The reaction energy A may be expressed in any units of energy which suit our convenience. In what follows, we shall always measure it in gram calories, taking for practical reasons as far as possible the 15° calorie as our unit of heat. However, the older data are often expressed without any precise definition of the calorie used, and the possibility even of a sure recalculation is not always afforded by the data given. The smallness of the difference between the mean, the zero and the 15° calorie makes this uncertainty less troublesome.

How composition Considering the composition of the gas as invariable during the reaction, we are at liberty to express this composition either in. concentrations—that is, in number of mols per unit of volume (litres)—or in partial pressures. The choice has a two-fold importance. If, for instance, we pass from conditions at one temperature to those at another, considering the composition as the same in both cases, it makes a difference whether we express the composition in concentrations or in partial pressures. In the first case the concentrations remain constant, and therefore, too, the volume of the gas mixture. It follows that, on heating, the pressure of the gas mixture will increase. In the second case, however, the partial pressures, and consequently their sum, the total pressure, remain constant, and the volume will therefore be increased. We deem it simpler to take the concentrations as the determining factors of the composition, and then to consider what small changes we must make in the formulæ we develop, in order to use partial pressures in their stead. There is also a second point of view. If we consider reactions in which a change in the number of molecules takes place, such as the formation of water or of ammonia from the elements, we find that it makes a difference whether we postulate a constant volume or a constant pressure, even when we restrict ourselves to the consideration of reactions at one temperature. If we denote the numbers of molecules by $\nu', \nu'', \nu''' \ldots$, considering the number of molecules of disappearing substances as negative, we have, for instance—

may be expressed.

$$2H_2 + O_2 = 2H_2O$$
$$\nu' = -2, \quad \nu'' = -1, \quad \nu''' = +2, \quad \text{and } \Sigma\nu' = -1$$
$$N_2 + 3H_2 = 2NH_3$$
$$\nu' = -1, \quad \nu'' = -3, \quad \nu''' = +2, \quad \text{and } \Sigma\nu' = -2$$

then at constant pressure the volume increases during the reaction by $\Sigma\nu' \times v$ (where v = volume of a gram molecule).[1] External work is therefore done on the atmosphere which at a constant external pressure, p, amounts to $(\Sigma\nu')pv$. At constant volume this expenditure of work does not occur. We shall see later that this expenditure of work may be eliminated from our formulæ, and consequently does not interest us. At the start

[1] In both examples the change of volume is, of course, negative.

[2] In both examples the work done is negative. The gas mixture then has work expended upon it.

we shall make its consideration unnecessary by defining the compositions by concentrations expressed in mols per litre, and always adhering strictly to the rule *that in every change which we may consider the volume shall remain constant.*

Reaction energy and temperature. The position of the equilibrium is dependent upon the temperature. A mixture of gases which is in equilibrium at one temperature will not in general be so at another temperature. The quantity A which measures the distance from equilibrium would have no definite meaning if we allowed the temperature to change in a gas mixture during the reaction phase we were considering when there was no perceptible change in composition. If, in a reaction such as the formation of water from a mixture of oxygen and hydrogen gas, we imagine that the process is resolved into an infinite number of very small stages, then the composition and the temperature too would remain constant in each separate stage, because only an infinitely small amount of gas would be transformed, and consequently only an infinitely small amount of heat liberated, so that the temperature would not perceptibly change throughout the whole mass of the gas. If, again, we consider the quantity of the gas mixture so great that the transformation of a mol produces no appreciable change in the composition, where consequently the value dx can be taken as equal to 1, we also have a condition in which the heat evolved in the transformation of a mol becomes vanishingly small, for the reaction heat is distributed over the whole mass of gas. *The reaction energy* A *then applies only when the composition and temperature of the reacting gas mixture remain constant.*

Reaction energy and reaction heat. If we compare the conditions under which the reaction energy A, as defined, is to be determined, with those conditions recognized in thermochemistry under which the reaction heat is determined, we are confronted with an important difference. In the case of the reaction heat, we make sure that the reacting substances shall be at the same or nearly the same temperature at the end as at the beginning of the reaction.[1]

[1] Two principles find general application in calorimetry of gas reactions. According to one, the reaction is made to take place in a vessel surrounded by a large heat reservoir, usually a water-bath of known heat capacity, and the rise of temperature in this, seldom amounting to more than a few degrees, is measured. Wide experience shows that reaction heat changes but slightly with

It does not matter if some time during the reaction the reacting mixture is heated temporarily to a higher temperature, for, according to fundamental principles, the heat evolved is only dependent on the initial and final stages, and in the case that the initial and final temperatures are identical, it amounts to the same in the end as if the temperature had remained constant all the while during the reaction. Changes in composition during the reaction are permissible in the determination of the reaction heat, but not in the measurement of reaction energy. From the experimental standpoint, the reaction heat when a very small quantity, dx, was transformed could not be actually determined, because it would be so small. The fact that the reaction heat, especially in the case of gases, is independent of the composition of the reacting mixture is decisive evidence that we can here neglect entirely the effect of changing composition. The reaction energy of the combustion of an oxygen-hydrogen mixture changes when we dilute the gas mixture with nitrogen, while the reaction heat remains unchanged.

Reaction heat and decrease of the total energy.

Let us consider a reaction at constant temperature and composition, in which neither mechanical nor electrical work is done, so that the heat evolved is a measure of the total energy change of the process. Since heat is liberated, the total energy of the reacting substances evidently decreases. If the reaction heat, which we measure in the calorimeter, is independent of the composition of the reacting mixture, then it represents without any further correction the decrease in the total energy. (With dissolved substances we must test this independence from case to case. This is easy to do, by simply testing in succession each substance taking part in the reaction to see whether a

the temperature. If, then, the temperature of the water-holder is raised from 10° to 20° by the reaction, there is no objection to assuming that the observed value corresponds with all desired accuracy to that which would be obtained had the calorimeter a heat capacity a thousand times as great, and the initial and final temperature had been 14·995° and 15·005° respectively. We have, therefore, measured the reaction heat for the reaction taking place isothermally at 15°. The second principle has been embodied by Junkers (Bueb, *Journal für Gasbel. u. Wasserversorgung*, 1893, p. 81; Haber, *idem.*, 1897, p. 751) in a calorimeter which is very convenient for the technical study of gases. Here the reaction products are cooled to exactly the initial temperature by flowing water, on the principle of counter-currents.

change in its concentration causes an absorption or evolution of heat. If no such heat change takes place with any of the substances, the reaction heat is independent of the composition. If such "heats of dilution" do appear, we have to determine to what extent they compensate one another. In the case of gases, the heats of dilution are very small, and for our purposes negligible.) The heat change corresponding to a unit amount transformed, as above defined, is called in thermochemistry the heat of reaction, or the reaction heat (*Wärmetönung*). For this thermochemical quantity we shall from now on use the symbol Q. For the decrement which the total energy undergoes when unit quantity is transformed at constant composition, we shall, according to the usual nomenclature, write the symbol U. The difference in the symbols will remind us that putting Q for U always involves the assumption that the change of the percentage composition does not alter the reaction heat.

Determination of the reaction energy.

We estimate the driving force of a reaction in synthetical chemistry by a sort of scientific instinct. We speak of stronger and weaker chemical reagents without measuring this strength in any specific units.

In natural science we usually measure forces by opposing them with other known forces. Whenever the opposing force is of the same nature as the force to be measured, this method of determination is easily comprehensible. But while we can easily measure mechanical forces by the use of a balance and weights, and electrical forces by means of compensating and opposing electrical forces, it is not possible for us to measure the driving force in chemical reactions with opposing forces of a chemical nature. We must therefore make use of dissimilar forces whose connection with the chemical forces is not self-evident, and consequently not so easy for us to comprehend. Such unlike forces can be either of a mechanical or an electrical nature. Mechanical opposing forces cannot be employed unless the reaction is accompanied by a change of volume.[1] For only in such cases

[1] Considered more exactly, opposed mechanical forces are not at all applicable to the measurement of the reaction energy of ideal gases, but only to the determination of the equilibrium point under various pressures. A gas reaction may be imagined to take place isothermically in a vessel with a movable piston, and the opposing pressure which at every instant it can compensate may be determined. If the vessel and the contained mass of gas

are reactions influenced by a change of pressure. Opposing forces of an electrical nature can be well employed in many gas reactions. For instance, the formation of water-vapour from oxygen and hydrogen, or of hydrochloric acid gas from chlorine and hydrogen, can be so carried on with the aid of heated glass [1] as electrolyte in the first case, or with fairly strong hydrochloric acid in the second case, that they will generate electrical energy. Arrangements which make this possible are called gas elements. The counter electromotive force which we must apply in order to just compensate the electromotive force of such a gas element is identical with the driving force of the chemical reaction taking place in the gas element. With most gas reactions, however, this procedure is impossible. Usually the only remaining alternative is to find by chemical methods at what composition of the gas mixture the reaction comes to a halt, regardless of the direction from which the equilibrium is reached. We then know that at this composition the driving force of the reaction is zero, and we may calculate its value at other compositions by means of well-known laws with which we shall shortly become acquainted.

If on the one hand we determine the reaction energy in the manner just described, and on the other the decrease of the total energy, then of course these two quantities might accidentally be equal. In general, however, these quantities have different values. The difference is most striking if we consider the last of the successive phases of the reaction—that is, the one at which the distance from the equilibrium, and consequently the driving force, is vanishingly small. The decrease of the total energy U is here just as large as ever. This decrease, however, can only give us heat, for the driving

The latent heat of reaction.

is so large that the composition is not altered by change of unit quantity, then the opposed force just compensated during this change remains constant. Then, however, the work which is done against the opposing force is for every newly formed gas mol $= pv$, and quite independent of the nature of the reaction and of the value of the opposing force, *i.e.* of the value of the pressure p. If we change the composition of the gas mass, and consequently its reaction energy, the pressure p becomes different; the work pv, however, is the same. This means that the experiment only gives the position of the equilibrium point at the pressure p.

[1] Till recently, it has been assumed that water containing acid or alkali could serve this purpose. However this so-called "Grove's element" has been proved to be incorrect, as will be seen in the Fifth Lecture.

force of the reaction is gone, and with it the possibility of generating electrical energy. If, in the element just mentioned, where hydrogen and chlorine combine to form hydrochloric acid, we choose such a concentration of chlorine, hydrogen, and hydrochloric acid gas that the system is infinitely near the equilibrium point, then the electromotive force is infinitely small, although the reaction heat (corresponding to the formation of two mols of HCl) has now, just as before, the high value of 44,000 cals. By choosing various relative concentrations, we are enabled to give all possible values to the difference between the reaction energy A and the reaction heat Q. Further, at given concentrations, this difference is dependent on the temperature we choose. Every isothermal chemical reaction which proceeds with the production of the maximum amount A of work, is connected, according to temperature and to the concentration, with the appearance and disappearance of definite quantities of sensible heat. The whole application of thermodynamics to chemical processes hinges on this point. The development of chemistry is responsible for the fact that this connection never occurs to us in the case of ordinary chemical reactions, while it appears as wholly self-evident in the very closely related phenomena of change in the state of aggregation. In chemical reactions we are always inclined to assume that the reaction energy (which we measure by the amount of work which the process can do, as above set forth), is equal to the total energy change which we measure in the calorimeter. On the other hand, in the case of a simple change in the state of aggregation, we consider the very reverse as self-evident. In principle, there is absolutely no difference between the two classes of phenomena.

Comparison with a change in state of aggregation.

Let us imagine a large vessel, closed by a movable piston, containing water at a temperature of 100°. There is water-vapour above the water, and this, acting against the inside surface of the movable piston, is in equilibrium with the external atmospheric pressure. An increase of this external pressure causes the piston to sink and the water-vapour to condense. If we lessen the pressure, water will be converted into steam, and the piston will rise indefinitely, provided only that there be a sufficient quantity of water present. In both the former and the latter case it is assumed that provision has been made for the abstraction or introduction of heat in such quantities that the temperature is kept constant at 100°. When a mol of water vapour condenses, an extraordinary quantity of heat, amounting to 9650 gram-cal., is set free. When a mol of water-vapour is formed from liquid water, the same quantity of heat is absorbed. The fact that in the formation of the water-vapour a small amount of work (*pv* for one mol, *i.e.* 746 cal. at 100° C.) is done by the piston rising against the pressure of the atmosphere is unimportant, for this amount is negligible in comparison with the 9650 cal. Overlooking, then, this small quantity, we must admit that the heat of vaporization corresponds exactly to the total energy change in a chemical reaction taking place at, or very near, the equilibrium, which gives no work. When a change of state takes place at constant temperature, and sensible heat is absorbed or evolved, we say in everyday phraseology, according to the direction of the change, that in vaporization heat becomes latent, and in condensation heat reappears out of the latent condition. In the same sense, we can, with Helmholtz,[1] speak of the latent heat of a chemical reaction, meaning thereby the difference between the maximum electrical or mechanical work A which we can get as the reaction progresses at constant temperature and constant composition, and the accompanying decrease in total energy, U. The latent heat of a chemical reaction, then, is that amount of sensible heat which appears or disappears when we cause the process to proceed furnishing, the largest possible yield of electrical or mechanical energy. On the other hand, it should be recalled that the decrease of total energy is equal to the heat which is evolved when we allow no electrical or mechanical work to be done. Both quantities are equal when the transformation is incapable of doing work (that is, in the case of chemical equilibrium). Otherwise they differ by the value of the reaction energy (work obtainable) of the reaction. This gives us the expression—

$$A - U = -q \quad . \quad . \quad . \quad . \quad . \quad . \quad (11)$$

Here we use $-q$ to designate the heat which is used up, or, in other words, "becomes latent."[2]

[1] Vorlesungen, vol. vi. "Theorie der Wärme" (edited by Richarz), p. 286 (Leipzig, 1903).

[2] In these lectures, heat or work taken up by the reacting system is

According to another mode of expression, likewise originating from Helmholtz, A is called the decrease in free energy, q the decrease in bound energy.[1]

Just as the latent heat of vaporization of water changes with the temperature, so the latent heat of a chemical reaction is also dependent on the temperature. The fact that the latent heat of a chemical reaction is also dependent on the composition of the reacting mixture is the only point of dissimilarity. A change in composition is evidently impossible in the case of a mere change of state of a simple substance.

Direct measurement of the latent heat.

We are able in some cases to directly measure the latent heat of reaction in a calorimeter. We can, for instance, allow the reaction to take place in a galvanic cell immersed in a calorimeter, and put the resistances in which we convert its electrical energy into heat in a second calorimeter. We can connect the cell with the resistances by means of thick wires almost devoid of resistance, so that no appreciable electrical energy shall be converted by them into heat. When the element acts, we obtain an evolution of heat in the first calorimeter which represents the quantity q. We must, however, make a small correction for the heating effect of the current in the cell itself, this correction being easily calculated from the resistance of the cell and the current which passes through it. In the second calorimeter, on the other hand, we measure the reaction energy A in heat units. If we place both cell and circuit in a single calorimeter, instead of in separate ones, the generation of heat is the same as though we let the reaction take place in the calorimeter without electrical assistance. It corresponds, therefore, to the value of Q or U. Jahn[2] has, indeed, carried out experiments of this kind with success. Gas reactions were not chosen as examples in these experiments. But so far as the principle is concerned,

considered negative, heat given off, or work done by the reacting system, is considered positive. According to (11)—

$$A + q = U$$

that is, the decrease, U, in the total energy is equal to the work done, A, plus the heat evolved, q.

[1] *Sitzungsberichte Berliner Akad.* (1882), 1, p. 22; and Ostwald's *Klassiker*, No. 124, edited by Planck.

[2] "Grundr. d. Elektrochemie" (Vienna, 1895), p. 180; *Z. phys. Chem.*, 18 (1895), 399.

there is nothing in the way of carrying out an experiment of this kind with gas reactions; the formation of hydrochloric acid from hydrogen and chlorine, for instance. For this purpose we should have to pass hydrogen and chlorine in at the electrodes of a suitable cell, and then remove gaseous hydrochloric acid from its concentrated solution, which forms the electrolyte, as fast as it was formed, so that the cell would lose as much hydrochloric acid by evaporation as it gained from the union of the gaseous elements of the electrodes. The result would be an isothermal formation of hydrochloric acid with the production of electrical energy, which in turn would revert to heat in the second calorimeter, and would be measured as such. The more completely we were able to avoid disturbances—such, for instance, as the polarization of the electrodes—so much nearer would the actual voltage of our cell approach its true electromotive force. But the electromotive force is equal to the driving force of the chemical reaction taking place. In the limiting case, where our electrodes are unpolarizable and our cell is of vanishingly small internal resistance, the electrical energy given off is equal to the reaction energy. If chlorine and hydrogen at atmospheric pressure be taken, the temperature kept at 30° C., and if the vapour pressure of the hydrochloric acid over its 12½ normal solution acting as electrolyte has the value corresponding to this temperature, then A, according to Dolezalek's[1] determination, equals, for the formation of one mol hydrochloric acid gas, +22,509 gram-cal. The heat change observed in the first calorimeter, or q, then amounts to -509 cal. The sum $A+q$, that is 22,000, represents the well-known heat of formation of one mol of hydrochloric acid gas.

The use of more dilute hydrogen or chlorine or a higher tension of hydrochloric acid above the solution, which would result, for instance, from the use of a hotter electrolyte, causes A to sink in value and q to rise. On the other hand, a higher pressure of the gaseous factors and a lower pressure of the resultant hydrochloric acid gas causes A to rise and q to fall. The sum $A + q$, however, is always equal to the reaction heat, which is not perceptibly affected by these changes.

Latent heat and temperature. Our problem, then, is to determine the connection between the latent heat q and the temperature and composition of a

[1] *Z. f. physik. Chemie*, 26 (1898), 334.

mixture of reacting gases. We will first solve that part of the problem connected with the temperature, and we will do this in two steps. First, in view of the connection between latent heat and obtainable work as expressed in equation (11), let us take the simplest possible case where heat becomes latent and work is simultaneously done, and find out what connection obtains between the latent heat and the temperature. The second step consists simply in showing that the same relation holds in all cases, because it is inherent in the nature of heat and work.

The porous plug experiment. Change in state of aggregation is not, as one might think, the simplest case in which latent heat and work occur together, for here the two states have different total energies. But the simple expansion of a gas, where, according to experiments of Gay Lussac,[1] the total energy does not change, answers our requirements. The total energy of a given mass of gas is made up of its chemical and heat energy. It is clear that as long as the gas does not change chemically, its chemical energy remains constant. The fundamental experiments of Gay Lussac have shown that the heat energy is not altered by changing the volume of the gas, provided no work is done. To prove this, he connected a large vessel (I.) full of gas by means of a stopcock with an evacuated vessel (II). When he opened the cock, the gas rushed over into the evacuated vessel. Expansion took place in vessel I., and the gas became colder, as it had to compress the first portion of the gas that had rushed over. On the other hand, the temperature in vessel No. II. rose, because these first portions of gas were compressed by the following portions. After the pressure of the gas had become equalized, the half of the gas in the first vessel was cooler than the initial temperature by the same amount that the other half in the second vessel was warmer. Therefore the mean temperature of the whole mass of gas had not changed, as it must have done had the thermal energy of the gas changed and heat been absorbed or set free.

The Gay Lussac experiment[2] was greatly refined by Joule and

[1] Mach, "Prinzipien der Wärmelehre," 2nd edit. (1900), pp. 198 and 310; also in the English translation.

[2] An illuminating criticism of this experiment and later ones by Gay Lussac on the same subject is to be found in an article by Regnault, *Memoirs de l'académie*, 26 (1852), 11. Compare also the Sixth Lecture.

Thomson.[1] They forced a constant stream of a compressed gas through a piece of boxwood, enclosed in a tube impermeable to heat. The pressure in front of the boxwood plug was P; behind it, p; that is, the gas expanded from a higher to a lower pressure. No heat was absorbed or given off during the process, and no work was done; therefore, had the total energy of the gas remained unchanged during the expansion, the temperature must have been the same behind as it was in front of the plug. The experiments showed that this was not strictly the case, although the more nearly the gas used adhered to the laws for an ideal gas, the smaller was the change in temperature on passing through the plug. We may therefore say with all rigour that the total energy of an ideal gas is independent of the temperature, but that with actual gases there are small deviations from this rule, especially at low temperatures.[2] Yet for our purposes the deviations are entirely negligible.

Isothermal expansion.

If we now let a gas expand and do work, keeping the temperature constant, we find that heat becomes latent. Remembering that the total energy of the gas does not change with changing volume, we conclude that the heat which becomes latent, that is, was used up, is the equivalent of the work done. To conclude otherwise would be equivalent to saying that energy had either vanished into nothing or had been created out of nothing, and that the principle of the conservation of energy had been violated. Now we can easily determine the amount of the work done. When a gas expands at the pressure p by the infinitely small volume dv, an infinitely small amount of work[3] is done, and—

$$dA = pdv \quad . \quad . \quad . \quad . \quad . \quad . \quad (12)$$

If, making use of equation (10), we replace p by $\frac{RT}{v}$, we have—

$$dA = \frac{RTdv}{v} = RTdlnv \quad . \quad . \quad . \quad . \quad (13)$$

[1] *Phil. Trans. Roy. Soc. Lond.*, 1853, 357; 1854, 321; 1862, 579.

[2] These deviations have the greatest interest in other connections. In the first place, Linde's beautiful method for the liquefaction of the air depends upon them. In the second place, a study of these deviations is the only means at our disposal for accurately determining the absolute zero of temperature, of which we spoke at the beginning of this lecture.

[3] Work = force × distance; force = pressure × area; volume = area × distance.

The continuation of the isothermal expansion which furnishes the work represents the sum of all these infinitely small quantities $RTdlnv$. If the initial volume be v and the final volume v', we can easily find this sum by means of the integral calculus (ln signifying natural logarithm) thus—

$$A = RT\int_{v}^{v'} dlnv = RTln\frac{v'}{v} \quad . \quad . \quad . \quad (14)$$

Remembering that here the work done is identical with the heat which becomes latent, we get—

$$A = RTln\frac{v'}{v} = -q \quad . \quad . \quad . \quad . \quad (15)$$

This important formula does not by any means say that the work A *must* always be done when a gas expands isothermally from the smaller volume v to the larger volume v'. We only need to call the Gay Lussac experiment to mind in order to see that the term A can become equal to zero. Every possible value of A between zero and $RTln\frac{v'}{v}$ is not only conceivable, but attainable. The value $RTln\frac{v'}{v}$ is, however, a maximum one. If we try to bring the gas back isothermally from the greater volume v' to the initial volume v, we need to expend more or less work depending on the efficiency of our compression machine. But here again the minimum value of the work required is A. The maximum production of work is therefore that which accompanies a process taking place reversibly. It represents the ideal limiting case of actual work production which only a frictionless compression machine could realize, and then only when we took care that the acting and opposing pressure never differed by a finite amount at any instant of the process. Let us consider the same process to be carried out, at a temperature higher by an infinitely smaller amount, in such a way that the initial and final volumes are the same; then—

$$A + dA = R(T + dT)ln\frac{v'}{v} = -(q + dq) \quad . \quad (16)$$

Subtracting (15) from (16)—

$$dA = RdTln\frac{v'}{v} \quad . \quad . \quad . \quad . \quad . \quad . \quad (17)$$

$$\text{or} \quad \frac{dA}{dT} = Rln\frac{v'}{v} \quad . \quad . \quad . \quad . \quad . \quad . \quad . \quad (18)$$

On the other hand, we get from (15), dividing through by T—

$$-\frac{q}{T} = Rln\frac{v'}{v}$$

It follows from this that—

$$-\frac{q}{T} = \frac{dA}{dT} \quad . \quad . \quad . \quad . \quad . \quad . \quad . \quad (19)$$

or, differently expressed—

$$-q = T\frac{dA}{dT} \quad . \quad . \quad . \quad . \quad . \quad . \quad . \quad (20)$$

Equation (20) gives us an expression in which the latent heat is equated to the product of the absolute temperature and the "temperature coefficient." This temperature coefficient represents the increment of work which we gain when we carry out the above process at a temperature higher by an infinitely small amount dT, divided by this temperature difference.

It now remains for us to show that this relationship is not conditioned by the properties of the gas, but is inherent in the nature of heat and work; that it applies to all latent heats so long as we restrict ourselves to isothermal processes yielding the maximum amount of work. To this end we will represent diagramatically in Fig. 1 the procedure we have just carried through with a gas. We will assume that the perpendicularly shaded area represents work which we gain when we let a mol of gas expand from v to v' at a temperature $T + dT$, and starting at the pressure $p + dp$. Then let us consider that the temperature is decreased to T by the abstraction of an infinitely small quantity of heat dq, and that the process is now made to go in the reverse direction with an expenditure of work represented by the obliquely shaded area. Raising the temperature by the amount dT brings us back to the starting-point. The heat used up at $T + dT$ we get back at T. The infinitely small amounts of heat absorbed and given off in the process of heating and cooling cancel. There remains only the infinitely small quantity of work dA, resulting from the falling of the heat q through the temperature interval dT. This is represented in the diagram by the narrow band containing only vertical shading. Now let

General applicability of equation (20).

us imagine the same process carried out, not with a gas, but with any desired system. Our sole condition is that the process shall produce the maximum amount of work, and that thereby the heat q shall be absorbed (become latent). Any desired changes of the total energy are permissible. Nor does the work done need to be mechanical, it may be electrical. Here, too, the final result is that everything returns to its original condition except that q has fallen in temperature by the amount dT. If in this second case a different yield of work were obtained, it would be possible to construct a perpetual-motion machine. If, for example, the work yielded by the second system were greater, then we should first execute

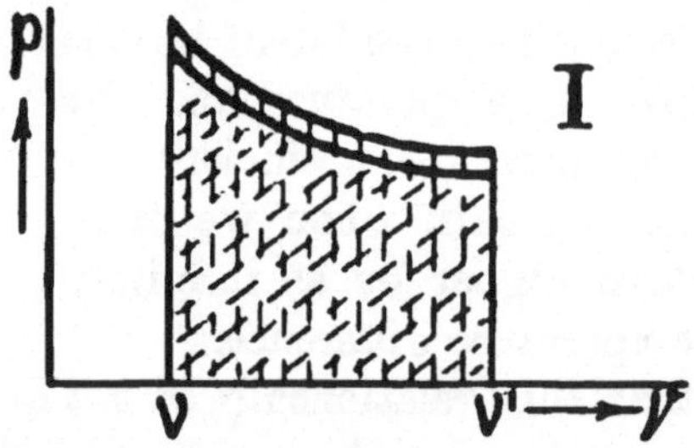

Fig. 1.

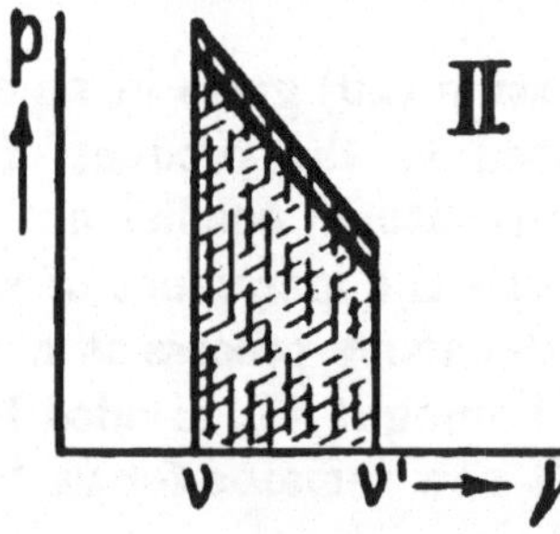

Fig. 2.

the cycle with it and obtain the work dA. We should then repeat the cycle in the reverse direction with the first system, and thus raise q back to its original temperature T + dT, with a net expenditure of work less than dA. The result would be that a certain amount of work would remain over without the temperature, or any other property of the substances concerned, having in any way changed. We should therefore have created work out of nothing.[1] It follows from this that the relation (20)

[1] The fact that formula 20 applies not only to gaseous processes, but also to all processes in which heat and work are coupled together, constitutes the so-called second law of thermodynamics. This second law cannot be deduced from the first law (conservation of energy) *alone*, though it may seem to have been done so in the above proof. Indeed, there is an assumption concealed in the deduction, which has nothing to do with the law of conservation of energy. This assumption, which will be more fully enlarged upon later, is that the temperature is always reckoned from the absolute zero point.

If the quantity of heat q falls in temperature by dT, as explained in the

applies to all systems, provided the change takes place at constant temperature and with the production of the maximum amount of work. This permits us to substitute equation (20) in our fundamental equation (11), and to write—

$$A - U = T\frac{dA}{dT} \quad . \quad . \quad . \quad . \quad . \quad (21)$$

We can make a slight change in the method of writing this equation, which will serve to remind us that the heating and cooling through dT is always considered as taking place at constant volume. This assumption of constant volume was made in the derivation of the expression, and is of importance, because it obviates any work being done by the heating or

$\frac{dA}{dT}$ is the partial differential of A with respect to T.

text, a part dq of it is used up and converted into work. This quantity of heat is the same in processes I. and II., since the values q, T, and dT are in both processes equal by supposition, and are connected by the expression—

$$q : dq = T : dT \quad . \quad . \quad . \quad . \quad . \quad . \quad . \quad . \quad . \quad (a)$$

This expression applies, as Mach has thoroughly demonstrated, to all forms of energy. In every case where a quantity of energy c at a constant value i of its "intensity factor" changes its form, the expression—

$$c : dc = i : di \quad . \quad . \quad . \quad . \quad . \quad . \quad . \quad . \quad . \quad (b)$$

holds good. Now, according to the first law, we have—

$$-dA = dq$$

From this formula 20 obviously follows without any consideration being given to the special properties of heat.

With all other forms of energy i may be reckoned from any arbitrary zero point, while with heat it is only possible from the absolute zero point, if the expression a is to agree with experimental evidence. If we change the zero point from which i is reckoned in equation b, dc will become different, though c and di remain the same. If it is only allowable to reckon T from the absolute zero point, dq will always have an unchanged value for same values of q and dT. An indirect proof of this fact is generally considered necessary. Suppose in processes I. and II. dq has the different values dq' and dq'', then the corresponding work must be dA' and dA''. If now processes I. and II. are combined, so that one always takes place forwards and the other backwards, any amount of heat we please may be changed into work without anything else undergoing permanent change. The impossibility of this, which will be enlarged upon in the Third Lecture, is the result of general experience. Thus, dq' and dq'' must be always equal. Since, however, from these indirect proofs nothing more than the necessity of reckoning T from $-273°$ C. is proved, it seems simpler to start here from this assumption, and later to explain its importance more fully.

cooling through the interval dT which could disturb the balance of heat and work in the cycle. That is, we can simply use the notation $\left(\frac{dA}{dT}\right)_v$ for the temperature coefficient. This shows us that A is dependent, not only on the temperature, but also on the volume, and that during the change dT the volume is kept constant. $\left(\frac{dA}{dT}\right)_v$, therefore, represents the *partial* differential of A with respect to T. So we can write—

$$A - U = T\left(\frac{dA}{dT}\right)_v \quad . \quad . \quad . \quad . \quad . \quad (21a)$$

Integration of equation (21). It will be useful in our later discussions to slightly rearrange this equation, transposing the members, and at the same time dividing through by T^2. We then have—

$$\frac{1}{T} \times \frac{dA}{dT} - \frac{A}{T^2} = -\frac{U}{T^2}$$

Further, by the elementary principles of the differential calculus—

$$\frac{d\left(\frac{A}{T}\right)}{dT} = \frac{1}{T} \times \frac{dA}{dT} - \frac{A}{T^2}$$

and therefore—

$$\frac{d\left(\frac{A}{T}\right)}{dT} = -\frac{U}{T^2} \quad . \quad . \quad . \quad . \quad . \quad . \quad . \quad . \quad (22)$$

Integrating this expression, we get—

$$\frac{A}{T} = f(v) - \int\frac{U}{T^2}dT \quad . \quad . \quad . \quad . \quad . \quad (23)$$

Here, instead of the constant which appears in the integration of a complete differential, we have a function $f(v)$ of the variable (volume), which was assumed constant in the differential. We can further transpose (23) into—

$$A = T \times f(v) - T\int\frac{U}{T^2}dT \quad . \quad . \quad . \quad . \quad (24)$$

If we replace the change in the total energy U by the reaction heat Q, it follows that—

$$A = T \times f(v) - T\int\frac{Q}{T^2}dT \quad . \quad . \quad . \quad . \quad (25)$$

Thus the quantity A is dependent on the heat of reaction, the temperature, and the volume, and because of the very nature of heat.

This expression is not immediately applicable, because we do not yet know anything further about the volume function $f(v)$ or the integral—

$$\int \frac{Q}{T^2} dT$$

We shall concern ourselves further with them in the Third Lecture. Here we shall simply consider a little more closely the relation we have just derived between heat and work.

If, remembering equation (15), we write— Peculiarities of latent heats.

$$-q = RT ln \frac{v'}{v}$$

and

$$-q' = RT' ln \frac{v'}{v}$$

it follows (for processes where $U = 0$ or independent of T) that—

$$q : q' = T : T' \quad . \quad . \quad . \quad . \quad . \quad . \quad (26)$$

This relation says that the latent heat of an isothermal process yielding the maximum amount of work is proportional to the intensity factor of its heat—that is, to the temperature. If, then, we leave everything as it is, and simply exchange the temperature 2T for the temperatue T, then the latent heat, that is the heat converted into work, would be twice as great. This exceedingly simple relation also exists in energy transformations of other kinds. When we use the energy of position which a lifted weight possesses to drive clockwork, and by it an electric motor, the height through which the weight falls represents the intensity factor of the position energy which we transform into electrical energy. We see immediately that if we let the weight fall twice as far, we consume twice as much position energy. When we send a certain quantity of electrical energy at the voltage E into a circuit where electrical energy is converted into heat or mechanical energy, evidently the doubling of E, the intensity factor, doubles the consumption of electrical energy, provided none of the other conditions are altered. The relation expressed in (26), therefore, represents an entirely general law. It has, nevertheless, a specific peculiarity. When the weight sinks, the intensity factor is represented by the *difference* of height (between initial and final position); in the electrical example, by the *difference* in potential at the ends of the circuit where the energy is used. But in the heat-work processes, where heat becomes latent, that is, where heat is used to do work, the *absolute temperature* must be taken as the intensity factor, and expression (26) becomes incompatible with experience when we choose any other zero point for our temperature scale. We can express this in another way by saying that, in the case of heat as well as electrical and mechanical energy, the difference of the intensities is the determining factor, but in the case of heat, we must reckon the difference between the given temperature and the *absolute* zero point, while in other energies we may arbitrarily choose any relative zero suitable to the conditions of the experiment. So far-reaching is this difference that we have absolutely no conception of an absolute zero of position energy or electrical potential. It is, of course, a pure convention to take the sea-level or earth potential as a zero point. Weight can sink below the level of the sea, and potential can be negative relative to the earth. Conversely, the absolute zero has acquired such importance in heat-work processes, because of the relationship expressed in equation (26), that temperatures are exclusively reckoned from it in all thermodynamical calculations.[1]

Difference of the fundamental laws for latent heats and for the utilization of heat.

The theory of practical heating processes often makes use of an expression similar to the one we have just considered—

$$w : w' = t : t'$$

where by w we understand the heat which a body contains at $t°$ C., and by w' that which it contains at $t'°$ C. The meaning of this expression is, however, entirely different. It has nothing to do with heat-work processes, but simply with heat transfer without the performance of work. For the purposes of this rule, therefore, temperatures can be just as well reckoned in degrees Celsius as from any other arbitrary zero point.

Conditions governing the isothermal

One observation still remains to be made with regard to the conditions under which sensible heat can be transformed

[1] See Mach, "Prinzipien der Wärmelehre" (Leipzig, 1900), p. 328: "Ueber die Konformität und die Unterschiede der Energien."

into work at constant temperature. It is that some process of dilution, such as we have just seen to occur in the case of the expansion of a gas, must always take place. Every such dilution is naturally limited, for it will only continue to progress so long as the pressure, steadily diminishing in the course of the dilution, is greater than that of the surroundings. If no pressure differences existed, the process of dilution and the consequent conversion of heat into work would be impossible. Now, where chemical reactions are taking place in gaseous, liquid, or molten mixtures, substances appear and disappear under certain concentration relationships. Every newly formed fraction of a reaction product must assume the concentration of the fraction already present. Every disappearing fraction of a reacting substance must relinquish a certain state of concentration as it ceases to exist. These unavoidable concentration changes are responsible for the existence and the importance of latent heats in chemical reactions.

Transformation of heat into work.

Returning now in conclusion to the expression—

$$A = U - q = U + T\frac{dA}{dT}$$

we shall be able to understand the significance of the member $T\frac{dA}{dT}$ if we consider a contention made by Berthelot. Berthelot maintained that U and A were equal if changes of state were excluded. He realized well enough that the latent heats accompanying change of state disturbed this relationship. But he erred in thinking that by excluding changes of state he prevented all latent heats from coming into play. Helmholtz therefore (*l.c.*) pronounced Berthelot's limitation as insufficient. Latent heats are only completely excluded in one case, and that is when the reaction takes place at absolute zero. In this case—

The principle of Berthelot, and Helmholtz's criticism of it.

$$A = U$$

In every other case a difference is possible, its size depending, according to equation (25), on the concentration relations ($f(v)$) in the reaction, and on the temperature. At given relative concentrations the deviations are greater the higher the temperature. At all temperatures relative concentrations can exist at which A and U are very different. Practically, however, concentrations vary only within quite narrow limits. If the substance disappearing in a reaction be so far used up that only $\frac{1}{100}$ per cent. of the original substance remains, we call the reaction complete, and do not concern ourselves with the great difference between reaction heats and reaction energies in the transformation of these last traces. At the ordinary temperature of about 20° C., which does not lie very high on the absolute scale (+293°), and in the really important regions of greater concentrations, the difference between A and U is not, as a rule, very great, and Berthelot's rule affords a useful approximation. This rule, however, cannot be applied to gas reactions, for these often take place at temperatures which are higher by many thousands of degrees.

SECOND LECTURE

ENTROPY AND ITS SIGNIFICANCE IN GAS REACTION

THE expression—

$$A = U - q \quad . \quad . \quad . \quad . \quad . \quad . \quad (1)$$

which we have treated in the preceding lecture, takes an especially simple form in the case just mentioned of a gas expansion, because the total energy of the gas does not change in isothermal expansion, that is, the quantity U is equal to zero.

In the case of chemical reactions we substitute the reaction heat Q for U. This statement, however, is not definite enough. In the first place, it is clear that what we mean is the reaction heat at the temperature of the reaction. Now, the reaction heat generally changes with the temperature. The extent of the change depends, as Kirchhoff has shown, on the specific heats of the substances formed and used up in the chemical reaction. If it were possible to carry out gaseous reactions at the absolute zero without having the gases suffer any change in their state of aggregation or losing any of their ideal properties, we should find a reaction heat which could be connected with the reaction heat at the temperature T by a simple consideration. That is, if we imagine the reaction to take place at the temperature T with a gain of heat Q_T, and the resulting product to be then cooled down to absolute zero, we should obtain the same net amount of heat as though we had first cooled the factors of the reaction to absolute zero and then allowed the reaction to take place with a heat gain of Q_0.

Reaction heat and temperature.

If $c_{\epsilon(0,T)}$ be the mean specific heat of the products between 0° and T°, then the net gain of heat in the first procedure would be $Q_T + Tc_{\epsilon(0,T)}$; in the second procedure it would be $Q_0 + Tc_{f(0,T)}$, where $c_{f(0,T)}$ is the mean specific heat of the factors. Since these two quantities are equal, it follows that—

$$Q_T = Q_0 + T(c_{f(0,T)} - c_{\epsilon(0,T)}) \quad . \quad . \quad . \quad (2)$$

Heat capacity; mean and true specific heats.

We must here say a word in regard to mean and true specific heats, and the relation between heat capacity and true specific heats. Heat capacity may be best defined as the amount of heat which must flow into a unit quantity of a substance in order to raise its temperature 1°. If we designate this quantity of heat as w, this definition is equivalent to the more precise statement that the heat capacity is $\frac{dw}{dT}$; that is, it is equal to the quotient of the very small increment of heat over the rise in temperature which it causes. The same quantity is also called the true specific heat of a unit mass at T°. Strictly speaking, the specific heat is, to be sure, only the *ratio* between the above-mentioned value, $\frac{dw}{dT}$, and the similar value, $\frac{dw'}{dT}$, of a standard substance. But since 1 gram of water at 15° is taken as the standard substance, for which $\frac{dw'}{dT}$ equals unity, it follows that the ratio and the quantity itself are identical. The two expressions are, in fact, used interchangeably.[1] If we represent the true specific heat by c_w, then—

$$c_w \times dT = dw \quad . \quad . \quad . \quad . \quad . \quad . \quad (3)$$

or, since we can here just as well use Celsius degrees, no heat-work process being involved, but only a pure transfer of heat from one substance to another, we can write—

$$c_w \times dt = dw \quad . \quad . \quad . \quad . \quad . \quad . \quad (3a)$$

If we integrate this expression between 0° C. and $t°$, or between 0° absolute and T° absolute, we get—

[1] For instance, see Wüllner, "Lehrbuch der Experimentalphysik," vol ii., 5th edit. (Leipzig, 1896), p. 169. Th. W. Richards (*Z. f. phys. Chemie*, 36 (1901), 358) lays stress on the rigorous definition of heat capacity (as calories divided by degrees), and at the same time makes the proposal to take as a unit of heat capacity the capacity which is warmed 1° C. by 1 joule (that is, 1 watt-second, or 10^7 erg). He calls this capacity 1 mayer, and a thousand times this capacity 1 kilomayer. This suggestion fits into a system of heat measurement which Ostwald has introduced in the third edition of his "Grundriss der allgemeinen Chemie," according to which the calorie will no longer be the unit quantity of heat, but will be replaced by the thermal equivalent of a joule. From a theoretical point of view, this system would result in some simplifications, but as yet it has met with no acceptance.

$$w_{(0,T)} = \int_0^T c_w \times dT = c_{m(0,T)} \times T \quad . \quad . \quad . \quad (4)$$

or

$$w_{(0,t)} = \int_0^t c_w \times dt = c_{m(0,t)} \times t \quad . \quad . \quad . \quad . \quad (4a)$$

Here $c_{m(0,T)}$ signifies the mean specific heat between $0°$ and $T°$ on the absolute scale; $c_{m(0,t)}$ the mean specific heat between $0°$ and $t°$ C. The two values are by no means always identical. They are identical among themselves, and with c_w when c_w has the same value at all temperatures. On the other hand, if c_w has a different value at different temperatures, all three quantities, c_w, $c_{m(0,T)}$, and $c_{m(0,t)}$, differ from one another. The simplest relation exists between the true specific heats and the temperature when the equation—

$$c_w = a + bt \quad . \quad . \quad . \quad . \quad . \quad . \quad . \quad (5)$$

holds. If we replace t in it by $(T - 273)$, we get—

$$c_w = a + bT - b273$$

If we call—

$$a - b \times 273 = a'$$

we can write—

$$c_w = a' + bT \quad . \quad . \quad . \quad . \quad . \quad . \quad . \quad (5a)$$

In the first case we find, from (4a) and (5)—

$$\int_0^t c_w dt = \int_0^t (a + bt)dt = (a + \tfrac{1}{2}bt) \times t \quad . \quad . \quad (6)$$

in the second, from (4) and (5a)—

$$\int_0^T c_w dT = \int_0^T (a' + bT)dT = (a' + \tfrac{1}{2}bT) \times T \quad . \quad (6a)$$

Comparing (5), (6), and (4a), it follows that—

$$c_{m(0,t)} = a + \tfrac{1}{2}bt$$

and

$$c_w = a + bt$$

On the other hand, from (5a) and (6a) and (4)—

$$c_{m(0,T)} = a' + \tfrac{1}{2}bT$$
$$c_w = a' + bT$$

If, then, the true specific heat is a linear function of the temperature, it differs from the mean specific heat between $0°$ and the temperature in question, simply in the fact that its second member is twice as great. We shall have frequent occasion, in the progress of our lecture, to make use of these formulæ. Just now, we will simply point out that according to (4) the mean specific heat between 0 and T may be considered as the heat, w, which is necessary to raise unit mass of a substance from absolute zero to T°, divided by the temperature T; that is $\frac{w}{T}$. In our later statements we shall not give the specific heats referred to in the usual unit of mass, the gram, but use the product of the specific heat per gram into the molecular weight. We call this the specific heat per mol. Ostwald calls it the "molecular heat."

Reaction heat at constant volume and pressure. If we return to what we were discussing at the beginning, we may say that in the second place the reaction heat of a gas reaction is dependent on whether the reaction takes place at constant pressure or constant volume. The difference existing between the reaction heat at constant volume and that at constant pressure may be ascribed either to work done against the atmosphere when the reaction is accompanied by an increase in volume, or to work done on the gases by the atmosphere when the reaction is accompanied by a decrease in volume. As we know from thermochemistry, and as we pointed out in the First Lecture, this work is $pv\Sigma\nu'$, or $RT\Sigma\nu'$, where $\Sigma\nu'$ in the equation representing the reaction represents the sum of all the numbers of molecules (the number of molecules of the factors being counted as negative). We have decided to consider work done by the *chemical reaction at constant volume* the basis of our calculations. Hence Q_T and the specific heats in equation (2) are to be taken at constant volume. The specific heat of an (ideal) gas at constant pressure, referred to one mol, is greater than the specific heat at constant volume by the quantity R (*i.e.* 1·98). If, for the sake of brevity, we substitute σ for the difference appearing in equation (2), first assuming that σ has the same value at all temperatures, and add the index v in order to show that this difference (σ) is taken at constant volume, we get—

$$U = Q_T = Q_0 + \sigma_v T \quad . \quad . \quad . \quad . \quad . \quad (7)$$

and from (1)—

$$A = Q_0 + \sigma_v T - q$$

We will now consider the "latent heat of reaction," q, from another standpoint. In the first lecture we discussed its relation to the temperature, and became familiar with the simple case (First Lecture, equation (26)) where— Latent heat and entropy.

$$\frac{q}{T} = \frac{q'}{T'}$$

If we denote the quotient $\frac{q}{T}$ by S, then, in general, we may write (1) and (7) as—

$$A = U - TS = Q_0 + \sigma_v T - TS \quad . \quad . \quad . \quad (7a)$$

Clausius, who introduced this term into the theory of heat, called it "entropy." Helmholtz[1] published a fundamental discussion of the *rôle* it plays in chemical reactions, and Planck[2] developed the subject systematically.

We found before that—

$$-q = T\frac{dA}{dT} \quad . \quad . \quad . \quad . \quad . \quad . \quad . \quad (8)$$

from which it immediately follows that—

$$-S = \frac{dA}{dT} \quad . \quad . \quad . \quad . \quad . \quad . \quad . \quad . \quad (9)$$

It therefore amounts to exactly the same thing whether we base the statements upon the peculiarities of the quantity S or of the quotient $\frac{dA}{dT}$, and only formal differences can incline us to use the one instead of the other. In this lecture we shall first make our calculation using the entropy S, because we can thus deduce the final formula in a more perspicuous manner. Later we shall obtain the same formula by another method, involving several steps and making use of the temperature coefficient of the maximum work, $\frac{dA}{dT}$.

At the outset, we will once more call to mind that $-q$ represents the heat taken up in an isothermal process yielding

[1] Helmholtz, "Zur Thermodynamik chem. Vorgänge," *Sitzungsberichte der Kgl. Preuss. Akademie, Berlin*, 1882, 1st half-vol., p. 22; and Ostwald, "Klassiker der exakten Wissenschaften," vol. 124; "Abhandl. zur Thermodynamik chemischer Vorgänge," von H. Helmholtz, edited with notes by M. Planck (Leipzig, 1902).

[2] Planck, "Thermodynamik." Leipzig, 1905.

its maximum amount of work as it progresses. If we imagine the heat to be resolved into an intensity factor and a capacity factor, S is the capacity factor of the heat in such a process. The quantity S, therefore, shares with the specific heat (mean specific heat between 0° and T°) the peculiarity that it is the quotient of a certain number of calories divided by the absolute temperature. But these two numbers of gram calories signify very different things. To determine the mean specific heat between 0° (absolute) and T°, we have, according to equation (4), merely to divide the heat required in raising the temperature by the temperature T. To determine the entropy, we must divide the heat used up in an isothermal process progressing with a maximum yield of work by the temperature T at which the process is taking place.

If a gas is made to expand isothermally to ten times its original volume—

$$-q = RTln10 = 1{\cdot}98 \times T \times 2{\cdot}3 = 4{\cdot}56T \text{ gram-cal.}$$

and consequently—

$$-S = 4{\cdot}56 \frac{\text{gram-cal.}}{\text{degrees}}$$

Since q represents heat given off by the system in changing, S also represents entropy given off, or a decrease of the entropy of the system. The above equation then says that under these conditions the entropy of a gas mol increases by 4·56 (S being negative). We are cognizant only of *changes* in entropy. We know nothing of the absolute amount of the entropy contained in a substance. We can merely ascribe some arbitrary value S_i to the substance before it undergoes a series of changes, then measure the entropy change, and so arrive at a value S_f which represents the entropy of the substance in the final state. We can also see, from the illustration just given, that a change in entropy does by no means involve a change in the specific heat, for the specific heat of the gas mol concerned is the same before and after the expansion to ten times the original volume.

To illustrate the conception of entropy, we may further observe that the entropy of a mol of water which is being converted into vapour at 100°, with an expenditure of 536 × 18

= 9650 gram-cal. in overcoming the pressure of the atmosphere, increases by $\frac{9650}{373} = 27 \cdot 8 \frac{\text{gram-cal.}}{\text{degrees}}$, while it would increase $\frac{10{,}926}{273} = 40 \frac{\text{gram-cal.}}{\text{degrees}}$ if the vaporization took place at 0°, where the heat of vaporization amounts to 18 × 607 gram-cal., and the maximum pressure to be overcome is only 4·5 mm.

After the foregoing explanations of entropy, we immediately see that in an infinitely small isothermal change, where the infinitely small maximum work dA is done, and the infinitely small quantity of heat dq is used up, the entropy change of the reacting system should have the value—

$$-dS = -\frac{dq}{T}$$

Let us now fix our attention upon a reversible, but not isothermal, process, where a maximum yield of work is secured from the heat consumed. We may consider the process to be resolved into an infinite number of infinitely small steps, in each of which T remains constant, or differs by only an infinitely small amount from its value in the preceding or succeeding step.[1] Then in every step the entropy change of the reacting system is—

$$-dS = -\frac{dq}{T}$$

and in all taken together—

$$-S = -\int\frac{dq}{T}$$

With this, the conception of entropy is explained sufficiently for our needs. Physics lends it still greater importance by considering what change the entropy experiences when a process takes place non-reversibly, that is, when it does not produce its

[1] It would be more accurate to say that the process may be resolved into a succession of two kinds of infinitely small steps. One kind consists in changes of temperature by expenditure or reception of work by the system, which are not connected with expenditure or reception of heat. Every one of these steps causes an infinitely small change of temperature without change of entropy. The second kind are isothermal steps, which fulfil the equation $-dS = -\frac{dq}{T}$. The entropy change of the whole process is given by $-S = -\int\frac{dq}{T}$, the first kind of steps being without influence upon the entropy change.

Entropy change in non-reversible processes.

maximum yield of work. As previously pointed out, it is a peculiarity of the maximum work that it is equal to the minimum work requisite to reverse the process. Physics, then, solves the problem of how much the entropy changes in the processes where the maxinum work is not done, by considering the minimum amount of work necessary to return the system to the initial condition. If this minimum work in the reverse change is coupled with an expenditure of heat q at T°, that is with the decrease of entropy $\frac{q}{T}$, we conclude that the entropy in the case of the forward process which did not yield a maximum amount of work has increased by the amount $-\frac{q}{T}$. There are many generalizations regarding changes of entropy. The best known of these is the principle of Clausius, which says that the entropy of the world tends toward a maximum. They are based, however, on non-reversible phenomena, and hence do not now concern us.

In order to find the entropy of a gas reaction by means of formula (7*a*), two points must be considered. This formula represents the maximum yield of work, not for any particular value of the temperature, but for all possible values. We must, therefore, first find out how the entropy of a single gas changes, if at all, with changes of its temperature and volume.

Further, at least two and usually several substances take part in a reaction, some of which are being used up, others formed. We must, therefore, determine the relation between the entropy of a mixture of gases and the entropy of the simple gases by themselves.

The entropy of a single gas and the entropy of a gas mixture.

In order to answer the first question, we revert to the idea that a process which is not isothermal may be imagined, so far as entropy change is concerned, as consisting of an infinite number of infinitely small steps, each one of which is isothermal. If the non-isothermal process takes place with a maximum of work, then in every single small step the infinitely small quantity of heat $-dq$ will be taken up by the system in changing, and, at the same time, its entropy will increase by the amount $\frac{-dq}{T} = -ds$. The entropy change in the total non-isothermal process will therefore be, as already mentioned—

$$-S = -\int \frac{dq}{T} \quad . \quad . \quad . \quad . \quad . \quad . \qquad (10)$$

Let us now imagine a reversible process of expansion to be carried out with an ideal gas, so that no heat is absorbed or given off. In such an "adiabatic" process every infinitely small amount of work done, pdv, must be accompanied by a corresponding decrease, dU, in the total energy of the gas. Since the gas does not change chemically, we are concerned only with the thermal and not the chemical part of its energy. If we at first assume that the specific heat of the gas at constant volume is independent of the temperature, this heat energy of the gas[1] equals c_vT, and the decrease dU is equal to $-c_vT$.[2] We thus get— **Adiabatic processes.**

$$dU = -c_v dT \quad . \quad . \quad . \quad . \quad . \quad . \qquad (11)$$

and since the work done, pdv, equals the decrease dU—

$$0 = c_v dT + pdv \quad . \quad . \quad . \quad . \quad . \qquad (12)$$

If the reversible process does not take place adiabatically, the quantities $-c_v dT$ and $+pdv$ are not equal. An infinitely small quantity of heat $(-dq)$ will then be taken up from the surroundings. We then get— **Non-adiabatic processes.**

$$-dq = c_v dT + pdv \quad . \quad . \quad . \quad . \quad . \qquad (13)$$

and dividing through by T—

$$-\frac{dq}{T} = c_v \frac{dT}{T} + \frac{p}{T} dv \quad . \quad . \quad . \quad . \quad . \qquad (14)$$

We replace p by $\frac{RT}{v}$, using the relation—

$$pv = RT$$

and writing for $\frac{dT}{T}$ and $\frac{dv}{v}$ the identical expressions $dlnT$ and $dlnv$ respectively, we get—

$$-\frac{dq}{T} = c_v dlnT + Rdlnv \quad . \quad . \quad . \quad . \qquad (15)$$

[1] Here we must take the specific heat at constant volume, c_v, and not at constant presure, c_p, for if we imagine the gas deprived of all its heat energy by being cooled to the absolute zero, only the amount of heat c_vT is taken out. In order to remove the amount c_pT we must allow the atmosphere to do work, compressing the gas to zero volume while it cools to the absolute zero of temperature; that is, we must introduce energy from without.

[2] It is to be remembered in this connection that, as pointed out in the first lecture, the experiments of Gay Lussac, Joule, and Thomson (First Lecture, p. 17) have demonstrated that the thermal energy of a gas does not depend on the volume.

If we now carry out this reversible non-isothermal process so that, starting from the initial temperature T_i and the initial volume v_i, we reach the final temperature T_f and the final volume v_f, the sum of all the infinitely small steps represents the difference between the final entropy S_f and the initial entropy S_i. That is—

$$S_f - S_i = -\int_{T_i v_i}^{T_f v_f} \frac{dq}{T} = c_v ln \frac{T_f}{T_i} + R ln \frac{v_f}{v_i} \quad . \quad (16)$$

If we introduce concentrations instead of the volumes occupied by a mol,

$$c = \frac{1}{v} \quad . \quad . \quad . \quad . \quad . \quad . \quad . \quad (17)$$

and we obtain— $S_f = S_i + c_v ln \frac{T_f}{T_i} - R ln \frac{c_f}{c_i}$

Choice of the initial condition.

We are free to define any state as the initial state. If we take as initial state the temperature $T = 1$[1] and the concentration $c = 1$, it follows that—

$$S = S_{\substack{T=1 \\ c=1}} + c_v ln T - R ln c \quad . \quad . \quad . \quad . \quad (18)$$

Here S signifies the entropy of a mol of gas at T° and the concentration c, both of which may have any value.

Separation of gases without expenditure of heat or work.

The second question may be settled by consideration of the fact that the separation of gases involves no expenditure of work, provided the concentration per unit of volume remains constant. This principle, first proved by Gibbs,[2] is a surprising one, because we know that gas mixtures, such as air, cannot be

[1] The fact that with this choice of an initial temperature we reach a higher temperature, not by expansion, but by compression, has no effect on the derivation. If one would convince himself of this, treat the non-isothermal reversible process as a compression and not as an expansion process. In this way, expression (13) with opposite signs is readily obtained, while at the same time the integral $-\int_{T_i v_i}^{T_f v_f} \frac{dq}{T}$ assumes the opposite sign, so that the result is the same as before.

[2] Gibbs, "Trans. Conn. Academy," vols. ii. and iii.; also "Thermodynamische Studien, Leipzig," translated by W. Ostwald (Leipzig, 1892).

separated into their components without the expenditure of work. But this is solely due to the fact that increase of volume under ordinary conditions involves expenditure of work against the atmosphere. If we imagine the atmospheric pressure to be removed, a suitable engine would carry out the separation of air into its constituents without expenditure of work or heat. **The separating apparatus.**

Thus, for instance, in a working space where vacuum would be maintained, a mixture of equal parts of hydrogen and oxygen could be separated into its component parts without the expenditure of heat or work, by means of the little arrangement here depicted.

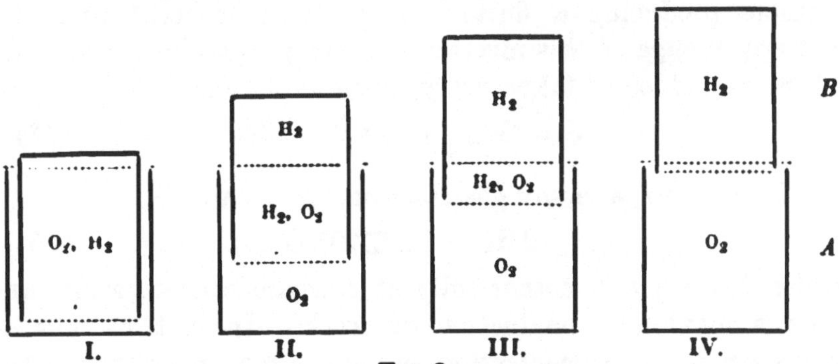

FIG. 3.

The arrangement consists of two boxes, A and B, whose walls are impermeable to gases on all sides except the cover of A and the bottom of B. The cover of A is permeable to hydrogen, but not oxygen, while the bottom of B is permeable to oxygen, but not to hydrogen.

The hydrogen exerts a lifting force on the cover of B, due to its partial pressure, equal to the force it exerts against the bottom of B. The oxygen has the same pressure on both sides of the bottom of B. If, then, the external pressure is zero, there is no position of the boxes where a force is acting in B which hinders or favours a displacement, and we can make the apparatus, which is assumed to be frictionless, pass through all four positions given in the diagram without losing or gaining a finite amount of work. Since a difference between pressure and counter-pressure never exists, it follows that the system will always be in equilibrium.

At the end of the operation, in position 4, each separate

gas has the same concentration as it had in the mixture in position 1. Since the process took place in the equilibrium condition and isothermally, and since no work was done either on or by the system, the latent heat of the process, and consequently the entropy change $\frac{q}{T}$, is zero. This amounts to saying that the entropy of the gas mixture is equal to the sum of the entropies of the separate components at the same temperature and concentration. It further follows from this that, when a mol of a gaseous component is abstracted from, or added to a very large quantity of gas mixture at constant temperature without producing a finite change in the concentration, the entropy change of this mixture is simply equal to the entropy of the gas added or taken away, and therefore is—

$$S = S_{\substack{T=1 \\ c=1}} + c_v lnT - Rlnc \quad . \quad . \quad . \quad (18)$$

Changes of entropy in gas reactions.

Let us now take any gas reaction, for instance—

$$2H_2 + O_2 \rightleftarrows 2H_2O \quad . \quad . \quad . \quad . \quad . \quad (19)$$

which takes place isothermally at constant concentrations and with a maximum production of work. In it two mols of hydrogen at the concentration c_{H_2}, and a mol of oxygen at the concentration c_{O_2}, disappear, while two mols of water-vapour are formed at a concentration c_{H_2O}. The entropy of the gas mixture thereby decreases by the entropy of $2H_2 + 1O_2$, and increases by the entropy of $2H_2O$. It therefore experiences the following changes:—

$$-2S_{H_2} = -2\left(S_{\substack{(H_2) \\ T=1 \\ c=1}} + c_{v(H_2)} lnT - Rlnc_{H_2}\right)$$

$$-S_{O_2} = -\left(S_{\substack{(O_2) \\ T=1 \\ c=1}} + c_{v(O_2)} lnT - Rlnc_{O_2}\right)$$

$$+2S_{H_2O} = +2\left(S_{\substack{(H_2O) \\ T=1 \\ c=1}} + c_{v(H_2O)} lnT - Rlnc_{H_2O}\right)$$

$$-S\,[1]) = -(2c_{v(H_2)} + c_{v(O_2)} - 2c_{v(H_2O)})lnT + Rln\frac{c^2_{H_2} \times c_{O_2}}{c^2_{H_2O}}$$

$$-\left(2S_{\substack{(H_2) \\ T=1 \\ c=1}} + S_{\substack{(O_2) \\ T=1 \\ c=1}} - 2S_{\substack{(H_2O) \\ T=1 \\ c=1}}\right) \quad . \quad . \quad . \quad . \quad (20)$$

[1] In regard to the signs, we should notice that, as pointed out before, q represents heat given off, and S a decrease in entropy. Therefore $-S$ is the increase in entropy in the reversible isothermal gas reaction. But this

The quantity enclosed in the brackets is nothing more than the difference of the specific heats of the disappearing and appearing substances, for which we have already used the symbol σ_v. If we use it here again, and collect in a single constant the entropies at T = 1 and $c = 1$ enclosed in parentheses, it follows from (20) that—

$$-S = -\sigma_v ln T + R ln \frac{c^2_{H_2} \times c_{O_2}}{c^2_{H_2O}} - k$$

If we substitute this value in equation (7*a*), we get—

$$A = Q_0 - \sigma_v T ln T + RT ln \frac{c^2_{H_2} \times c_{O_2}}{c^2_{H_2O}} + (\sigma_v - k)T. \quad (21)$$

Except, then, for the constant k, we know how the ability of our gas reaction to do work depends upon the reaction heat, the specific heats of the gases concerned, and the concentration at all temperatures and percentage compositions. **General formulation.**

The relation which we have deduced for this specific case of the formation of water may be similarly deduced for any desired gas reaction. It is only necessary to substitute the general expressions $c'_{\text{disappearing}}$, $c''_{\text{disappearing}}$, . . . , $c'_{\text{appearing}}$, $c''_{\text{appearing}}$, . . . , together with the number of molecules ν', ν'', . . . , for the special concentrations of hydrogen, oxygen, and water, and for the particular number of molecules which appear in the member—

$$RT ln \frac{c^2_{H_2} \times c_{O_2}}{c^2_{H_2O}}$$

The number of molecules of the disappearing substances must be taken as negative in this general mode of expression. We then obtain—

$$A = Q_0 - \sigma_v T ln T - RT \times \Sigma \nu' ln c' + (\sigma_v - k)T \quad (22)$$

Let us return to the process of the formation of water from the elements, and accordingly in this equation put for $c'_{\text{disappearing}}$ the value c_{H_2}, and for ν' its value -2, because, in accordance with the equation, $2H_2$ disappear in the reaction. For $c''_{\text{disappearing}}$ we put the value c_{O_2}, and for ν'' its value -1, because 1

increase consists of the entropy of two mols H_2O taken as positive, and the entropy of two mols H_2 plus one mol O_2 taken as negative.

mol O_2 is used up; for $c_{\text{appearing}}$ the value c_{H_2O}, and finally for ν''' we put $+2$, because 2 mols H_2O are formed. We then have—

$$A = Q_0 - \sigma_v T ln T - RT ln \frac{c^2_{H_2O}}{c^2_{H_2} \times c_{O_2}} + (\sigma_v - k)T$$

This is identical with equation (21).

The expression of Helmholtz.

Formula (22) has already been in part obtained by Helmholtz (*l.c.*). That is, in his fundamental paper on the subject, Helmholtz developed an expression which, when written in our notation, runs as follows :—

$$A - A_1 = (\sigma_v - S_1)(T - T_1) - \sigma_v T ln \frac{T}{T_1} \quad . \quad (23)$$

In it he made no further assumptions than that the process in question yielded its maximum amount of work, that the difference in specific heat of the substances appearing and disappearing had the same value σ_v at all temperatures, and that the ratio of the volumes or the concentrations could be regarded as unchangeable. If in this equation we put $T_1 = 1$, and understand by A the maximum work at 1° abs., and by S_1 the change of entropy under these conditions, we get from (23)—

$$A = A_1 + S_1 - \sigma_v - \sigma_v T ln T + (\sigma_v - S_1)T$$

But now, if Q_1 signifies the reaction heat at 1° abs.—

$$A_1 + S_1 = Q_1$$

and according to (7)—

$$Q_1 - \sigma_v = Q_0$$

that is—

$$A = Q_0 - \sigma_v T ln T + (\sigma_v - S_1)T$$

If we compare formula (22) with this, we see that both become identical when—

$$S_1 = k + R\Sigma\nu' ln c'$$

S_1 signifies the change in entropy at T = 1° and at any given concentration of the gases taking part in the reaction, *k* the same quantity where both T and the concentrations equal 1 and the gases obey the laws for ideal gases. By using the general term S_1, Helmholtz avoided any special assumption regarding the laws which govern changes of concentration. He thus gave his expression a more general form, but at the same time one less directly applicable to our particular problems. Helmholtz certainly knew it was possible to derive the expression—

$$k + R\Sigma\nu' ln c'$$

from S_1 by means of the gas law—

$$pv = RT$$

In his "Dritter Beitrag zur Thermodynamik chemischer Vorgänge,"[1] he makes a special application of this, and in his lectures he develops in a more precise way (on the basis of the gas law) the influence of concentration. Helmholtz did not give our expression (22), simply because it had no immediate value in the consideration with which he was busied. The extraordinary stimulus given by Van't Hoff to the study of the influence of concentration in systems which obeyed the gas law, directed general attention to those phenomena whose treatment requires our more specialized formula (22).

The indeterminate thermodynamic constant *k*.

If, now, we examine the constant k, we can, in the first place, make its meaning somewhat clearer if we consider that the latent heat of the reaction is identical with TS. This product, however, when $T = 1°$ abs., is equal to the entropy. Now, we have seen that the constant k is equal to the change in entropy of a reaction proceeding reversibly at concentrations which are all equal to 1, and at the temperature of 1° abs. Therefore we can also call the constant k the latent heat of the reaction under these conditions. Thermodynamics tells us nothing more about this constant. Consequently, the latent heat of a gas reaction taking place reversibly, at concentrations all equal to 1, and at the temperature $T = 1$, remains theoretically indeterminate. Let us see what further information we can gather about this quantity.

Use of Berthelot's principle to determine *k*.

To take a short cut, calling the principle of Berthelot to our assistance, we may reason somewhat as follows: At absolute zero, Berthelot's principle is recognized as correct, but the higher the temperature the less does it suffice. However, at 1° abs. it ought to be very nearly true. We may, therefore, write—

$$A_{T=1} = Q_{T=1}$$

[1] Ostwald, Klassiker 124, "Abhandl. zur Thermodynamik chem. Vorgänge von H. Helmholtz," p. 63.

Since at 1° abs.—

$$Q_{T=1} = Q_0 + \sigma_v$$

it follows that—

$$A_{T=1} = Q_0 + \sigma_v \quad . \quad . \quad . \quad . \quad . \quad . \quad (24)$$

In formula (22) we now put the term $\Sigma\nu' lnc'$ equal to zero, and T equal to 1. The first substitution simply means that all concentrations are equal to 1. It then follows that—

$$A_{\substack{T=1\\c=1}} = Q_0 + \sigma_v - k \quad . \quad . \quad . \quad . \quad (25)$$

We now compare (24) with (25), and conclude that k is zero. This is, however, a very arbitrary procedure. Equation (24) will be approximately true, whether we choose the concentrations as all just equal to 1, or somewhat different, for only at extreme differences in concentrations does the term $R\Sigma\nu' lnc'$ attain a value at all comparable with Q_0. On the other hand, a ratio x of concentrations must exist at which—

$$A_{\substack{T=1\\c=x}} = Q_0 + \sigma_v$$

But x may have a value other than 1. We cannot, therefore, decide in this way whether a perceptible difference between A and Q does not exist, though small in comparison with Q_0. But, unfortunately, this is a crucial point. For although we can certainly neglect k at a temperature of 1° abs., this does not help us at all, for we never observe gas reactions at such low temperatures. But if the observations are carried out at 2000° abs., kT is two thousand times greater than k at 1°, and we do not know whether we can then neglect kT or not.

Other attempts at a determination of k.

If we examine the matter more closely, we shall be inclined to assume rather that A does not change greatly between 0 and 1°, and to write—

$$A_1 = A_0 = Q_0$$

from which it follows that $k = \sigma_v$.

If we picture the substances as containers of heat, and imagine that a diminution in the heat capacity σ_v, in the course of a reaction, arises from the fact that a part of the total energy of the reaction is used up in compressing those substances and in squeezing heat out of them, it follows that only the part of this total energy remaining after the atomic compression will be available for the outside work.[1] In the simplest case where $T = 1$, this remaining part, which is the reaction energy, may have the value—

$$A_1 = Q_1 - \sigma_v$$

that is Q_0. Yet even this third assumption is not very satisfactory. Let us, therefore, examine the subject still more closely.

The first point to be noted is that the numerical value of the constant k is affected by the choice of the unit of concentration, if $\Sigma\nu'$ is not equal to zero. It is easy to see the connection between the two, for the term $RT \times \Sigma\nu' lnc'$ in (22), as well as k, depends upon this same choice. If we express the ratio of concentrations in a given mixture of gases, first in mols per litre, c, and then in mols per cubic centimetre, C, then— The relation of the constant k to the size of the temperature degree.

$$(R \times \Sigma\nu' lnc' + k)T = (R \times \Sigma\nu' lnC' + k')T$$

for otherwise our formula for the reaction energy of a gaseous mixture would give different results depending on how we reckoned our concentrations, which is of course impossible. If we call s the ratio of the two units of concentration—

$$k_{(\text{mol : litre})} = k' + R \times \Sigma\nu' lns$$

If $\Sigma\nu'$ equals zero, the constant k is independent of the unit of concentration. We will study this case more minutely. Let us here remember that while thermodynamics fixes the location of the absolute zero, it does not fix the value of the interval of temperature which we call a degree. It is mainly due to chance that thermometry has adhered to the suggestion made by Celsius in 1742, that the distance between the boiling-point of water, and the freezing-point of ice be divided into 100 parts. Réaumur had previously (1730) suggested that we divide the same interval into 80 parts. As far as the principle is concerned, it could just as well be divided into some other number of parts. Let us suppose that we have a degree a thousand times smaller than the one now in use, and let us designate temperature measured in these new degrees by τ. In the first place, the gram-calorie would become a thousand The degree interval of the temperature scale considered for the case that $\Sigma\nu' =$ zero.

[1] It seems to me that this idea forms the basis of T. W. Richards' study of the relation between atomic volume, compressibility, and free energy (*Z. f. phys. Chemie*, 40 (1902), 169 and 597; also 42 (1903), 129).

times smaller than the one now used, for it would represent the quantity of heat necessary to heat one gram of water from 15° to 15·001° on the old scale of temperature. Reaction heats would for this reason be represented by a number of the new calories a thousand times greater than that of the old. The value of the specific heats would not be altered, because it is the quotient of calories over degrees, and the numerator and the denominator would change equally. The quantity R in absolute units (erg × T^{-1}) would be a thousand times smaller, but in heat units would remain the same, because the work-equivalent of the calorie by which we divide in order to convert R into heat units, is likewise a thousand times smaller. The concentration is not affected by this change in the unit of temperature.

Influence of the size of a degree on the constant *k*.

Now, of course, we see that the energy of a reaction cannot depend on what value we may choose to call a degree interval. Any given reaction which, for instance, takes place at the boiling-point of sulphur must have one and the same reaction energy whether we use the τ or T scale for reckoning this temperature. We must always bear in mind that both scales differ from one another only by a proportionality factor, and become identical at absolute zero. We have just used $\frac{1}{1000}$ as an example of such a proportionality factor. In general we may call it *a*. Then—

$$T = \tau \times a$$

If we use the τ scale instead of the T scale, the reaction energy remains identical only when we measure it in absolute units (ergs). Expressed in calories it is different, because the calorie is *a* times as small. We can reduce the value of the reaction energy expressed in heat units on the τ scale to heat units on the T scale by the relationship—

$$A_{(T)} = aA_{(\tau)}$$

Similarly—

$$Q_{(T)} = aQ_{(\tau)}$$

If we write our equation (22) first in units of the T scale, and then in units of the τ scale, we have—

$$A_{(T)} = Q_{0(T)} - \sigma_v T ln T - RT \times \Sigma\nu' lnc' + (\sigma_v - k_{(T)})T \quad (22)$$

$$A_{(\tau)} = Q_{0(\tau)} - \sigma_v \tau ln \tau - R\tau \times \Sigma\nu' lnc' + (\sigma_v - k_{(\tau)})\tau \quad (26)$$

Let us now reduce (26) to (22) by the help of the relationship, just considered, between $A_{(T)}$ and $A_{(\tau)}$, $Q_{(T)}$ and $Q_{(\tau)}$, and between τ and T themselves. To do this we multiply (26) by a, and replace $aA_{(\tau)}$ by $A_{(T)}$; $aQ_{0(\tau)}$ by $Q_{0(T)}$, and $a\tau$ by T, and get—

$$A_{(T)} = Q_{0(T)} - \sigma_v T ln \frac{T}{a} - RT \times \Sigma\nu' lnc' + (\sigma_v - k_{(\tau)})T$$

If we compare this with (22), it follows that—

$$k_{(T)} = -\sigma_v lna + k_{(\tau)}$$

Thus we see that the value of the indeterminate thermodynamic constant depends on the value of the degree we choose. If we choose it a thousand times smaller than the Celsius interval, the constant is greater by $-\sigma_v ln 0{\cdot}001$, *i.e.* $+6{\cdot}9\sigma_v$. It would be especially advantageous if we could so determine a that the constant became just equal to σ_v. The term $(\sigma_v - k)$ would then completely disappear, and our expression would take the form—

$$A = Q_0 - \sigma_v ln\tau - R\tau \times \Sigma\nu' lnc' \quad . \quad . \quad (27)$$

Perhaps the meaning of the τ scale will be still clearer from the following discussion of the formula (27). According to equation (27), A becomes equal to Q_0 at the absolute zero. But it would be equal to Q_0 a second time when $\Sigma\nu' lnc'$ equalled zero and τ equalled 1. Now, by a suitable choice of concentrations, the action of mass, represented by the term $R\tau\Sigma\nu' lnc'$, may be eliminated, so that this term becomes zero. We then determine at what temperature other than absolute zero the reaction energy becomes equal to Q_0, and call the interval between this temperature and the absolute zero one degree.

T. W. Richards' proof that k is zero in galvanic cells.

The question first arises whether, after all, such a procedure is in general possible. If we take the case where there is no difference between the specific heats of factors and products, $\sigma_v lna$ is also equal to zero, and the thing we have attempted to do seems impossible unless k, in this case is equal to zero. This question cannot be definitely settled from a study of gas reactions, because there is no gas reaction known where σ_v is beyond all question equal to zero at all temperatures. We can, however, glean the desired facts from another field. T. W. Richards has shown[1] that the heat of reaction and the

[1] Th. W. Richards, *Z. f. phys. Chemie*, 42 (1903), 137 ff.

reaction energy in galvanic cells are equal when the concentrations of factors and products are equal, and where no change in heat capacity ensues during the reaction. We conclude from this that in gas reactions too, for the ideal case where $\sigma_v = 0$, equation (22) would become—

$$A = Q - RT \times \Sigma\nu' lnc'$$

without any term kT making its appearance. Gas reactions do not contradict this conclusion, as we shall see in the Fourth Lecture, although our knowledge may not be sufficient to demonstrate that they support it. Thus, if no differences of heat capacity exist, and mass action is excluded by choosing concentrations such that—

$$R \times \Sigma\nu' lnc' = 0$$

we can see no good reason why heat and work should be different. Yet if k and the term $\sigma_v ln$T either both have a finite value or both equal zero, the conclusion is justified that everything depends on the units in which we measure the term $\sigma_v ln$T. We saw before that for every isothermal and reversible gas reaction where σ_v had the same value at all temperatures, we could so define our temperature interval that $(\sigma_v - k)$ would become equal to zero. We now conclude that for all reactions of this class where the number of reacting molecules does not change during the reaction, one and the same temperature interval or degree would fulfil equation (27). We will call this temperature interval a "chemodynamic" degree.

The chemodynamic and the Celsius degree.

Starting from the deductions of T. W. Richards, Van't Hoff[1] has recently discussed similar relations. In those galvanic cells, and consequently those reactions, where the number of mols on each side of the reaction equation is the same ($\Sigma\nu' = 0$), he puts the term $(\sigma_v - k)$ equal to zero, or, speaking more accurately, the term in his very different derivation corresponding to it. He did not consider the effect of the magnitude of the temperature degree on the quantity k. But since, by choosing the Celsius degree, he got an agreement between calculation and observation, it follows that the "chemodynamic" degree, which makes $(\sigma_v - k)$ equal to zero, does not differ very much from a

[1] "Festschrift," dedicated to Ludwig Boltzmann (Leipzig, 1904), p. 233.

Celsius degree. If this be the case, using the Celsius degree, we may write as an approximation, at least—

$$A = Q_0 - \sigma_v T ln T - RT \times \Sigma \nu' ln c' \quad . \quad . \quad (28)$$

It will be recalled that it is the logarithm of the ratio a between the two temperature intervals or degrees, and not the ratio itself, which influences the value of k. The approximation (28) can therefore well be a very close one, even though the two kinds of temperature degrees are distinctly different. Since, further, the effect of the temperature interval chosen depends on the product $\sigma_v ln a$, the approximation would be especially close when the σ_v concerned was small.

Possibility of experimentally determining the value of the temperature interval.

It would be of great value if we could accurately evaluate k and σ_v in a single favourable instance. If the number of reacting mols did not change during the reaction, we could determine from it the value of the chemodynamic degree. If we then investigated a case where the number of reacting mols did change during the action, we could tell by determining k and σ_v in what units we must reckon concentrations in order to make $(\sigma_v - k)$ disappear. Were we once in possession of this value, we could predict the value of the reaction energy for any given mixture of gases, knowing the reaction heats and the specific heats. There are, to be sure, certain tacit assumptions made in this statement. We assume, in the first place, that the gases follow the ideal gas laws with rigour, as is surely the case if the temperature be high enough. In the second place, we assume that the difference in the specific heats of factors and products is the same at all temperatures. The latter assumption is certainly by no means fulfilled.

Alteration of the formula for $\sigma_v = f(T)$

We must, therefore, now undertake to alter our formula so that the change of the difference of the specific heats with the temperature will be taken into account.[1] Here we are con-

[1] The following is a very short derivation of the relationships so far obtained. Let k be the change of entropy in a reversible isothermal reaction when T equals 1 and all the concentrations equal 1; that is, let k be the latent heat under such conditions. Then—

$$A_1 = Q_0 + \sigma_v - k$$

Let us imagine the factors to be brought adiabatically from $T = 1°$ to a higher temperature, and then allowed to react isothermally. The products of the reaction could then be cooled down adiabatically to $T = 1°$. The entropy

fronted with the difficulty that we are only able to deduce the empirical formula

$$c = a + bT + cT^2 + \ldots$$

to express the effect of temperature on the specific heats, and that our observations are hardly accurate enough to allow us to determine more than the first two constants, a and b.

Assumption regarding the mean specific heats.

We must therefore content ourselves with representing the mean specific heats of each gas taking part in the reaction by an expression—

$$c_v = a + bT$$

We can then write for the mean specific heats of all the factors (all at constant volume)—

$$c_{\text{factors}} = \Sigma a + T\Sigma b$$

and similarly for the products—

$$c_{\text{products}} = \Sigma a^{\cdot} + T\Sigma b^{\cdot}$$

It therefore follows, from analogy with (2), that the reaction heat at absolute zero—

$$Q_T = Q_0 + T(\Sigma a - \Sigma a^{\cdot}) + T^2(\Sigma b - \Sigma b^{\cdot}) \quad . \quad (2a)$$

change at the higher temperature is still k, for the entropies of all gases are constant during adiabatic changes. The energy at T is then—

$$A = Q_0 + \sigma_v T - kT$$

On the other hand, the difference σ_v of the specific heats is connected with the values of the concentrations at the higher temperature T by the law governing adiabatic changes, namely—

$$-\sigma_v lnT = R \times \Sigma\nu' lnc'$$

It follows from this that—

$$A = Q_0 + (\sigma_v - k)T - \sigma_v T lnT - RT \times \Sigma\nu' lnc'$$

Any desired isothermal and reversible change of concentrations of the gases only affects the value of A on the left-hand side of this equation and the value of lnc' in the last term on the right-hand side. Q_0 and σ_v are considered independent of concentration and temperature, while k depends on the value of the temperature interval and on the concentration unit. If now $\Sigma\nu'$ equals zero, and the temperature degree is so chosen that σ_v equals k at 1° abs., $(\sigma_v - k)$ drops out, and we get—

$$A = Q_0 - \sigma_v \tau ln\tau - R\tau\Sigma\nu' lnc'$$

The choice of τ has nothing to do with thermodynamics, but rests on the experience or assumption that (for $\Sigma\nu' = 0$) σ_v and k appear and disappear at the same time, and on the deduction from this that k only depends on the value of the temperature degree if $\Sigma\nu'$ equals zero.

As before, we put—

$$\Sigma a - \Sigma a' = \sigma'_v,$$

and—

$$\Sigma b - \Sigma b' = \sigma''$$

and get instead of (7) the expression [1]—

$$U = Q_T = Q_0 + \sigma'_v T + \sigma'' T^2 \quad . \quad . \quad . \quad (7a)$$

The effect of our changed assumption regarding specific heats upon the changes of entropy in chemical reactions may easily be seen from the results of our previous discussion. In the first place, we should now represent the heat energy of a gas by—

$$aT + bT^2$$

The decrease in the total energy which we represented in equation (11) now takes the form—

$$dU = -d(aT + bT^2) = -adT - 2bTdT \quad . \quad (11a)$$

Similarly, from equation (12) we get—

$$0 = adT + 2bTdT + pdv \quad . \quad . \quad . \quad (12a)$$

from (13)—

$$-dq = adT + 2bTdT + pdv \quad . \quad . \quad . \quad (13a)$$

and finally from (14)—

$$-\frac{dq}{T} = a\frac{dT}{T} + 2bdT + \frac{p}{T}dv \quad . \quad . \quad . \quad (14a)$$

Here let us pause a moment to consider whether we are justified in introducing the gas law (at this point), and so to pass to equation (15*a*), and at the same time consider the specific heats as variable with the temperature where thermodynamics elsewhere assumes them to be constant. But, according to all we know, actual gases obey the gas law very well at high temperatures, although it is precisely there that their specific heats are

[1] The term σ'' does not need the index v, since it has the same value whether we mean specific heat at constant volume or constant pressure. This is due to the fact that the difference of the specific heats has the constant value R, which is independent of the temperature, or more precisely, independent of it so long as the gases behave ideally, that is, obey the expression—

$$pv = RT.$$

E

known to vary with the temperature. We need not hesitate, therefore, to make use of the relationship—

$$pv = RT$$

Continuing, then, (15) becomes—

$$-\frac{dq}{T} = adlnT + 2bdT + Rdlnv \quad . \quad . \quad (15a)$$

and (16)—

$$S_f - S_i = aln\frac{T_f}{T_i} + 2b(T_f - T_i) + Rln\frac{v_f}{v_i} \quad . \quad (16a)$$

But then (18) becomes—

$$S = S_{\substack{T=1 \\ c=1}} + alnT + 2b(T - 1) - Blnc \quad . \quad (18a)$$

The further development remains the same, and leads to the final result—

$$A = Q_0 - \sigma'_v TlnT - \sigma''T^2 - RT \times \Sigma\nu' lnc' + (\sigma'_v + 2\sigma'' - k)T \quad (22a)$$

Relation of equation (22a) to Helmholtz's more general formula.

We could have obtained the same expression from a general formula given in Helmholtz's fundamental paper. This formula, expressed in our notation, is—

$$A = Q_1 - TS_1 + \int_{T_1}^{T} xdT - T\int_{T_1}^{T} xdlnT$$

Here x is the difference between the true specific heats of factors and products at T°.

If we substitute for it—

$$x = \sigma'_v + 2\sigma''T$$

we get, putting $T_1 = 1$—

$$A = Q_1 - TS_1 - \sigma'_v TlnT - \sigma''T^2 + (\sigma'_v + 2\sigma'')T - \sigma'_v - \sigma''$$

If, now, we replace the reaction heat at the temperature 1° absolute by $Q_0 + \sigma'_v + \sigma''$, corresponding to the circumstance that according to our definition $\sigma'_v + \sigma''$ represents the mean specific heats between absolute zero and 1°, we get, after a slight transformation—

$$A = Q_0 - \sigma'_v TlnT - \sigma''T^2 - TS_1 + (\sigma'_v + 2\sigma'')T$$

When we introduce the gas law as before, and put—

$$S_1 = k + R \times \Sigma\nu' lnc'$$

this goes directly over into (22a).

We can repeat here (22a) all the observations made in

connection with equation (22). If the effect of mass is eliminated by a suitable choice of concentrations, so that the term $\Sigma\nu' lnc'$ becomes equal to zero, and if T is then taken as 1, we get—

$$A_1 = Q_0 + \sigma'_v + \sigma'' - k$$

We can again show that the value of k depends upon the temperature interval chosen, which in this case is connected with σ'_v in the same way as explained above. We conclude, on the basis of our former considerations, that the Celsius degree is nearly enough equal to a chemodynamic degree to make $\sigma'_v + \sigma'' - k$, when the number of molecules does not change during the reaction, if not exactly equal to zero, still very near it. We shall, therefore, in general use the expression—

Again the influence of the temperature interval.

$$A = Q_0 - \sigma'_v T lnT - \sigma'' T^2 - RT \times \Sigma\nu' lnc' + \text{const.}\, T \quad (28a)$$

in our further discussion, assuming for the particular case where $\Sigma\nu'$ is equal to zero that the constant, though not equal to zero, is still very small. For the case in which the number of reacting molecules changes during the reaction, we shall make the constant assume a small value by writing the formula in the way explained on p. 53.[1] We thus free ourselves, as well as the present state of our knowledge permits, from the troublesome constants which hinder our passing from quantities of heat to quantities of energy. The way in which we can know and eliminate them more completely has already been pointed out. We will append a brief consideration which increases the probability that the constant vanishes in case $\Sigma\nu' = 0$. Le Chatelier[2] made the noteworthy observation that the true specific heats of nearly all gases and vapours under constant pressure can be represented by an expression of the form—

Relation to Le Chatelier's law of the true specific heats.

$$c_p = 6{\cdot}5 + \alpha T$$

at least as a first approximation. As far as the gases follow the simple gas law, we can expect the specific heats at constant volume, smaller by R (1·98) than the specific heats at constant pressure, to show a similar convergence. We conclude, from this rule, that in those reactions where the number of reacting molecules does not change, σ'_v, if not equal to zero, is still more

[1] This, of course, is purely formal, the importance of k, relative to the reaction energy, not being influenced by any method of writing the formula.

[2] *Compt. Rend.*, 104 (1887), 1780.

nearly equal to it than it is in the reactions where the number of reacting molecules does change. In case $\Sigma\nu' = 0$, we get as a first approximation—

$$A_1 = Q_0 + \sigma'' - k$$

Now, σ'' is always a small number. If k on the τ scale equals σ'', and if the "chemodynamic" τ scale does not differ greatly from the T scale, k will differ only slightly from σ''. So on this ground too, using the Celsius degree instead of the chemodynamic one, we are justified in calling the quantity $(\sigma_v + \sigma'' - k)$ equal to zero.

Use of partial pressures instead of concentrations.

It still remains for us to transform our expressions applying at constant concentrations into such as shall apply at constant partial pressures. It was shown in the First Lecture that partial pressures expressed in atmospheres, and concentrations expressed in mols per litre, are connected by the equation—

$$p = 0{\cdot}0821 \mathrm{T} c$$

Substituting p for c in (28a) on this basis—

$$\Sigma\nu' ln c' = \Sigma\nu' ln \frac{p'}{0{\cdot}0821 \, . \, \mathrm{T}} = \Sigma\nu' ln p' - \Sigma\nu' \, . \, ln \mathrm{T} - \Sigma\nu' \, . \, ln 0{\cdot}0821$$

and therefore—

$$-\mathrm{RT}\Sigma\nu' ln c' = -\mathrm{RT}\Sigma\nu' ln p' + \mathrm{RT}\Sigma\nu' \, . \, ln \mathrm{T} + \mathrm{RT}\Sigma\nu' \, . \, ln 0{\cdot}0821$$

But now, because specific heats at constant pressure and at constant volume differ by R—

$$\sigma'_v - \mathrm{R}\Sigma\nu' = \sigma'_p$$

At first sight the sign in this expression seems strange. We should expect that the quantity $\mathrm{R}\Sigma\nu'$ would add itself to σ_v to give the difference, σ_p, of the specific heats at constant pressure. But according to our definition, we must call the number of molecules, ν, of the factors, that is, of the disappearing substances, *negative*, while, on the other hand, we have defined the difference of the specific heats, σ'_v, in such a way that the specific heats of factors are positive and those of the products are negative. This difference of signs necessitates subtracting $\mathrm{R}\Sigma\nu'$ from σ'_v. Therefore (28a) becomes—

$$A = Q_0 - \sigma'_p \mathrm{T} ln \mathrm{T} - \sigma'' \mathrm{T}^2 - \mathrm{RT} \times \Sigma\nu' ln p' + \text{const.}'' \times \mathrm{T} \quad (29)$$

The term const." is an abbreviation for (const. $-4{\cdot}95\Sigma\nu'$).[1]

[1] The quantity $-4{\cdot}95$ is nothing but the product R(ln0·0821), that is, 4·56 log 0·0821. Had we reckoned concentrations in mols per cubic

We shall later use formula (29) in all cases where the number of reacting molecules changes during the reaction, and avail ourselves of the device of choosing the unit of reacting molecules as small as possible, in order to make the change in the number of molecules during the reaction as small as possible. For instance, instead of—

$$2H_2 + O_2 = 2H_2O$$

we write—

$$H_2 + \tfrac{1}{2}O_2 = H_2O$$

This method has a certain practical significance which will be discussed at the beginning of the Fourth Lecture. As a result of this, the constant (const.") of formula (29) in every case is expressed by a small number.

The main point to observe is that, on the basis of the gas law, partial pressures have been substituted in an equation deduced for a reaction taking place at constant volume. *This merely formal transformation changes nothing in the derivation.* As before, the external pressure plays no *rôle*. **Significance of A when partial pressures are used.**

The quantity A which we obtain by means of formula (29) represents, as before, the work done at constant volume.

But is this the work which we actually wish to determine? In view of the fact that most of our gas reactions take place at the constant pressure of one atmosphere, must we not make a corresponding change in our formula? Examining an actual case, and seeing what the maximum work is which we wish to determine, we can answer this question in the negative. Two typical examples will illustrate this. We will first take the formation of water at the equilibrium point. According to our fundamental conception, the reaction energy is here zero. But this is only true when the external work is *not* considered, for the formation of water at the equilibrium point under atmospheric pressure is accompanied by a decrease in volume. The atmosphere does the work pv on our system for every mol $(-2H_2-1O_2+2H_2O)$ which disappears. The work done by our system at equilibrium and under constant pressure therefore has the (negative) value $-pv$. If, then, we are to hold strictly to our definition according to which the maximum work at the

centimetre, we should then have obtained const. + $4{\cdot}56\Sigma\nu'$ log $82{\cdot}1$ = const.", that is, const. + $8{\cdot}71\Sigma\nu'$.

equilibrium equals zero, it follows as an immediate consequence that work against the constant pressure of the surroundings cannot be counted; that is, that the reaction must be considered as taking place at constant volume.

Change produced in formula by a different definition of A.

But suppose we did count the work done against a constant external pressure, what effect would it have on our formula? A glance at our fundamental equation $A = U - q$, which at equilibrium becomes $0 = U - q$, shows us the answer to this question. If the work done against the constant external pressure is to be counted, we must, as we have already seen, add the term $pv\Sigma\nu'$ to the left-hand member. But, as we saw at the beginning of this lecture, the change in the total energy is also greater, by $pv\Sigma\nu'$, at constant pressure than at constant volume. If, therefore, we choose to change our definition of the equilibrium and add to reaction energy A the quantity $pv\Sigma\nu'$, we must also add to the right-hand member of expression (29) for the reaction energy, the equivalent of $pv\Sigma\nu'$, that is, $RT\Sigma\nu'$.

If we are not actually determining the equilibrium conditions of the gases, but are measuring their reaction energy, we make use of gas elements, as above mentioned. The oxyhydrogen cell described in the Fifth Lecture belongs to this class. It furnishes the maximum electrical energy A_e, from the combination of hydrogen with oxygen. If it is working at constant pressure, the atmosphere will do, in addition to this, the work pv on the system, since one mol disappears during the reaction

$$(-2H_2 - O_2 + 2H_2O)$$

Counting this in, the maximum work equals $A_e - pv$, or in general equals $A_e + pv\Sigma\nu'$. If this term is defined as the maximum work of the cell, then in equation (29) the right-hand side must be increased by $RT\Sigma\nu'$. It is, however, much simpler to call A_e the maximum work, and to use the equation without change.

All known cases can be grouped under these two examples, and we are therefore justified in always neglecting the work done against the atmosphere. Formulæ (28*a*) and (29) become identical when $\Sigma\nu'$ equals zero, that is, when the number of molecules on both sides of the chemical equation is the same; for in that case $\Sigma\nu' lnc'$ is equal to $\Sigma\nu' lnp'$.

THIRD LECTURE

ANOTHER DERIVATION OF THE FORMULA PREVIOUSLY OBTAINED, AND ITS BEARING ON REACTIONS BETWEEN SOLIDS

WE became familiar in the first lecture with the expression—

$$\frac{d\left(\frac{A}{T}\right)}{dT} = -\frac{Q_T}{T^2} \quad . \quad . \quad . \quad . \quad . \quad . \quad (1)$$

Q_T was the reaction heat at constant volume and at the temperature T. From it we can now derive the formulæ which we deduced from the entropy principle in the second lecture, provided we can succeed in expressing the maximum work A as a function of the concentrations.

Van't Hoff[1] has solved this problem with the help of a reaction space which we shall call the "equilibrium box" (Fig. 4).

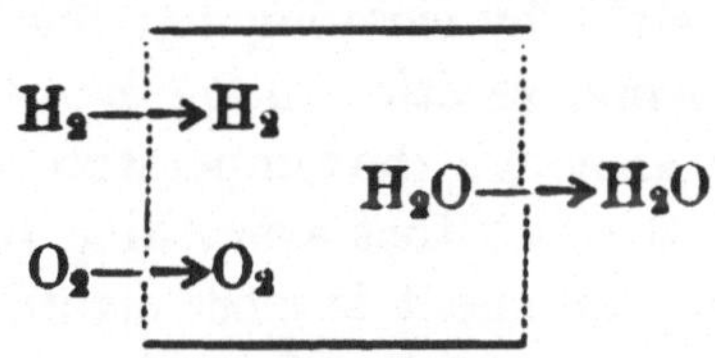

FIG. 4.

Chemical equilibrium is assumed to always prevail in this box. The substances disappearing in the reaction are introduced through the side wall on the left, while the products of the reaction are removed through the wall on the right. That is, the left-hand wall is only permeable to the factors, the right-hand wall to the products of the reaction—as illustrated in Fig. 4. In

[1] "Die Gesetze des chemischen Gleichgewichts," by J. H. van't Hoff, Ostwalds *Klassiker*, No. 110, edited by G. Bredig. Leipzig, 1900.

this way we can imagine any desired quantity of the gases reacting without gain or loss of work, provided only that the substances be put in or taken out at the same concentration as they had in the equilibrium mixture. If we know the concentrations at the equilibrium point, it is not difficult to determine what the reaction energy would be at any other concentrations.

The roundabout way through the equilibrium box as a means of determining A.

Thus we need only imagine that a substance which is to react is brought by an isothermal change of volume, with the production of the maximum amount of work, to the concentration which it has in the equilibrium box. We then imagine it to be introduced into the equilibrium box, and to react there without gain or loss of work. The products of the reaction are similarly and simultaneously removed from the equilibrium box, and are brought back to the original concentration by a similar isothermal and reversible process.

The net result of this process is the same as if we had carried out the reaction reversibly at the given temperature without any equilibrium box and without any changes of concentration.

The maximum work is independent of the path.

The maximum work cannot depend on the path we take in passing from the initial to the final state. If this were the case, it would be possible, by carrying out the direct reaction in one way and the reverse reaction in another, to have some work left over at the end, although the temperature had not changed. And this in spite of the fact that everything returned to its original condition. This statement is most significant. But it does not have its roots, as we might at first think, in the law of the conservation of energy, for we see, on closer inspection, that this work need not necessarily have been created out of nothing. It could have been generated from heat which was transformed into work at constant temperature without any other accompanying change. Were such a process possible, it would by no means invalidate the law of the conservation of energy. Let us see why, in spite of this, our experience declares it to be impossible.

The perpetuum mobile of the second kind.

For this purpose let us admit for the moment that heat can be turned into work at constant temperature without any permanent change. Such an assumption would mean that it should be possible to invent some sort of a motor boat which could derive its heat from the ocean, convert it into work, and then, through the agency of friction, return it again to the surrounding ocean. Such a boat would be just as much a perpetual motion machine as any apparatus which created work out of nothing. Ostwald[1] has called this a perpetual-motion machine of the second kind.

Experience tells us that this is just as impossible as a perpetual-motion machine of the first and better-known kind, in which work is created out of nothing. If the impossibility of all forms of perpetual-motion machines of the first kind is referred back to the law of the conservation of energy, we may, with Boltzmann,[2] recognize the source of our conviction that perpetual motion of the second kind is impossible in the *invariable direction of time*. That is, our inner consciousness can never conceive of time as progressing backwards. If we question why, on waking from sleep, we are so sure that we always are older and never younger, we can only answer that, in our world, things transpire in a fixed direction as far as time is concerned. Direction in all spacial things requires a special testing. If we fall asleep in a coach facing forward, we are by no means sure on waking that we have gone forward. The coach may have stood still, or may even have turned about and gone in the opposite direction. A fixed direction does not, then, exist in spacial things. The consciousness of a fixed direction of time is a result of experience based chiefly on the observation that whatever happens in the world of reality does not of itself completely revert to its earlier condition; it never retraces its steps. A swinging pendulum, though it be ever so finely hung, comes slowly to rest, and does not again begin to swing unless it is moved from its position of rest. Friction never becomes negative. If it did, a machine might of itself start running, cooling its bearings, and converting the heat thus obtained into work. Gases or liquids, once mixed, never spontaneously separate and thus return to the higher concentrations existing before the mixing. Warm water, which we have prepared by mixing hot and cold water, never spontaneously separates into a hot and cold portion. Every one of these processes, which appear so impossible to us as to be almost absurd, if they could take place, would make possible a perpetual-motion machine of the second kind. Every perpetual-motion machine of the second kind would make it possible for us to turn backwards the hands of

[1] Ostwald, "Lehrbuch d. allgem. Chem.," iii. 474 (Leipzig, 1893).

[2] *Wied. Ann.*, 60 (1897), 392.

time. A cinematograph running backwards furnishes to the eye a picture of the world in which time is retrogressing. We can picture the same thing in our imagination if we think of ourselves as being projected from the earth with a velocity greater than that of light. Images of terrestrial happenings would then strike our eyes, directed toward the earth, in a reverse order of time, just as in the reversed cinematograph.[1]

The reaction energy deduced from conceptions of Van't Hoff.

We will now return to the work-process which gave rise to these considerations, and we shall first see what result may be obtained in the case of the formation of water from the elements, making use of the law that the maximum work is independent of the path. For the sake of convenience, we will start from the assumption that the three substances taking part in the equilibrium have the concentrations c'_{H_2O}, c'_{H_2}, and c'_{O_2} at T°. The energy obtainable from their transformation at T° on the absolute scale in case the concentrations are c_{H_2O}, c_{H_2}, and c_{O_2}, may then be represented as the algebraic sum of three quantities of work, namely those required to bring the H_2 and the O_2 from the concentration c to the equilibrium concentration c', and that required to return the H_2O from the equilibrium concentration to the concentration c. The amounts of the three gases to be compressed or expanded are connected by the equation—

$$2H_2 + O_2 \rightleftarrows 2H_2O$$

According to our earlier conclusions (p. 18), these quantities of work can be represented as follows—

$$A_{(H_2)} = 2RTln\frac{c_{H_2}}{c'_{H_2}} \text{ for 2 mols of } H_2 \text{ from } c_{H_2} \text{ to } c'_{H_2}$$

$$A_{(O_2)} = RTln\frac{c_{O_2}}{c'_{O_2}} \text{ ,, 1 ,, } O_2 \text{ ,, } c_{O_2} \text{ ,, } c'_{O_2}$$

$$A_{(H_2O)} = 2RTln\frac{c'_{H_2O}}{c_{H_2O}} \text{ ,, 2 ,, } H_2O \text{ ,, } c'_{H_2O} \text{ ,, } c_{H_2O}$$

[1] The entropy theory, when applied to actual non-reversible processes, teaches that in all the examples adduced above (friction, diffusion, and conduction of heat) the total entropy of all participating systems increases. On the other hand, for the reversible processes which we shall alone consider in the future, while single systems or parts of systems may increase or decrease in entropy, the total entropy of all the systems concerned, including that of the surrounding calorimeter or atmosphere, does not change. From this we derive the law of Clausius already mentioned on p. 34. W. Thomson (*Phil. Mag.*, (4) 4, 304) had previously given expression to a popularised conception of the law.

Their sum is—

$$A = 2RTln\frac{c_{H_2}}{c'_{H_2}} + RTln\frac{c_{O_2}}{c'_{O_2}} + 2RTln\frac{c'_{H_2O}}{c_{H_2O}}$$

or arranged—

$$A = RTln\frac{c'^2_{H_2O}}{c'^2_{H_2} \times c'_{O_2}} - RTln\frac{c^2_{H_2O}}{c^2_{H_2} \times c_{O_2}}$$

The quotient—

$$\frac{c'^2_{H_2O}}{c'^2_{H_2} \times c'_{O_2}} = K_c$$

which is characteristic of the equilibrium at T°, is called the *equilibrium constant*. We may now generalize this expression by noticing that—

$$ln\frac{c^2_{H_2O}}{c^2_{H_2} \times c_{O_2}} = \Sigma\nu' lnc'$$

We then obtain—

$$A = RTlnK_c - RT\Sigma\nu' lnc' \quad . \quad . \quad . \quad . \quad (2)$$

Partial pressures may be used in the preceding deduction just as well as concentrations, and therefore we may write—

$$A = RTlnK_p - RT\Sigma\nu' lnp' \quad . \quad . \quad . \quad . \quad (3)$$

Yet the numerical value of K_p is, in general, not equal to K_c, unless $\Sigma\nu'$ is accidentally equal to zero, and hence the term $\Sigma\nu' lnp'$ is identical in value with the term $\Sigma\nu' lnc'$.

It follows from (2) that— Combination with Helmholtz's equation.

$$\frac{A}{T} = RlnK_c - R\Sigma\nu' lnc'$$

This can be directly substituted in equation (1), and we get—

$$\frac{dRlnK_c}{dT} - \frac{dR\Sigma\nu' lnc'}{dT} = -\frac{Q_T}{T^2} \quad . \quad . \quad . \quad (4)$$

The term—

$$\frac{dR\Sigma\nu' lnc'}{dT}$$

however, here equals zero, for neither R nor $\Sigma\nu'$ nor the gas concentrations can be changed by a change of temperature dT at constant volume. Taking R out, we then obtain—

$$\frac{RdlnK_c}{dT} = -\frac{Q_T}{T^2} \quad . \quad . \quad . \quad . \quad . \quad . \quad (5)$$

This expression has acquired the highest importance because of the work of Van't Hoff. Integrated, it gives—

$$RlnK_c = \text{const.} - \int \frac{Q_T}{T^2} dT \quad . \quad . \quad . \quad . \quad . \quad (6)$$

or—

$$RTlnK_c = \text{const.}\, T - T\int \frac{Q_T}{T^2} dT \quad . \quad . \quad . \quad (7)$$

On comparing with equation (2), it follows that—

$$A = RTlnK_c - RT\Sigma\nu' lnc' = -T\int \frac{Q_T}{T^2} dT - RT\Sigma\nu' lnc' + \text{const.}\, T$$

Introduction of Kirchhoff's law.

If now we introduce the value which Kirchhoff's law gives for Q_T (see (1*a*), p. 49), it follows that—

$$A = -T\int \frac{Q_0 + \sigma'_v T + \sigma'' T^2}{T^2} dT - RT\Sigma\nu' lnc' + \text{const.}\, T \quad (8)$$

Or, finally—

$$A = Q_0 - \sigma'_v TlnT - \sigma'' T^2 - RT\Sigma\nu' lnc' + \text{const.}\, T \, . \quad (9)$$

Substitution of partial pressures instead of concentrations.

If we wish to substitute partial pressures for concentrations in this expression, we first find that—

$$\frac{dRlnK_p}{dT} - \frac{dR\Sigma\nu' lnp'}{dT} = -\frac{Q_T}{T^2}$$

(Here again Q_T is the reaction heat at constant volume and T°.) Now—

$$\frac{dlnp'}{dT}$$

is not equal to zero, for the pressure changes when we raise the temperature at constant volume. But we can readily find out the magnitude of this change from the gas law—

$$\frac{dlnp'}{dT} = \frac{dlnRTc'}{dT} = \frac{dlnR}{dT} + \frac{dlnT}{dT} + \frac{dlnc'}{dT}$$

All of the three terms—

$$\frac{dlnR}{dT}, \frac{dlnT}{dT}, \text{ and } \frac{dlnc'}{dT}$$

except the second, are equal to zero under these conditions, and it therefore follows that—

$$\frac{dRlnK_p}{dT} - \frac{dR\Sigma\nu' lnT}{dT} = -\frac{Q_T}{T^2}$$

This expression, however, is the same as—

$$\frac{dRlnK_p}{dT} - R\Sigma\nu'\frac{1}{T} = -\frac{Q_T}{T^2} \quad . \quad . \quad . \quad . \quad (10)$$

Or, if we arrange the terms more suitably, and add the index (v) to Q_T to show more conspicuously that we mean the reaction heat at constant volume, we get—

$$\frac{dRlnK_p}{dT} = -\frac{Q_{(v)T} - RT\Sigma\nu'}{T^2} = -\frac{Q_{(p)T}}{T^2}$$

Integration gives us—

$$RlnK_p = \text{const.} - \int\frac{Q_{(p)T}}{T^2}dT \quad . \quad . \quad . \quad (11)$$

which, on multiplication with T, becomes as before—

$$RTlnK_p = \text{const.}\ T - T\int\frac{Q_{(p)T}}{T^2}dT \quad . \quad . \quad (12)$$

It further follows, just as in the previously discussed case, that—

$$A = -T\int\frac{Q_{(p)T}}{T^2}dT - RT\Sigma\nu' lnp' + \text{const.}\ T \quad . \quad (13)$$

On the basis of Kirchhoff's law, we now introduce the difference of the specific heats at constant pressure—

$$Q_{(p)T} = Q_0 + \sigma'_p T + \sigma'' T^2 \quad . \quad . \quad . \quad . \quad (14)$$

and obtain—

$$A = Q_0 - \sigma'_p TlnT - \sigma'' T^2 - RT\Sigma\nu' lnp' + \text{const.}\ T \quad (15)$$

We have thus deduced again the same equations which we found in our Second Lecture. It is very evident in this new method of deduction that, in spite of the introduction of the reaction heat at constant pressure ($Q_{(p)T}$), the external work done by the atmosphere should not be taken account of. For the walls of the equilibrium box, in which the reaction is imagined to take place, although partially permeable, are nevertheless rigid, and neither allow the atmosphere to do work of compression upon our system, nor our system, in expanding, to do work upon the atmosphere.

If we examine somewhat more closely the expression which we deduced in the Second Lecture from the conception of entropy, the integration constant obtained here will appear in a new light. According to our earlier discussion, in the case The indeterminate thermodynamic constant.

of formula (22*a*) (p. 50), this constant was made up of the difference in the true specific heats of the factors and products of a reaction and of the latent heat, provided the reaction took place at T = 1° and at equalized concentrations ($\Sigma\nu' lnc' = 0$). We represented this particular value of the latent heat by the letter k. This relationship no longer appears in our altered method of deduction, but we can easily recognize it if we insert the value 1 for T and 0 for $\Sigma\nu' lnc'$, obtaining in this way from equation (9)—

$$A_1 = Q_0 - \sigma'' + \text{const.} \quad . \quad . \quad . \quad . \quad (16)$$

According to the fundamental equation (First Lecture, p. 14), $A = U - q$. If, as before, we call k that value of q which corresponds to T = 1 and $\Sigma\nu' lnc' = 0$, it follows that $A_1 = U_1 - k$. We can here substitute for U_1 the reaction heat at the absolute temperature 1°, and thus obtain—

$$A_1 = Q_0 + \sigma'_v + \sigma'' - k \quad . \quad . \quad . \quad . \quad (17)$$

If we compare this with (16), we immediately get the former value of our constant (Second Lecture, p. 50, (22*a*) and (28*a*))—

$$\text{Const.} = \sigma'_v + 2\sigma'' - k \quad . \quad . \quad . \quad . \quad (18)$$

Another method of combination.

Finally, we may carry through the discussion in a somewhat different way. Starting from the expression—

$$A = U + T\frac{dA}{dT} \quad . \quad . \quad . \quad . \quad . \quad (19)$$

and substituting—

$$A = U + T\frac{dA}{dT} = Q_T + T\frac{dA}{dT} = Q_0 + \sigma'_v T + \sigma'' T^2 + T\frac{dA}{dT} \quad (20)$$

we differentiate the resulting equation with respect to T at constant volume, and thus obtain [1]—

[1] We shall later (p. 80) draw a particular conclusion from this relation, for which we will now take the preliminary steps. Thus, if $T\frac{d^2A}{dT^2} = 0$, the difference of the specific heats of factors and products is also equal to zero, and *vice versâ*, as is evident from equation (20) and (21). Now, $\frac{d^2A}{dT^2}$ is the change of the temperature coefficient with variable temperature. We conclude, therefore, that wherever the specific heats of factors and products are equal, $\frac{dA}{dT}$ has the same value at all temperatures. But where the specific

$$\frac{dA}{dT} = \sigma'_v + 2\sigma''T + T\frac{d^2A}{dT^2} + \frac{dA}{dT}$$

or—

$$-T\frac{d^2A}{dT^2} = \sigma'_v + 2\sigma''T \quad . \quad . \quad . \quad . \quad (21)$$

If $\sigma'_v + \sigma''T$ (see p. 49) is the difference of mean specific heats of factors and products between 0 and T° at constant volume, $\sigma'_v + 2\sigma''T$ is the difference of the true specific heats at constant volume and at T°. Rearranging—

$$\frac{d^2A}{dT^2} = -\frac{\sigma'_v}{T} - 2\sigma'' \quad . \quad . \quad . \quad . \quad . \quad (22)$$

and integrating, it follows that—

$$\frac{dA}{dT} = -\sigma'_v lnT - 2\sigma''T + f(v)^1 \quad . \quad . \quad . \quad (23)$$

Now it is easy to show, on the basis of certain considerations due to Van't Hoff, that the function of the volume ratios $f(v)$ in this equation is identical, except for a constant x, with—

$$-R\Sigma\nu' lnc'$$

We then obtain—

$$\frac{dA}{dT} = -\sigma'_v lnT - 2\sigma''T - R\Sigma\nu' lnc' - x$$

Substituting this in (20), it finally follows that—

$$A = Q_0 - \sigma_v'TlnT - \sigma''T^2 - RT\Sigma\nu' lnc + (\sigma_v' - x)T \quad (24)$$

heats (at constant volume) of factors and products are equal, the reaction heat (at constant volume) is constant and independent of the temperature. Consequently, constancy of reaction heat at constant volume signifies constancy of $\frac{dA}{dT}$ at all temperatures. But $-T\frac{dA}{dT}$ is equal to the latent heat q (see p. 19). If $\frac{dA}{dT}$ is constant, $\frac{q}{T}$ is also constant at all temperatures, and therefore, finally, if the reaction heat is constant, $\frac{q}{T}$ is identical at all temperatures.

[1] A function of those variables which were taken as constant during the integration make its appearance here in place of the integration constant, just as in the corresponding derivation of equation (23) in the First Lecture. It is here a function of the volume or concentration ratio.

Comparing with (9) and (18), we see that $x = k - 2\sigma''$. The expression $(\sigma'_v - x)$ is the same one which we previously called $(\sigma'_v + 2\sigma'' - k)$, and which we obtained in equation (9) under the simple guise of "const." That it has a different formal appearance depending on the method of derivation is due to the circumstance that in every integration there must always remain an indeterminate constant.

The latent heat is, then, as we have already seen in the Second Lecture, the only remaining unknown factor in a gas reaction which progresses, or, more precisely, which can be considered to be progressing, at the absolute temperature $T = 1°$ and at equalized concentrations $(\Sigma\nu' lnc' = 0)$. We need not again consider how this unknown quantity is related to the value of a degree on the temperature scale.

The approximation formulæ. It now remains for us to learn what modifications may be made in our formulæ to adapt them for special cases. Since our knowledge of the reaction heat or of the specific heats is often most unsatisfactory, it is of the very greatest importance to know what thermodynamic conclusions we can draw without their aid.

Van 't Hoff's approximation. The most important of these conclusions is one which Van't Hoff has derived from the formula—

$$R\frac{dln K_c}{dT} = -\frac{Q_{(v)T}}{T^2} \quad . \quad . \quad . \quad . \quad . \quad . \quad (5)$$

He points out that the reaction heat (Q_T) at constant volume always changes very slowly with the temperature. We may therefore assume, without hesitation, that between any two fairly adjacent temperatures, T' and T'', it would have the constant average value Q_m. Integrating (5) between these limits, it follows that—

$$Rln K_{c(T'')} - Rln K_{c(T')} = Q_m\left(\frac{1}{T''} - \frac{1}{T'}\right)$$

or—

$$Rln\frac{K_{c(T'')}}{K_{c(T')}} = Q_m\left(\frac{1}{T''} - \frac{1}{T'}\right) \quad . \quad . \quad (25)$$

Knowing the experimentally determined value of K_c at the temperatures T' and T'', we can easily find the mean reaction heat Q_m between T' and T'' (at constant volume).

If T' and T'' are high temperatures, not very far apart, the value of Q_m computed in this way may be very different from the value of the reaction heat measured calorimetrically at ordinary temperatures and at constant volume. By comparing both values, we may see how the reaction heat depends on the temperature, and estimate the difference between the specific heats of the factors and products and its change with the temperature. This method is of great theoretical importance in determining or checking other determinations of the change of the specific heats with the temperature. We shall see how Bodenstein actually made use of it in studying the dissociation of hydriodic acid. In practice the method has the drawback that $K_{c(T')}$ and $K_{c(T'')}$ must be determined very accurately. A numerical example will best illustrate this. If T' and T'' are respectively 2000° and 1800° on the absolute scale—

$$\frac{1}{T''} - \frac{1}{T'} = \frac{1}{1800} - \frac{1}{2000} = 5{\cdot}55 \times 10^{-5}$$

That is, $\frac{1}{T''} - \frac{1}{T'}$ is a very small number. The mean reaction heat, which in this case we may consider the reaction heat at 1900° abs., is then—

$$Q_m = \frac{R}{5{\cdot}55 \times 10^{-5}} ln \frac{K_{c(1800)}}{K_{c(2000)}}$$

or, substituting Briggsian logarithms for the natural logarithms, and the value 1·98 for R—

$$Q_m = 0{\cdot}829 \times 10^5 \log \frac{K_{c(1800)}}{K_{c(2000)}}$$

Now, suppose that the ratio of the two constants as found in a series of experimental determinations, which are in most cases not simple, was really 10 per cent. too great. Q_m would then come out wrong by $0{\cdot}829 \times 10^5 \log (1{\cdot}1)$, or in round numbers, 2500 cals. An error of this magnitude makes it idle to attempt to calculate the change in the specific heats with the temperature from the heat of reactions at ordinary temperatures and Q_m, the heat of reaction at 1900° abs.

In this state of affairs it is natural that determination of Q_m should only be used as a sort of check to see whether the values of K as found for any two temperatures stand in approximately the right relation. We usually know enough about the specific heats to be able at least to estimate how great the reaction heat could be at another temperature. If we get approximately the same result from the equilibrium constants of equation (25), then these constants must certainly bear nearly the right ratio to one another.

The approximation usual in theoretical physics.

While we may unhesitatingly neglect the change of the reaction heat within narrow temperature intervals, we should possibly introduce a considerable error into our calculations if we neglected it at all temperatures. We do not, therefore, get such a good approximation using the equation—

$$A = Q_v - RT\Sigma\nu' \ln c' + \text{const. } T \quad . \quad . \quad (25a)$$

This approximate equation serves chiefly for calculation of equilibria, A being put equal to zero, and $\Sigma\nu' \ln c'$ equal to K_c. It is widely used in theoretical physics. Planck, for instance, bases all his examples on it.[1] Its basis is a very interesting one. It depends on the supposition that the specific heats of gaseous compounds are made up additively of the specific heats of the gaseous components.

The fundamental assumption of Buff and Clausius.

This supposition was first made and defended by Buff[2] (1860), and also by Clausius[3] (1861). Both based their arguments on the experimental results of Regnault. Clausius gives a table (*l.c.* and later corrected in accordance with Regnault's final values and published as Appendix B to his sixth paper on thermodynamics[4]), in which he compares the experimental values with those calculated on the assumption that all diatomic gases have the same specific heat at constant volume, and that, further, the triatomic gases have specific heats half again as great, the tetratomic gases twice as great as these, and so on. The agreement is very imperfect. The deviations occasionally exceed 30 per cent. Regnault,[5] in his critical discussion of the experimental facts, therefore denies the view championed by Buff and Clausius, or is at least very sceptical about it. It is primarily an extension to gases of the rules of

[1] For a more detailed discussion see the Fifth Lecture.

[2] Buff, Liebig's *Ann. d. Chemie u. Pharm.*, 115 (1860), 301.

[3] Clausius, Liebig's *Ann. d. Chemie u. Pharm.*, 118 (1861), 106.

[4] "Abhandlungensammlung," Bd. ii. (Braunschweig, 1864), S. 286.

[5] *Mem. Inst. de France*, 26 (1862).

Neumann (1831) and of Kopp (1864). These rules state that the specific heats of solid compounds are made up additively of the specific heats of the solid constituents. Viewed from the standpoint of the mechanical theory of heat, it represents an application of Berthelot's principle, which we discussed at the close of the first lecture. To be sure, the assumption of Buff and Clausius (as expressed in 25*a*) is by no means a sufficient condition that heat (Q_v) and work (A) (when mass action is equalized—that is, when $\Sigma\nu' lnc' = 0$) are identical. For, as we see in the equation, they differ by the term const. T. But the assumption of Buff and Clausius, which makes the terms $\sigma'_v T ln T$ and $\sigma'' T^2$ (which do not appear in equation 25*a*) equal to zero, is one of the conditions necessary for the identity of Q_v and A as maintained by Berthelot.

From the atomistic point of view which Clausius took, the theory proposed by Buff and himself had a still further significance. Clausius asked himself what effect the heat which one added to the gas had upon the molecules. He divided it into a portion which exerted a heating effect only, and another portion which changed the internal nature of the molecule, exercising, in particular, a loosening power upon the forces which held the atoms together. Because of this distinction, Clausius naturally came to investigate the limiting case where heat worked solely upon the thermal condition of the gas, and had no effect on the molecular properties of the gas particles. He was able to deduce mathematically for this limiting case the view proposed by Buff and himself. Clausius therefore considered the additive nature of the specific heats as a characteristic of perfect gases.

The importance of atomistic considerations in thermodynamic questions is very differently evaluated. Many important discoveries have undoubtedly been made by their help, and Boltzmann[1] has shown that the mechanical theory of heat can be completely represented on an atomistic basis with the aid of the theory of probability. On the other hand, Gibbs,[2] Helmholtz, Planck,[3] and Mach[4] have shown that the theory of heat becomes simpler and more convincing when stripped of its atomistic

[1] "Vorlesungen über Gastheorie," ii. Teil (Leipzig, 1898).
[2] "Thermodyn. Studien," trans. by Ostwald (Leipzig, 1892).
[3] "Thermodynamik," 2 Aufl. (Leipzig, 1905).
[4] "Prinzipien der Wärmelehre," 2 Aufl. (Leipzig, 1900).

clothing. This especially applies to the work of Mach. If we abandon the atomistic basis and no longer consider the peculiarities and movements of molecules imperceptible to all our senses, and consider instead those qualities of heat and work—things directly accessible to our senses corresponding to molar quantities—the view of Clausius and Buff loses the nature of a postulate, which it possessed with Clausius, and becomes simply a statement of experimental fact. It is in this sense that Planck introduces it into his thermodynamic treatment, while we, because of the poor confirmation, which experiment affords it, do not use it in our general considerations.

Historically, it is interesting to note that the first application of thermodynamics to chemistry resulted from the use of the Buff-Clausius approximate assumption.

Clausius called attention only to the general applicability of the theory of heat to chemical reactions. We have Horstmann[1] to thank for the fundamental advance from this incidental observation to a fruitful thermodynamic treatment of chemical problems. His development lacked the completeness which was later attained by Gibbs, Helmholtz, and Van't Hoff.[2] It followed directly the line of reasoning adopted by Clausius, and practically resulted in equation (25a). He does not fail to call attention to the variability of the reaction heat with the temperature and to its effects. But, standing within the sphere of the thought and influence of Clausius, he naturally ascribes only a subordinate importance to this possibility. It is of considerable interest to compare Horstmann's treatment with that of Le Chatelier, which we mentioned at the beginning of the First Lecture. Both treat the same relationships, and in part the same examples. But almost a generation lies between the two, during which Helmholtz and Van't Hoff, particularly, had shown the way to apply systematically thermodynamics to chemistry. The view of Buff and Clausius is no longer entertained by Le Chatelier.

[1] "Abhandlung zur Thermodynamik chemischer Vorgänge," edited by J. H. Van't Hoff (Leipzig, 1903; Ostwald's Klassiker, No. 137). Compare the historical observations of Van't Hoff on p. 73 of this paper.

[2] "Die Gesetze des chemischen Gleichgewichts." Translated into the German, and edited by Bredig (Leipzig, 1900; Ostwald's Klassiker, No. 110).

If we substitute for c in equation (25a) as before, the expression $\frac{p}{RT}$, it follows that— Roughest approximation.

$$A = Q_v + RT\Sigma\nu' lnT - RT\Sigma\nu' lnp' + const.' T$$

If we consider the term lnT as constant, which is permissible within narrow temperature limits, or if $\Sigma\nu'$ equals zero, we may rewrite the formula as follows—

$$A = Q - RT\Sigma\nu' lnp' + const.'' T \quad . \quad . \quad (26)$$

This expression is the one used by Bodländer.[1] Applying it to the whole range of temperature involves an approximation which, again, is less complete than the former one, except when the number of molecules on each side of the chemical equation is the same.

Bodländer's formula calls Q the reaction heat at ordinary temperatures. It does not matter whether we take reaction heat at constant pressure or constant volume, for we are only concerned with the roughest kind of an approximation formula, and the two sorts of reaction heats are always nearly the same at ordinary temperatures. The fundamental idea in the approximation, namely, that the specific heats of gases at constant pressures are additive, is older and less complete than the view of Buff and Clausius expressed in equation (25a), as Clausius himself pointed out. It can neither be explained in terms of the atomistic conception, nor is it compatible with Berthelot's principle.

Equation (26) has an important empirical bearing on certain phenomena related to gas reactions, particularly dissociations in which solids and a single gas take part. If it were not for this, equation (26) would only be of value as a rough approximation formula with which to compute, from values of the free energy and the equilibrium constant observed at known temperatures, values at intermediate temperatures. To make this clear, we shall briefly discuss the connection between the reactions of solids and of gases.

Where both solid and gaseous substances take part in a reaction, we can deduce in the same way the formulæ which we found for simple gas reactions. Suppose, for instance, that the reaction— When solid substances join in the reaction.

$$a + b + c = d + e + f$$

where c and f are solids, were to be studied instead of a gas

[1] *Zeitschr. f. Elektrochemie*, viii. (1902), 833.

reaction. We can imagine an equilibrium box containing all six substances in a state of equilibrium, just as we did at the beginning of this lecture. Here, as before, we introduce the gaseous substances a and b at their concentrations in the equilibrium mixture, and draw out the gaseous substances d and e in a similar way. The reaction then progresses in the equilibrium box without any expenditure of work so long as the supply of c and f holds out, and we can, of course, consider the supply as indefinitely large. The concentrations a, b, d, and e are the sole variables. Only these substances, therefore, can be brought to other concentrations by isothermal expansion. We therefore find exactly the same expression (2) or (3) for the energy of our reaction, excepting only that the gaseous substances and not the solids appear in the term $\Sigma\nu' lnc'$, or $\Sigma\nu' lnp'$. The subsequent treatment is the same, step for step, as that of gaseous systems. Only the reaction heats and the specific heats are affected by the presence of the solid substances.[1] Among the cases here included there is one which is of especial interest to us. It is the case of dissociations, where but a single gaseous component is present, as, for instance—

The dissociation of calcium carbonate.

$$CaCO_3 = CaO + CO_2$$

[1] There is an observation which ought to be made regarding the specific heat of solids. It is as easy to determine the specific heats of solids at constant pressure as it is difficult to determine their specific heats at constant volume. Fortunately, it is of no importance what this specific heat at constant volume is, so long as we are only dealing with simple substances, and not with solid solutions of variable composition. In deducing our formulæ (p. 22), we assumed the volume to remain constant when we raised the temperature dT degrees, for we wished the heating to take place without doing work. In this way we obtained expressions referring to constant volume or constant concentrations, and in which the specific heat at constant volume appeared. Now, solid substances have such a small coefficient of thermal expansion, in comparison with gases, that we may neglect without hesitation the work which they would do in expanding against any small external pressure (1 atmosphere, for instance). We may therefore put the specific heats of solids at constant pressure, except when very high pressures are considered, as identical with the specific heats, when the solids are heated without their doing any work against external pressure. These latter specific heats, however, are implied in thermodynamical formulæ. Besides, by determining the values of the specific heat of solids in a vacuum, we could easily realize the case where practically no work was done against an external pressure. It is naturally supposed that no vaporization of the solid substances takes place and interferes with this determination of their specific heats in the vacuum.

We may apply our conclusions to such a reaction in the following way.[1] Let us imagine a box containing both calcium oxide and calcium carbonate to be heated to a high temperature. A certain pressure, K_p, of carbon dioxide will establish itself in the box, corresponding to the equilibrium of the three substances. If, now, we imagine one of the walls of the box to be permeable to carbon dioxide, and that through it we draw out this carbon dioxide at the pressure prevailing in the box, the reaction will progress steadily in the box at the equilibrium point without doing any work. It will only come to an end when the supply of carbonate is exhausted, and we may imagine this to be as large as we please. If we wish to know what the reaction energy of the decomposition will be, at some other pressure of carbon dioxide p_{CO_2}, we may first carry out the decomposition at the pressure K_p, as above described, and then bring the carbon dioxide thus obtained from the pressure K_p to the pressure p_{CO_2} by an isothermal and reversible process. This would yield the work—

$$A = RTlnK_p - RTlnp_{CO_2}$$

which must be identical with the reaction energy of the direct decomposition sought, for the maximum work is independent of the path by which we obtain it. Just as before, we can now deduce the relation between this reaction energy and the heat of decomposition at constant pressure—

$$R\frac{dlnK_p}{dT} = -\frac{Q_p}{T^2} \quad . \quad . \quad . \quad . \quad (27)$$

If, as a first approximation, we assume the heat of decomposition to be independent of the temperature, we further obtain by integration—

$$RlnK_p = \frac{Q_p}{T} + \text{const.} \quad . \quad . \quad . \quad . \quad (28)$$

This formula is perfectly comparable with the expression (26) which we obtained before. This becomes evident as soon as we rearrange it to—

$$0 = Q_p - RTlnK_p + \text{const. } T \quad . \quad . \quad . \quad (29)$$

[1] This case, so frequently discussed, is somewhat complicated by the fact that, according to Raoult (*Comp. Rend.*, 92 (1881), pp. 189, 1110, and 1457), the alkaline carbonates form basic carbonates with carbon dioxide at high temperatures. We shall not consider this aspect of the problem.

where the value 0 tells us that the reaction energy vanishes at the pressure K_p.

The importance of dissociation equilibria in gas reactions.

We can make very profitable use of this equation if we know the value of the constant it contains. For with its help we can pass over from gas reactions proper to reactions between gases and solids. Suppose, for example, that we wished to predict whether a certain mixture of carbon monoxide and dioxide would reduce a metallic oxide, say FeO, or oxidize the corresponding metal, Fe, at a temperature of T°. Using the formula, we can calculate the partial pressure of oxygen which is in equilibrium with iron and ferrous oxide at T°. If, then, on the other hand, we know enough about the equilibrium $CO + O \rightleftarrows CO_2$ to tell what pressure of oxygen can exist together with carbon monoxide and dioxide at their given pressures, the problem is solved. That is, if the equilibrium pressure of the oxygen in the given mixture of carbon monoxide and dioxide is greater than that which can exist in equilibrium with iron and ferrous oxide, the iron will be oxidized and oxygen used up. This oxygen is furnished by the carbon dioxide, which is thereby reduced to carbon monoxide. If the gas mixture is steadily replenished, the reaction will continue till all the iron is consumed. If it is not replenished, then the reaction comes to a halt when the partial pressure of the dioxide has diminished, and the partial pressure of the monoxide increased so far that the partial pressure of the oxygen in equilibrium with them shall have become equal to the equilibrium pressure of the oxygen over the ferrous oxide and iron. If from the start we had such partial pressures of monoxide and dioxide that the oxygen partial pressure in equilibrium with them was less than the dissociation pressure of the ferrous oxide, the latter would dissociate, giving up oxygen to the carbon monoxide. If the gas mixture is steadily renewed, this would continue till all the ferrous oxide has been used up, otherwise it would cease when the partial pressure of the oxygen in equilibrium with the monoxide and dioxide was equal to the dissociation pressure of the ferrous oxide.

We can apply the same reasoning to the transformation of ferrous oxide into ferric oxide and the corresponding transformation of ferrous oxide into ferrous-ferric oxide. The more accurate investigation of this case which Bauer and Glaessner[1]

[1] *Z. f. phys. Chemie*, 43 (1903), 354.

have carried out naturally disclosed certain deviations, because the approximate assumption regarding the reaction heats is not sufficiently exact. Yet it does not invalidate formula (28) for purposes of illustration. Very similar considerations apply to the oxidizing or reducing action of moist hydrogen.[1]

Other examples

All oxidations by carbon monoxide and hydrogen (or technically by water-gas or half-water-gas), and by carbon dioxide and water-vapour, may be similarly treated. Indeed, if we look more closely, we see that most reactions of solid substances with gases can be referred in this way to a gaseous equilibrium, and one or more dissociation equilibria. Thus we may consider the action of carbon monoxide on molten caustic soda[2]—

$$CO + 2NaOH \rightleftarrows CO_3Na_2 + H_2$$

to consist of a competition between the dissociation of sodium carbonate into carbon dioxide and sodium oxide, and of sodium hydroxide into sodium oxide and water, and the "water-gas" reaction, as illustrated in the following reaction:—

$$\begin{array}{ccccc} & 2NaOH & & Na_2CO_3 & \\ & \downarrow\uparrow & & \uparrow\downarrow & \\ CO + & H_2O & \rightleftarrows & CO_2 & + H_2 \\ & + & & + & \\ & & Na_2O & & \end{array}$$

[1] We find cases, such as the action of hydrogen on silver chloride and bromide, which belong there, and which Jouniaux (*Compt. Rend.*, 129 (1889), 883; 132 (1901), 1270; 133 (1901), 228) has studied. Here the dissociation of the halogen salt into halogen and metal is linked with the reversible formation of the halogen acid from halogen and hydrogen. If the pressure of hydriodic acid over any aqueous solution of hydriodic acid between 0·015 and 0·167 normal were known, we could easily calculate from Danneel's electrical measurements (*Z. f. phys. Chemie*, 33 (1900), 442) the corresponding equilibrium between silver iodide, hydrogen, and hydriodic acid. We could further, with the help of Bodenstein's measurements of the hydriodic acid equilibrium (to be discussed in the next lecture), calculate the dissociation equilibrium of the silver iodide.

We should also mention in this connection Pélabon's investigations (*Ann. Chim. Phys.*, vii. 25 (1902) 365), of the behaviour of hydrogen towards the sulphides of the heavy metals, where the dissociation of the sulphide into metal and sulphur is correlated with the equilibrium between sulphur vapour, hydrogen sulphide, and hydrogen. A comparison could here be made with Bernfeld's measurements (*Z. f. phys. Chemie*, 25 (1898), 46).

[2] Haber and Bruner, *Z. f. Elektrochem.*, x. (1904) 708.

We might similarly resolve the well-known sulphate process of Hargreaves—

$$2NaCl + SO_2 + O + H_2O = Na_2SO_4 + 2HCl$$

into the two dissociation equilibria—

$$2NaCl \rightleftarrows Na_2 + Cl_2$$
$$Na_2SO_4 \rightleftarrows Na_2O + SO_3$$

and the gaseous equilibria—

$$SO_2 + O \rightleftarrows SO_3$$
$$Cl_2 + H_2O \rightleftarrows 2HCl + O$$

One of these gaseous equilibria is the basis of sulphuric anhydride manufacture, the other of the Deacon process.

Two solid substances, such as lime and calcium carbonate, which exist in a dissociation equilibrium with a gas, act on the gas like a pressure regulator. They take up the gas when the pressure tries to rise, and give it off when it tries to fall. Thus they exert the same influence on the pressure as a condensation product of the gaseous substance (for instance, solid CO_2 in presence of gaseous CO_2) would do. In this case the vapour pressure of the gas corresponds exactly to the dissociation pressure in the above sense. A gaseous equilibrium linked with a vaporization is consequently entirely analogous to one linked with a dissociation. Pélabon[1] has studied an example of this kind. It consists in the reversible formation of gaseous hydrogen selenide from solid selenium and gaseous hydrogen. Here the equilibrium—

$$Se_{(solid)} \rightleftarrows Se_{(vapour)}$$

is coupled with the equilibrium—

$$SeH_{2(gas)} \rightleftarrows Se_{(vapour)} + H_2$$

We are also in possession of a similar study of the formation of hydrogen sulphide from liquid sulphur and hydrogen gas.[2] Dissociation equilibria, therefore, deserve just as full a treatment as gaseous equilibria. Yet we cannot undertake it here. We must content ourselves with considering but one side of the matter, namely, the constant of the approximation formula (28), where the reaction heat is taken as unchanging.

[1] *Z. f. phys. Chemie*, 26 (1898), 659.

[2] Bodenstein, *Z. f. Phys. Chemie*, 29 (1899), 315, and Pélabon (*l.c.*).

We have justification for the remarkable statement that this constant has a similar value, about 32, for a great number of simple processes of dissociation where but a single gas is present besides solid substances. This being so, formula (28) becomes— The constant of formula (28).

$$-\frac{Q}{T} = \text{about } 32 \quad . \quad . \quad . \quad . \quad . \quad (30)$$

when the dissociation pressure of the gases represented by K_p equals one atmosphere. This number has been found to vary some seven units in each direction. Le Chatelier,[1] who first called attention to this surprising relation, found values in the few cases at his disposal which were generally a trifle below 32. Forcrand,[2] who has recently concerned himself with the matter, frequently finds somewhat higher values.

The first question that one naturally asks, in view of this fact, is whether the agreement is due to certain regularities among the specific heats or to the nature of the dissociation process itself. It therefore seems advisable to examine dissociation from a somewhat different point of view. We take as an example the dissociation of calcium carbonate.

Let us consider this dissociation as the passage of carbon dioxide from the solid, bound condition to the gaseous, free condition. It is then, as we have already pointed out, completely similar to the process of vaporization. Indeed, we can obtain equation (27) directly by treating the reaction from this point of view. According to Clapeyron and Clausius, every vaporization is governed by the thermodynamic relation— Dissociation viewed as a vaporization.

$$q = T(v_a - v_b)\frac{dp}{dT} \quad . \quad . \quad . \quad . \quad . \quad (27a)$$

Here q signifies the latent heat of vaporization, that is the heat used up in the formation of vapour, v_a the volume in the gaseous condition, v_b the volume in the solid (or liquid) condition, and p the pressure at the temperature T. The quantities of heat and the volumes all refer to one mol of substance. Applying this equation to our process, which we may write as—

$$CO_{2(\text{solid, bound})} \rightarrow \text{vaporization} \rightarrow CO_{2(\text{gaseous, free})}$$

[1] *Ann. des Mines* [8], 13 (1888), 157.

[2] *Ann. Chim. Phys.*, vii. 28 (1903), 384, 531.

we ought to obtain the same result as we did from our other conception of the process, as embodied in equation (27). On comparing the two, we note that the heat liberated by the splitting off of the carbon dioxide was called Q_p in (27), while the heat absorbed in the same process was called q in (27*a*). It follows that $q = -Q_p$. The equilibrium pressure, which we called K_p in (27) we called p in (27*a*). We notice, further, that the volume of solid, bound carbon dioxide (that is, the difference between the volume of one mol solid calcium carbonate and of one mol solid lime) is thousands of times smaller than the volume v_a of a mol of gaseous carbon dioxide under the equilibrium pressure of the dissociated mixture. The difference $v_a - v_b$ is therefore not perceptibly different from v_a. Then, too, the carbon dioxide obeys the gas laws almost exactly under the conditions of the dissociation equilibrium where we have either very low pressures at low temperatures or moderate pressures at high temperatures. We may therefore substitute $\frac{RT}{p}$ for v_a, or, as well, for $(v_a - v_b)$; and, further, $\frac{RT}{K_p}$ for $\frac{RT}{p}$ and K_p for p. Therefore (27*a*) becomes—

$$-Q_p = \frac{RT^2}{K_p} \times \frac{dK_p}{dT}$$

or, when transposed—

$$R\frac{dlnK_p}{dT} = -\frac{Q_p}{T^2}$$

which is identical with (27).

Application of Kirchhoff's law to the phenomena of vaporization.

Now, in order to pass over to the integrated form of our equation, we will replace the inexact assumption of a constant heat of reaction by the more exact expression afforded by Kirchhoff's law that—

$$Q_p = Q_0 + \sigma'_p T + \sigma'' T^2$$

The difference of the specific heats which here appears is the difference between the specific heats of a mol of carbon dioxide in the solid fixed condition and in the gaseous free condition. We can easily determine what the specific heat of the solid fixed carbon dioxide is by measuring the amounts of heat given off, respectively, when we allow a mol of calcium carbonate and a mol of lime, each heated to T°, to fall into an ice-calorimeter. In this way we determine the mean specific heat of the calcium carbonate and of the lime between 273° abs. and T° abs. The difference between the two represents the mean specific heat of solid, fixed carbon dioxide over this interval of temperature. By repeating these experiments at other temperatures we may determine this specific heat over a wider temperature interval. In this way we can derive an expression $a + bT$ which will represent the mean specific heat of solid, fixed carbon dioxide between 0° abs. and T° abs. There might possibly be still a third term, cT^2, in the expression, but we will neglect it here for the sake of simplicity. If we subtract from this the expression $a' + b'T$, representing the mean specific heat of gaseous free carbon dioxide at constant pressure, we obtain the value σ_p' and $\sigma''T^2$. Substituting these values in (27) and integrating, we get the expression—

$$R ln K_p = \frac{Q_0}{T} - \sigma_p' ln T - \sigma'' T + \text{const.}$$

instead of (28).

We may again make use of Kirchhoff's law and replace Q_0 by the heat of reaction at ordinary temperatures (about 290° abs.) which appears in equation (30), writing—

$$Q_{p(290)} = Q_0 + \sigma'_p \times 290 + \sigma'' \times 290^2$$

If we substitute this expression in the preceding equation, choosing at the same time such a temperature that $K_p = 1$ and consequently $R ln K_p = 0$, we obtain—

$$-\frac{Q_{p(290)}}{T} = -\sigma'_p\left(ln T + \frac{290}{T}\right) - \sigma''\left(T + \frac{290^2}{T}\right) + \text{const.} \quad (30a)$$

The left-hand members of equations (30) and (30*a*) are identical. Consequently the right-hand member of equation (30*a*) must be very generally equal to 32. If we examine the examples upon which this fact is usually based, we find that they are either carbonates which split off carbon dioxide, solid hydrates which give off water-vapour, ammonia compounds which lose ammonia, or finally oxides where oxygen is set free by heating.

Now, the specific heats of the gaseous products of dissociation are, according to Kopp,[1] as follows :—

O_2	CO_2	H_2O
8·0	9·8	8·6

The atomic heat of nitrogen in the solid fixed condition is uncertain, and the value for ammonia is omitted on this account. The other three values lie sufficiently near the mean specific heats of the same substances in the gaseous state (per one mol at constant pressure between 0° and T°) to allow us to assume that σ_p' and σ'' and the terms containing these quantities in (30*a*) are very small. Looked at in this light, the fact expressed in equation (30) simply means that in the reaction—

$$CO_{2(\text{solid, fixed})} \rightarrow \text{vaporization} \rightarrow CO_{2(\text{gaseous, free})}$$

and in analogous ones, the constants of the equation governing it have a nearly identical numerical value, and that the approximate formulæ (28) and (30) give the same constant just because the terms neglected are small in the examples at our disposal.

But if the numerical value 32 is correlated to the change from solid to gaseous state, the suggestion immediately presents itself that we might find the same number in the case of an ordinary vaporization, such as—

$$CO_{2(\text{solid, free})} \rightarrow \text{vaporization} \rightarrow CO_{2(\text{gaseous, free})}$$

representing the quotient of the heat of vaporization and that temperature at which the vapour pressure of the solid substance equals an atmosphere. As Forcrand (*l.c.*) has shown, this is really the case.

To make the matter clearer we may proceed in yet a different way, making use of a principle which has to do with the reaction of solid substances with other solid substances to form solid products. For brevity's sake, we will call such reactions "solid" reactions. We can then say that in "solid" reactions the heat of reaction and the maximum work are very nearly identical.

This principle is easily understood, because, in the first place, according to the Kopp-Naumann rule, the specific heat of a solid molecule is equal to the sum of the specific heats

[1] See compilation in Ostwald's "Lehrbuch der allgem. Chem.," i. (1891) 982.

of the component atoms. Now, every chemical reaction is simply a rearrangement of the atoms to form new compounds. If the atoms retain their original specific heats in all the compounds, then initial and final substances have the same specific heats, and all differences between the specific heats of the substances formed and the substances used up disappear. But, in the second place, the effect of variable concentration, that is of mass action, is absent in the case of solid reactions. Bodländer[1] has tested the principle in a most patent way on the basis of the following considerations. It ought to require the same amount of work to break up a substance in a saturated solution as in the solid state. Now, the saturated solutions of many substances are very dilute. Take, for example, the saturated solution p of silver chloride, bromide, and iodide in water. Here the salts are almost exclusively present as ions. If we call the solubility of a salt S, meaning the number of salt mols contained in a litre of the saturated solution, then one litre of saturated silver iodide solution will contain S mols of AgI, S being very small. The concentrations of the two ions ($Ag^{\cdot}$ and I') would also be S. The work required to bring the ions from the concentration S to the concentration 1 (one mol per litre) is known. The work required to bring the ions from this concentration 1 into the solid elementary condition by discharging them electrically (decomposition voltage) is also known. According, therefore, to the principle with which we started, the sum of these two amounts of work must be equal to the negative heat of formation, that is, to the heat of decomposition of the solid substance into its solid components.

In other words, we ought to be able to obtain the same quantity of work whether we convert the solid components directly into the solid compound, or whether we first bring them (by help of a suitable reversible cell) into solution as ions at a concentration of 1 mol per litre, and then dilute this solution in a reversible way (by help of an osmotic machine) to the concentration these ions have in a saturated solution of the compound. This work, according to the principle under discussion, must equal the heat of reaction of the solid substances. In this way, Bodländer was actually able to determine

[1] *Z. f. Phys. Chemie*, 27 (1898), 55.

with success the solubility of many solid substances from their heats of formation, on the one hand, and the electromotive forces of suitable electrodes on the other.

Haber and Tolloczko[1] have lent further support to this principle by showing that the counter electromotive force of polarization in the electrolysis of solid, non-molten substances corresponds very nearly to the heat of reaction.

Dissociation and vaporization are connected quantitatively through the principle of "solid" reactions.

If we assume the principle of "solid" reactions as true, we can compute the degree of dissociation and the constant 32 appearing in equation (30) in the following way. We will not discuss the general case, but instead take some specific case of dissociation like—

$$CaCO_3 \rightleftarrows CaO + CO_2$$

The line of reasoning is as follows: We produce a mol of carbon dioxide at atmospheric pressure from the calcium carbonate, first at the boiling point of solid carbon dioxide, and then at the temperature where the dissociation pressure of the carbon dioxide over the mixture of burnt and unburnt lime equals one atmosphere. In both cases, according to our initial assumption—

$$A = U - q$$

We now introduce the approximate assumption that U, that is the decrease in the total energy, has the same value at the two temperatures. We then know, from a former discussion of the matter (see p. 50, note 1) that $\frac{q}{T}$ would have the same value in both cases. Now, we can very easily determine the latent heat q of the reaction at the boiling-point of the solid carbon dioxide, by resolving the reaction into a solid reaction where the heat and work are identical, and a simple vaporization. That is, we imagine that the solid carbonate is first converted into solid lime and solid carbon dioxide, and then that the solid carbon dioxide vaporizes under atmospheric pressure (boiling-point = 194·8° abs.). The heat of vaporization is equal to the total latent heat of the reversible formation of carbon dioxide from the carbonate, at atmospheric pressure and 194·8° abs. According to Favre,[2] this amounts to −6257 cal.

[1] *Z. f. Anorg. Chem.*, 41 (1904), 407.

[2] *Ann. Chim. Phys.*, (5) 1 (1874), 250; Behn (*Drudes Ann.*, 1 (1900), 270) finds 142·4 cal. per gram, from which we get the value of 6265·6 cal. per mol.

($\pm$ 2 per cent. error). On the other hand, A is zero at the temperature where the carbon dioxide at atmospheric pressure is in equilibrium with the carbonate, and hence the latent heat q is equal to the change of the total energy of dissociation, or the directly measured heat of dissociation. This amounts to $-42,000$ cal.

Since $\frac{q}{T}$ is equal in both cases, $\frac{6257}{194 \cdot 8} = \frac{42,000}{T_x}$ where T_x signifies the temperature at which the pressure of carbon dioxide equals one atmosphere. The external work done against the atmosphere is neglected on the ground of our previous inquiry, which showed that A and U underwent an equal, and hence for our purposes immaterial, increment of $RT\Sigma\nu'$. Yet it should be observed that we have assumed the gas law to hold for carbon dioxide vapour at its boiling-point under atmospheric pressure, and this is certainly but a rough approximation at so low a temperature.

It makes no difference what carbonate we take, provided only the requirement is adhered to that the specific heat of the carbon dioxide shall be the same at constant pressure both when bound to the base as a carbonate or when existing free as a gas. Without this assumption we could expect no agreement, for it is only true in this single case that $\frac{q}{T}$ has theoretically the same value at all temperatures for any given pressure of carbon dioxide. The quantity $\frac{6257}{194 \cdot 8}$ has the value 32·12. Assuming that the heat of dissociation divided by 32 equals the temperature T_x at which the pressure of carbon dioxide reaches the value of one atmosphere, we compute the following values of T_x for several carbonates :—

	Heats of diss.	$\frac{\text{Heats of diss.}}{32} = T_x$	$T_x - 273° = t_x$ °C.
$AgCO_3$	16,500 (?)	513	240
$CaCO_3$	42,000	1312	1039
$BaCO_3$	55,900	1747	1474
Na_2CO_3... ...	89,960	2811	2538

The first of these numbers agrees with the observations of Joulin,[1] who found t_s to be about 225°. The second agrees with the observations of Debray, who found t_s to be about 1000°. The third agrees with the statement of Forcrand that barium carbonate is decomposed at the highest heat of a gas oven. In regard to the fourth, Forerand states that soda is extremely stable even when subjected to the very highest heat. Still, this is not as certain as one would infer from Forcrand's statement, for loss in weight certainly takes place when sodium carbonate is ignited. As is well known, this is the reason for not heating sodium carbonate to a glow in preparing it for use as a standard substance in acidimetry. Le Chatelier found t_s for calcium carbonate to be 812°, quite different from Debray's value. The heat of dissociation of silver carbonate is rather uncertain. The agreement is therefore not very pronounced, but depends on very interesting and instructive relationships, and seems natural enough. In an analogous way, Forcrand showed that for numerous double salts containing ammonia, which splits off when heated, the heat of dissociation, divided by the temperature at which the ammonia has a pressure of one atmosphere over the salt, gives the same number as does the heat of vaporization of solid ammonia under atmospheric pressure. Forcrand showed that the same held true for the heat of dissociation of a salt containing water of crystallization, and the temperature at which the water-vapour pressure from the salt is one atmosphere, compared with heat of vaporization of ice and the boiling-point of water.

Bearing on Trouton's Rule. The fact that the constant of the approximation formula (30) lies so near the quotient of the heat of vaporization and the boiling-point of the substance appearing as a gas in the dissociation reaction (under atmospheric pressure), indicates that this constant has its source chiefly in the change in the state of aggregation which the substance experiences when it becomes gaseous. It also teaches us that the change from the solid to the gaseous state generally involves the consumption of some $30 \times T_s$ cal. of heat in the most various substances; T_s being the boiling-point under atmospheric pressure.

This fact reminds us of a principle which Desprez and Pictet stated in an incomplete way, and which Trouton later

[1] *Ann. Chim. Phys.* (4), 30 (1873), 276.

developed into a rule. According to this rule, the quotient of the heat of vaporization (in gram calories) of homogeneous liquids, divided by their absolute boiling-point under atmospheric pressure, is likewise a constant, and equals about 21. Van't Hoff[1] has shown that this rule finds an explanation in the atomistic theory of corresponding states, enunciated by Van der Waals. Forcrand's extension of the Desprez-Pictet-Trouton rule to the transition solid-gaseous, shows us that the relation of latent heat and state of aggregation possesses a still more general regularity, and makes it seem possible to extend Van der Waal's theory to the solid state. Perhaps, too, these facts serve to render the idea of entropy more comprehensible, for the number 32, or in the case of the Desprez-Pictet-Trouton rule, the number 21, is nothing but the change of entropy involved in the change of the state of aggregation under a pressure of one atmosphere. The significance of entropy in the simplest changes of substances is therefore patent.

APPENDIX TO LECTURE III.

Nernst has recently put forward a new point of view for the calculation of chemical equilibria from thermal measurements.[2] His starting-point is the assumption that for "solid" reactions (p. 78), not only does the relation [3]—

Nernst's fundamental hypothesis.

$$[A_0] = [Q_0]$$

hold, according to which heat and work are the same for all reactions at absolute zero, but also—

$$[S_0] = 0$$

This means that not only does the product TS become zero when $T = 0$, but for this class of reactions at 0° abs. the change of entropy itself is zero. Whether the assumption is correct or not

[1] "Vorlesungen," 2nd edit., 1903, p. 52. Compare also Nernst, "Theoretische Chemie," 4th edit., p. 328. Numerous examples are to be found in both places. For other examples see Forcrand (*l.c.*) and Traube, *Berl. Ber.*, 31 (1898), p. 1562. Forcrand's objections to Van't Hoff's views do not seem to me grave ones.

[2] "Göttinger Nachrichten," 1906. Also Nernst, "Thermodynamics and Chemistry" (Chas. Scribner's Sons, New York, 1907).

[3] All terms applying to solid reactions may be put in square brackets.

can be determined solely by experiment. Since, however, direct observations are not possible at or near the absolute zero point, the deductions for higher temperatures which follow from this idea must be tested. Now, no single deduction of this nature can be reached without the help of new hypotheses, so that if a deviation occurs the subsequent hypotheses may be incorrect, and not Nernst's assumption. Conversely, if experiment should seem to confirm the assumption, it still remains unproved, for inexactness of the assumption and of the auxiliary hypotheses may compensate one another. Since, however, Nernst's fundamental hypothesis[1] is a very simple one, it appears allowable to accept it as a working hypothesis. A more detailed discussion will be given in the latter part of this appendix.

If we apply Helmholtz's equation to "solid" reactions—

$$[A] = [Q] + T\frac{d[A]}{dT}$$

and put—

$$[Q] = [Q_0] + [\sigma']T + [\sigma'']T^2 \quad . \quad . \quad . \quad . \quad (1)$$

Nernst's own way of expressing the hypothesis.

[1] Nernst expresses it somewhat differently. He puts—

$$lim\left(\frac{d[A]}{dT}\right)_{T=0} = lim\left(\frac{d[Q]}{dT}\right)_{T=0}$$

Now, [Q] denotes, according to Nernst, heat put in, while we define it as heat given out. Therefore, in our notation—

$$-lim\left(\frac{d[A]}{dT}\right)_{T=0} = lim\left(\frac{d[Q]}{dT}\right)_{T=0}$$

Now—

$$-\frac{d[A]}{dT} = [S]$$

and therefore—

$$-lim\left(\frac{d[A]}{dT}\right)_{T=0} = [S_0]$$

On the other hand, it follows from—

$$[Q_T] = [Q_0] + [\sigma']T + [\sigma'']T^2$$

that—

$$\frac{d[Q]}{dT} = [\sigma'] + 2[\sigma'']T$$

and—

$$lim\left(\frac{d[Q]}{dT}\right)_{T=0} = [\sigma']$$

Therefore—

$$[S_0] = [\sigma']$$

and from the later explanations in the text—

$$[S_0] = 0 = [\sigma']$$

we obtain, according to the explanations in Lectures II. and III.—

$$[A] = [Q_o] - [\sigma']T\,ln\,T - [\sigma'']T^2 + [const.]T \quad . \quad (2)$$

The specific heats are those obtained by heating the solid substances without production of work—that is to say, by heating in a vacuum. The usual values of specific heats of solids determined at the constant pressure of 1 atmosphere are not sensibly different from those in a vacuum, and may be taken in their stead. Equation (2), therefore, is identical with that which we obtained directly from Helmholtz's deductions (p. 50)—

$$[A] = [Q_o] - [\sigma']T\,ln\,T - [\sigma'']T^2 + ([\sigma'] + 2[\sigma''] - [S_1])T \quad (2a)$$

Hence—

$$[\sigma'] + 2[\sigma''] - [S_1] = [const.] \quad . \quad . \quad . \quad . \quad (3)$$

Substituting in (2)—

$$[Q_o] = [Q_T] - [\sigma']T - [\sigma'']T^2$$

we get—

$$[A_T] = [Q_T] - [\sigma']T(1 + ln\,T) - 2[\sigma'']T^2 + [const.]T \quad (2b)$$

Now, observing that—

$$\frac{[A_T] - [Q_T]}{T} = -[S]$$

we obtain—

$$[S] = [\sigma'](1 + ln\,T) + 2[\sigma'']T + [const.]$$

Hence, if T = 0—

$$[S_o] = [\sigma'](1 + ln\,0) + 2[\sigma'']0 + [const.]$$

If $[S_o]$ is to be zero, then in the first place $[\sigma']$ must also be zero, for, since $ln\,0$ equals $-\infty$, the product $[\sigma'](1 + ln\,0)$ would be $-\infty$ for every finite value of $[\sigma']$. Further, for $[S_o] = 0$ the constant must also be zero.

If we take $[\sigma'] = 0$, it is the same as saying that at the absolute zero point the hypothesis of Neumann and Kopp holds good, the sum of the atomic heats being exactly equal to the molecular heat.

Then it follows from (3) that—

$$0 = 0 + 2[\sigma''] - [S_1]$$

hence—

$$[S_1] = 2[\sigma''] \quad . \quad . \quad . \quad . \quad . \quad . \quad . \quad . \quad (4)$$

and from (2a)—

$$[A_T] = [Q_o] - [\sigma'']T^2 \quad . \quad . \quad . \quad . \quad . \quad (2c)$$

and from (1)—

$$[Q_T] = [Q_o] + [\sigma'']T^2 \quad . \quad . \quad . \quad . \quad . \quad (1b)$$

According to Nernst these exceedingly simple equations hold good for solid reactions, though of course he does not disregard the fact that higher terms with T^3 or T^4 may be of influence. If, for example—

$$[Q_T] = [Q_0] + [\sigma'']T^2 + [\sigma''']T^3 + [\sigma'''']T^4$$

then—

$$[A_T] = [Q_0] - [\sigma'']T^2 - 0{\cdot}5[\sigma''']T^3 - 0{\cdot}33[\sigma'''']T^4$$

Because of our insufficient knowledge of such higher terms as $[\sigma''']T^3$ and $[\sigma'''']T^4$, the question whether $[\sigma']$ equals zero is very difficult to decide experimentally.

The thermodynamically indeterminate constants of gas reactions and of vaporization processes.

Now, the connection with gas reactions appears from the following considerations. Let the vapours be in a state of equilibrium with the solid substances from which they are derived. Taking as an example the reaction—

$$2H_2 + O_2 \rightleftarrows 2H_2O \quad .$$

let us imagine a reversible isothermal cycle with this system, whereby, during the first step, 2 mols H_2 and 1 mol O_2 are vaporized; in the second, 2 mols of water-vapour are formed out of these 3 mols; in the third, these 2 mols of vapour are condensed to ice; finally, the 2 mols ice are decomposed into solid oxygen and solid hydrogen.

Then, if we write values which refer to vaporization in round brackets, and values which refer to gas reactions without brackets—

$$+2(A_{H_2}) + (A_{O_2}) + A_T - 2(A_{H_2O}) - [A_T] = 0$$

and—

$$[A_T] = A_T - \Sigma\nu'(A) \quad . \quad . \quad . \quad . \quad . \quad (5)$$

If, now, the equation for $[A_T]$, that is—

$$[A_T] = [Q_o] - [\sigma'']T^2 \quad . \quad . \quad . \quad . \quad . \quad (2c)$$

be true, and if the thermodynamically indeterminate constant in this equation is 0, as Nernst suggests, we can calculate the value of this constant in the equation—

$$A_T = Q_0 - \sigma'_p T ln T - \sigma'' T^2 - RT\Sigma\nu' ln p' + \text{const. } T \quad (6)$$

provided we know the value $\Sigma\nu'(A)$ completely.

Let us put, according to p. 77—

$$2(A_{H_2}) = 0 = 2(Q_{0(H_2)}) - 2(\sigma'_{p(H_2)})T ln T - 2(\sigma''_{(H_2)})T^2 \\ - 2RT ln p_{m(H_2)} + 2\gamma_{(H_2)}T$$

for the work of vaporization[1] of 2 mols solid hydrogen under the equilibrium pressure p_m, and similarly for 1 mol O_2—

$$(A_{O_2}) = 0 = (Q_{0(O_2)}) - (\sigma'_{p(O_2)})T\ln T - (\sigma''_{(O_2)})T^2 - RT\ln p_{m(O_2)} + \gamma_{(O_2)}T$$

and for 2 mols H_2O—

$$2(A_{H_2O}) = 0 = 2(Q_{0(H_2O)}) - 2(\sigma'_{p(H_2O)})T\ln T - 2(\sigma''_{(H_2O)})T^2 - 2RT\ln p_{m(H_2O)} + 2\gamma_{(H_2O)}T$$

then we obtain on summation—

$$\begin{aligned} -\Sigma\nu'(A) = {} & 2(Q_{0(H_2)}) + (Q_{0(O_2)}) - 2(Q_{0(H_2O)}) \\ & - \{(2\sigma'_{p(H_2)}) + (\sigma'_{p(O_2)}) - 2(\sigma'_{p(H_2O)})\}T\ln T \\ & - \{(2\sigma''_{H_2}) + (\sigma''_{O_2}) - (2\sigma''_{H_2O})\}T^2 \\ & + RT\ln\frac{p^2_{m(H_2O)}}{p_{m(O_2)}p^2_{m(H_2)}} \\ & + (2\gamma_{(H_2)} + \gamma_{(O_2)} - 2\gamma_{(H_2O)})T \end{aligned}$$

Instead of this, we can put—

$$-\Sigma\nu'(A) = -\Sigma\nu'(Q_0) + \Sigma\nu'(\sigma'_p)T\ln T + \Sigma\nu'(\sigma'')T^2 + RT\Sigma\nu'\ln p'_m - \Sigma\nu'\gamma T \quad . \quad . \quad . \quad (7)$$

Now it may readily be seen that—

$$-\Sigma\nu'(\sigma'_p) = [\sigma'] - \sigma'_p \quad . \quad . \quad . \quad . \quad . \quad (8)$$

or since, on Nernst's assumption—

$$[\sigma'] = 0$$

therefore—

$$\Sigma\nu'(\sigma'_p) = \sigma_p T_1 \quad . \quad . \quad . \quad . \quad . \quad . \quad (9)$$

Further—

$$-\Sigma\nu'(\sigma'') = [\sigma''] - \sigma'' \quad . \quad . \quad . \quad . \quad (10)$$

If we replace $[A_T]$ by its value in (2c) and A_T by (6), then (5) becomes—

$$-\Sigma\nu'(A) = [Q_0] - [\sigma'']T^2 - Q_0 + \sigma'_p T\ln T + \sigma''T^2 + RT\Sigma\nu'\ln p'_m - \text{const.} \times T$$

or from (9) and (10)—

$$-\Sigma\nu'(A) = -\Sigma\nu'(Q_0) + \Sigma\nu'(\sigma'_p)T\ln T + \Sigma\nu'(\sigma'')T^2 + RT\Sigma\nu'\ln p_m - \text{const. } T \quad . \quad . \quad . \quad (11)$$

and from (7) and (11)

$$\text{const.} = \Sigma\nu'\gamma \quad . \quad . \quad . \quad . \quad . \quad . \quad (12)$$

[1] The letter γ stands for the thermodynamically indeterminate constant of the vaporization process.

The relation $[Q_0] - Q_0 = -\Sigma\nu'(Q_0)$ has been made use of in this derivation. It is readily obtained when the cycle is imagined to be carried out at the absolute zero, and the heats given out are summed up. This sum must be zero, as in every reversible isothermal cycle.

The thermodynamically indeterminate gas-reaction constant is therefore found to be the sum of the vaporization-constants γ of the gases concerned.

Our notation compared with that of Nernst.

Nernst defines his values somewhat differently. According to him, the value which we call γ is $R(i + ln R)$. Our value (σ'_p), which represents the difference between the specific heat of 1 mol in the solid state at 0° abs. and the specific heat of the same mol in the gaseous state at 0° abs. and constant pressure, *i.e.*—

$$(\sigma'_p) = C^{\circ}_{(solid)} - C^{\circ}_{p(gas)}$$

does not appear in Nernst's equations. He puts for $C^{\circ}_{p\,(gas)}$ the value $a + R$, and for $C^{\circ}_{(solid)}$ the symbol a_0, thus—

$$-(\sigma'_p) = R + a - a_0$$

Similarly, according to Nernst's notation—

$$-(\sigma'') = \beta - \beta_0$$

Lastly, he uses $-\lambda_0$ instead of our symbol Q_0.

In the course of his work Nernst defines another constant C—

$$2{\cdot}302C = (i + ln R) \quad . \quad . \quad . \quad . \quad . \quad (13)$$

Our value γ is accordingly given by the expression—

$$\gamma = 4{\cdot}571C \quad . \quad . \quad . \quad . \quad . \quad . \quad . \quad (14)$$

(The value 4·571 is employed for $R/ln 10$. Nernst prefers this value to the value 4·56 used in this book.)

The basis of Nernst's deduction.

The validity of Nernst's deduction depends entirely upon our decision whether or not we can employ the expression—

$$0 = (Q_0) - (\sigma'_p)T ln T - (\sigma'')T^2 - RT ln p_m + \gamma T \quad . \quad (15)$$

for the vaporization of a solid. As was shown on p. 76, this formula is derived from that of Clapeyron and Clausius, which holds in all cases, when—

(*a*) the saturated vapour obeys the laws of perfect gases, and

(*b*) the volume of a mol in the solid or liquid state is practically vanishingly small compared with that of a mol of the saturated vapour.

Now, it is a well-known fact that both these suppositions are not in general fulfilled by saturated vapours. Consequently the author has in the Third Lecture used formula (15) with great care. Nernst now puts forward the very interesting proposition that formula (15) also holds when both suppositions are but very imperfectly realized. He establishes his theory so that he puts forward the empirical relation—

$$p_m(v_a - v_b) = \mathrm{RT}\left(1 - \frac{p_m}{p_k}\right) \quad . \quad . \quad . \quad . \quad (16)$$

for the relation of the molecular volume v_a in the gaseous state and v_b in the liquid state to the vapour pressure p_m and the temperature T of vaporization.

p_k denotes the critical pressure.

On the other hand, he gives the expression—

$$(Q_T) = (Q_0) - 3{\cdot}5\mathrm{T} + (\sigma'')\mathrm{T}^2\left(1 - \frac{p_m}{p_k}\right) \quad . \quad . \quad (17)$$

for the heat set free during vaporization (negative heat of vaporization). When both these equations hold, then the expression of Clapeyron and Clausius—

$$-(Q_T) = \mathrm{T}(v_a - v_b)\frac{dp_m}{d\mathrm{T}} \quad . \quad . \quad . \quad . \quad (18)$$

becomes converted into formula (15) without the necessity of using the law $pv = \mathrm{RT}$. Then by introducing (16) and (17) in equation (18) we obtain—

$$-(Q_0) + 3{\cdot}5\mathrm{T} - (\sigma'')\mathrm{T}^2 = \frac{\mathrm{RT}^2}{p_m} \times \frac{dp_m}{d\mathrm{T}} \quad . \quad . \quad (19)$$

But the integral of (19) is identical with (15) when we admit that—

$$(\sigma'_p) = -3{\cdot}5$$

We obtain from (16) the general gas equation for perfect gases when p_m becomes very small compared with the critical pressure, and v_b consequently very small in comparison with the gas volume v_a. At the critical pressure itself, where $v_a = v_b$, then $\mathrm{RT}\left(1 - \frac{p_m}{p_k}\right)$ becomes zero. As far as the relations at those temperatures where p_m is a considerable fraction of p_k are concerned, Nernst shows that for benzene fluoride, which Young has studied with special accuracy, the deviations of the expression $\mathrm{R}\left(1 - \frac{p_m}{p_k}\right)$ from the observed value $p\frac{(v_a - v_b)}{T}$ amount to only one to two per cent. of the whole, the temperatures being 0·66 to 0·84 times the critical temperature (559·6° abs.). For higher temperatures the agreement is much worse, for lower temperatures even better. Nernst tests equation (17) on ammonia (critical temperature 404° absolute), and finds at 313° absolute, *i.e.* at $0{\cdot}77T_k$, a deviation of 2·7 per cent., which becomes greater at higher temperatures and vanishes at decreasing temperatures.

No more examples are given. In both cases which have been considered, the vapour pressures which correspond to the temperatures of $0{\cdot}84T_k$ for benzene fluoride and $0{\cdot}77T_k$ for ammonia are 13 and 15½ atmos. respectively. Above 20 atmos. the formulæ are thus in both cases insufficient. But if it should be found that saturated vapours of all substances up to 20 atmos. obey with equally good approximation (16) and (17), whose introduction into (18) gives formula (15), then an important advance has been made. According to the interpretation of the author, this is a cardinal point of Nernst's theory.

Nernst's assumption concerning the numerical values of the specific heats at the absolute zero point.

Equation (17) contains the idea that the value (σ'_p) is $-3{\cdot}5$ for all substances. This means that the specific heats of all gases at absolute zero exceed those of their condensation products by 3·5, whilst, as is well known, the specific heat of saturated vapours is usually less than that of their condensation products. Nernst characterizes this as a preliminary assumption, the proof of which is wanting. It seems that up to now only Dewar[1] has determined the specific heat of gases in the solid state at very low temperatures. He found the specific heats of—

	CO_2	NH_3	SO_2
at	−130°	−145°	−143° C.
to be	9·68	8·5	14·7

The specific heats of these substances in the gaseous state at the same temperature are probably smaller.

The value (σ'') which occurs in equation (17) is deduced by Nernst in the following manner. Let T in (17) be a temperature for which p_m is only a small fraction of p_k, then the terms $3{\cdot}5T\frac{p_m}{p_k}$ and $(\sigma'')T^2\frac{p_m}{p_k}$ are so small that they may be neglected, and (17) can be written—

[1] *Chemical News*, 92 (1905).

$$(Q_T) = (Q_o) - 3{\cdot}5T + (\sigma'')T^2 - (Q_o)\frac{p_m}{p_k}$$

On differentiating with respect to T, we obtain—

$$\frac{d(Q_T)}{dT} = -3{\cdot}5 + 2(\sigma'')T - \frac{(Q_o)}{p_k} \times \frac{dp_m}{dT}$$

Now, the last term on the right-hand side is small, and in order to obtain an approximation, we may replace $\frac{dp_m}{dT}$, which is equal to $\frac{-(Q_T)}{T(v_a - v_b)}$ according to (18), by $\frac{-(Q_T)}{Tv_a}$; or, using the gas equation for v_a—

$$v_a = \frac{RT}{p_m}$$

$$\frac{dp_m}{dT} = -\frac{(Q_T)p_m}{RT^2}$$

whence it follows that—

$$\frac{d(Q_T)}{dT} = -3{\cdot}5 + 2(\sigma'')T + \frac{(Q_0)(Q_T)}{RT^2} \times \frac{p_m}{p_k} \quad . \quad (20)$$

If (Q_T) and $\frac{d(Q_T)}{dT}$ are known for any temperature T, as well as p_m and the critical pressure p_k, the values of (σ'') and (Q_o) can be obtained with the aid of both equations (17) and (20). When (Q_o) and (σ'') have been found in this manner, γ can be easily calculated from formula (15). The reliability of this γ value, to be sure, depends entirely upon whether the value $-3{\cdot}5$ for (σ'_p) can be accepted as correct.

Nernst still further extends the hypothesis based on this assumption regarding relationship at the absolute zero. He concludes, with the help of the kinetic theory of gases, that the monatomic gases show at all temperatures the same ratio of the specific heats at constant pressure and constant volume, namely 1·66. Therefore the value of their molecular heats at constant pressure and absolute zero is the same as at all other temperatures, namely 5·0. Now, since he puts $(\sigma'_p) = -3{\cdot}5$, the molecular heat of a monatomic substance in the solid state at absolute zero must be 1·5, because—

$$C^o_{(solid)} - C^o_{p(gas)} = (\sigma'_p) = -3{\cdot}5$$

and—

$$C^o_p = 5$$

therefore—

$$C^o_{(solid)} = 5 - 3{\cdot}5 = 1{\cdot}5$$

But if all monatomic substances at 0° (abs.) have the specific heat 1·5, it then follows from the previous supposition, according to which Kopp's law holds for all solid substances at $T = 0$, that the molecular heat of an n-atomic solid substance ($T = 0$) is[1]—

$$n \times 1{\cdot}5$$

and it follows that the molecular heat of an n-atomic gas at the absolute zero and under constant pressure is—

$$3{\cdot}5 + n \times 1{\cdot}5$$

Accordingly, for example, the specific heat of a diatomic gas such as O_2 and H_2 at constant pressure and $T = 0$ would be—

$$3{\cdot}5 + 2 \times 1{\cdot}5 = 6{\cdot}5$$

and for a triatomic gas, *e.g.* water-vapour—

$$3{\cdot}5 + 3 \times 1{\cdot}5 = 8{\cdot}0$$

The results of this conception are very wide-reaching. If the true specific heat of a gas at any temperature and constant pressure has been measured, we can deduce the mean specific heat of the gas between 0° abs. and any temperature which we please with the aid of the value—

$$C^{\circ}_{p} = 3{\cdot}5 + n \times 1{\cdot}5$$

assuming that the mean specific heat is a linear function of the temperature—

$$(3{\cdot}5 + n \times 1{\cdot}5) + bT$$

Hereby, however, one meets deviations from experiment either due to the fact that the expression—

$$C^{\circ}_{p} = 3{\cdot}5 + n \times 1{\cdot}5$$

is not exact, or to the fact that the specific heat is not a linear function of the temperature. On the other hand, we can obtain values $pf(\sigma'')$ from equations (20) and (17), and, knowing the specific heats of the saturated vapours, we can deduce the specific heats of the condensation products. For the same reasons we should also expect to find deviations in this case, the extent of which could only be determined after more accurate investigation.

Now, the points of view which have been developed are not

[1] For views and experimental facts concerning the specific heats of solids at very low temperature, see Van't Hoff, *Vorlesungen über theoretische und physikalische Chemie*, 1899, vol. iii. § 4, 8; and Forch and Nordmeyer, Drude's *Ann.*, 20 (1906), 426.

sufficient, even presupposing their correctness, for the calculation of γ in the cases which are of most importance. It is precisely those gases which chiefly take part in actual gas reactions, the so-called permanent gases, whose critical points lie so low that up to the present sufficient experimental information regarding (Q_T), $\frac{d(Q_T)}{dT}$, and p_m has not been gathered for us to calculate (Q_0) and (σ''), and these are essential, according to formulæ (17) and (20), to determine γ in formula (15). The whole deduction can only be serviceable for the numerical evaluation of the thermodynamically indeterminate constants when a further relation is found, which will render γ accessible in this case also. The method which Nernst adopts for this purpose is a very interesting one. It is generally acknowledged that the vapour pressure p_m at a temperature T could be calculated from the critical data T_k and p_k in an identical way for all substances provided we possessed a sufficiently exact characteristic equation of the state of universal applicability. The characteristic equation of Van der Waals—

$$\left(p + \frac{a}{v^2}\right)(v - b) = RT$$

is not sufficiently exact, and a non-arbitrary deduction from a theoretical consideration of a more exact expression is not possible. Under these circumstances we must have recourse to empirical approximations.

We shall, however, simply suppose that an exact formula does exist, which would take theoretically the form—

$$\frac{p_m}{p_k} = f\left(\frac{T_k}{T}\right) \quad . \quad . \quad . \quad . \quad . \quad . \quad (21)$$

Here p_k denotes the critical pressure, and T_k the critical temperature, f a function whose exact nature is unknown, as explained above. If $f\left(\frac{T_k}{T}\right)$ were known, we could express p_m in terms of $f\left(\frac{T_k}{T}\right)$ and p_k. But now, with the aid of (15), we can, according to Nernst, calculate p_m, at least for values of T which do not lie too near the critical temperature. This means that the equations (21) and (15) would give the same values of p_m. Consequently the values (Q_0), (σ'_p), (σ''), and γ, which determine the value of p_m in (15), would be connected with p_k and $f\left(\frac{T_k}{T}\right)$ in such a way that they could be expressed by them. For this fictitious case γ could also be calculated from the critical constants.

The empirical approximation of Van der Waals.

Now, Van der Waals has incidentally proposed—

$$\log \frac{p_k}{p_m} = a\left(\frac{T_k}{T} - 1\right) \quad . \quad . \quad . \quad . \quad . \quad (22)$$

as an empirical approximation. a denotes a constant. If we assume this approximation as sufficiently exact, then γ can be calculated from a, T_k, and p_k according to (22) and (15). This is Nernst's interpretation. He puts—

$$1 \cdot 1a = C = \frac{\gamma}{4 \cdot 571}$$

or (rounding off slightly)—

$$\gamma = 5a$$

This simple numerical relation between a and γ is, however, not directly capable of proof, because the formulæ (15) and (22) cannot be brought into entire agreement. This is readily seen on transforming (22) into—

$$lnp_m = -2 \cdot 302a\frac{T_k}{T} + 2 \cdot 302(a + \log p_k)$$

or written otherwise into—

$$0 = -4 \cdot 571 \times a \times T_k - RTlnp_m + 4 \cdot 571(a + \log p_k)T$$

If we now compare this with (15) we come to the conclusion that, were (σ'_p) and (σ'') zero, *i.e.* if the vaporization heat were independent of the temperature, (Q_o) would have the value $-4 \cdot 571 \times a \times T_k$, and γ the value $4 \cdot 571(a + \log p_k)$. Then the two expressions would be the same. But experimentally the values (σ'_p) and (σ'') are not zero, and any comparison at all of (15) and (22) is difficult, since (15) agrees better with experiment the further we are below the critical temperature, while (22) is most accurate in the immediate neighbourhood of the critical temperature.

The relation of (15) to (22) becomes still clearer after another algebraic transformation. If we refer (15) to the critical point, thus—

$$0 = (Q_0) - (\sigma'_p)T_k lnT_k - (\sigma'')T_k^2 - RT_k lnp_k + \gamma T_k \qquad (15a)$$

and subtract (15) from (15a), after dividing these expressions by RT and RT_k respectively, it follows that—

$$\log \frac{p_k}{p_m} = -\frac{(Q_0)}{4 \cdot 571T_k}\left(\frac{T_k}{T} - 1\right) - \frac{(\sigma'_p)}{R} \log \frac{T_k}{T} - \frac{(\sigma'')T_k}{R}\left(1 - \frac{T}{T_k}\right)$$

If we put—

$$(Q_0) = -4 \cdot 571aT_k$$

then the expression becomes—

$$\log \frac{p_k}{p_m} = a\left(\frac{T_k}{T} - 1\right) - \frac{(\sigma'_p)}{R} \log \frac{T_k}{T} - \frac{(\sigma'')}{R} T\left(1 - \frac{T}{T_k}\right)$$

But this is identical with (22) only when (σ'_p) and (σ'') are zero.

Still, an approximate agreement is at least conceivable when the terms $\frac{(\sigma'_p)}{R} \log \frac{T_k}{T}$ and $\frac{(\sigma'')T_k}{R}\left(1 - \frac{T}{T_k}\right)$ have nearly the same value, but opposite signs. It is then to be observed that (σ'_p) and (σ'') either cannot in the least depend on the nature of the substance, or if they do so, they must be functions of T_k, p_k, and a which have the same form for all substances. Nernst now puts $(\sigma'_p) = -3{\cdot}5$ tentatively for all substances, as has been formerly mentioned, and also—

$$\frac{\sigma'' T_k}{R} = \frac{a}{2{\cdot}36}$$

and finds from the resulting formula—

$$\log \frac{p_k}{p_m} = a\left(\frac{T_k}{T} - 1\right) + 1{\cdot}75 \log \frac{T_k}{T} - \frac{a}{2{\cdot}36}\left(1 - \frac{T_k}{T}\right) \quad . \quad (23)$$

approximately the same values for $\log \frac{p_k}{p_m}$ as from (22) when a in (23) is chosen only slightly different from a in (22).

Now, the values of a in général are somewhat uncertain. According to Van der Waals, who brings forward much direct and indirect evidence in favour of his view, a should have the same value, namely 3, for all substances. Ramsay[1] and Guye and Mallet[2] share in this view when polymerization does not interfere. Nernst represents the results of observation graphically, $\log \frac{p_k}{p_m}$ being taken as ordinate, and $\left(\frac{T_k}{T} - 1\right)$ as abscissa. But he does not obtain by any means the same line for all substances. He finds, on the contrary, that a is larger for higher molecular weight or for a greater number of atoms in the molecule than for smaller molecular weight or a smaller number of atoms in the molecule. The values of a vary from about 2 to 4, and therefore lie around 3, the value which Van der Waals thinks ought to apply. But it is further to be observed that the lines for one and the same substance are not straight, but are a little curved, so that formula (22) is by no means an exact

[1] *Zeitschrift f. physic. chem.*, 15 (1894), 106.

[2] *Archive des sciences physiques et naturelles, Genève*, (IV.) 13 (1902), 66.

expression of the facts.[1] Under these circumstances Nernst makes use of the following expedient. He calculates a from (22) for temperatures between $0{\cdot}8T_k$ and $0{\cdot}71T_k$, where this formula agrees comparatively well with experiment. On the other hand, he calculates γ (that is to say, his constant C, which corresponds to our γ) from (15) in the way previously described, and obtains in this manner the relation $\gamma = 5a$, which he accepts as valid in all cases. The exact details of this calculation have not yet been communicated. The following table for various values of C contains the results. The author has placed the values of γ beside corresponding values of C.

Numerical values of the thermodynamical indeterminate constants of vaporization processes.

	C	γ		C	γ
H_2	2·2	10·0	CO_2	3·2	14·6
CH_4	2·5	11·4	CS_2	3·1	14·2
N_2	2·6	11·9	NH_3	3·3	15·1
O_2	2·8	12·8	H_2O	3·7	16·9
CO	3·6	16·5	CCl_4	3·1	14·2
Cl_2	3·0	13·7	$CHCl_3$	3·2	14·6
I_2	4·0	18·3	C_6H_6	3·1	14·2
HCl	3·0	13·7	C_2H_5OH	4·1	18·7
NO	about 3·7	16·9	$C_2H_5OC_2H_5$	3·3	15·1
N_2O	3·3	15·1	CH_3COCH_3	3·7	16·9
H_2S	3·0	13·7	$CH_3COOC_3H_7$	3·8	17·4
SO_2	3·3	15·1			

Comparison of Nernst's views with those explained in the former lectures.

If we now consider the relation of Nernst's interpretation to that in the Second Lecture, the following may be noted :—

The hypothesis (for solid reactions) that

$$[S_0] = 0$$

is of prime importance. If in a system of rectangular co-ordinates we plot T as abscissæ and [A] and [Q] as ordinates, we obtain two lines, according to Nernst's equations (2*c*) and (1*b*), which converge at the value $[A_0] = [Q_0]$, without intersecting. It is important to observe that the directions of both curves become more nearly parallel to the axis of abscissæ the nearer we approach the absolute zero. However, if we take for [A] equation (2), and for [Q] equation (1), where $[\sigma]$ is not zero, we then obtain two lines, which at higher temperatures run similarly to those in the case above, and also meet one another at the absolute zero. But they show a difference in the neighbourhood

[1] Happel (*Drude's Ann. d. Physik*, 13 (1904), 340) finds, too, that a is dependent upon the temperature. H. v. Jüptner (*Zeitschr. f. physik. Chemie*, 55 (1906)) comes to the same conclusion.

of the absolute zero. The curve for [A] intersects the curve for [Q] very near the absolute zero, then bends and meets the curve for [Q] at the absolute zero itself. In the former case the angle of inclination of the two curves at the meeting-point, T = 0, is zero; in the latter case it is about 90°. That is to say—

$$\lim\left(\frac{d[Q]}{dT}\right)_{T=0}=[\sigma']$$

and

$$\lim\left(\frac{d[A]}{dT}\right)_{T=0}=\text{const.}-[\sigma'](ln0+1)=-\infty$$

Nernst's further supposition regarding the specific heat of gases renders it necessary that the difference of the specific heats at the absolute zero (σ'_p or σ'_v) be zero in all those gas reactions where the number of substances which are formed is equal to the number of those which disappear. Now, since the size of the degree interval of the thermometer scale has no effect upon the thermodynamically indeterminate constants of gas reactions except in those reactions where σ'_v does not differ from zero, we can no longer retain the deduction developed on p. 45 if we accept Nernst's hypothesis. The idea of the influence of the degree interval is based on the evidence produced by Th. W. Richards and Van't Hoff that the thermodynamically indeterminate constant becomes zero when the appearing and disappearing substances possess the same specific heat. This proof does not appear to be conclusive in the light of Nernst's interpretation.

If we return to the usual approximation (p. 66) formerly made in theoretical physics, according to which the assumption of Buff and Clausius holds, and the heat of gas reactions at constant volume is independent of the temperature, then the reaction energy of gas reactions may be written (see p. 69)—

$$A=Q_0+RT\Sigma\nu'lnT-RT\Sigma\nu'lnp'+\text{const. }T$$

When the number of disappearing mols exceeds the number of those formed by 1, then—

$$A=Q_0-RTlnT-RT\Sigma\nu'lnp'+\text{const. }T \qquad (24)$$

Nernst's expression, on the contrary, is as follows :—

$$A=Q_0-3{\cdot}5TlnT-\sigma''T^2-RT\Sigma\nu'lnp'-\Sigma\nu'\gamma T$$

Now, since σ'' is small, but generally not sufficiently well known, Nernst recommends the following approximation :—

$$A=Q_0-3{\cdot}5TlnT-RT\Sigma\nu'lnp'-\Sigma\nu'\gamma T$$

This means that the constant of the formula (24) is approximately—

$$\text{const.} = -\Sigma\nu'\gamma - 1{\cdot}5 ln\text{T}$$

The calculations of former physicists can in this way be conveniently compared with those of Nernst. The theoretical basis, however, is quite different, because Nernst's approximation does not agree with the assumption of Buff and Clausius.

The cardinal point of Nernst's theory lies in the hypothesis that the vapour-pressure curve can be represented by—

$$0 = (\text{Q}_0) - (\sigma'_p)\text{T}ln\text{T} - (\sigma'')\text{T}^2 - \text{RT}lnp_m + \gamma\text{T} \quad . \quad (15)$$

although the saturated vapours do not obey the gas law, which is all that has been used till now in deriving this formula from the strictly valid formula of Clapeyron and Clausius. The vaporization constant γ, and further the thermodynamically indeterminate constant for gas reactions, can only be deduced with the help of this hypothesis from the vaporization constants of each of the participating gases. The starting-point of this is the empirical formula—

$$p_m(v_a - v_b) = \text{RT}\left(1 - \frac{p_m}{p_k}\right) \quad . \quad . \quad . \quad . \quad (16)$$

On differentiating formula (15), we obtain—

$$\frac{\text{T}}{p_m} \times \frac{dp_m}{d\text{T}} = \frac{-(\text{Q}_0)}{\text{RT}} - \frac{(\sigma'_p)}{\text{R}} - \frac{(\sigma'')\text{T}}{\text{R}} \quad . \quad . \quad (15a)$$

But according to Clausius and Clapeyron (see (18), p. 89)—

$$\frac{-(\text{Q}_\text{T})}{(v_a - v_b)p_m} = \frac{\text{T}}{p_m} \times \frac{dp_m}{d\text{T}}$$

From this it follows, if (15) and consequently (15a) are correct—

$$\frac{(\text{Q}_\text{T})}{(v_a - v_b)p_m} = \frac{\text{Q}_0}{\text{RT}} + \frac{(\sigma'_p)}{\text{R}} + \frac{(\sigma'')\text{T}}{\text{R}}$$

On applying formula (16), it follows that—

$$(\text{Q}_\text{T}) = \{(\text{Q}_0) + (\sigma'_p)\text{T} + (\sigma'')\text{T}^2\}\left(1 - \frac{p_m}{p_k}\right) \quad . \quad (25)$$

If these empirical expressions (16) and (25) are further confirmed by experiment, an important perfection of our knowledge regarding the vapour-pressure curve has been attained.

The vaporization constant γ takes us back finally to the statements of Trouton and Forcrand.

Nernst puts forward the same formula which is given as formula (30a) on p. 77 of this book. But he does not assume that the terms containing (σ_p) and (σ'') compensate one another, giving zero; according to him, the constant of this expression, therefore, does not become about 30. He puts $-3{\cdot}5$ for (σ'_p), according to his former hypothesis. Further, he puts for (σ'') a number obtained from the observations of the specific heats of gases, and for the constant he puts his γ value according to the table given above, based on his hypothesis regarding the values at the absolute zero. In this way he actually obtains, for example, for the ammonia compounds of the metallic salts the observed value of 33 for $\frac{Q}{T}$. He then puts generally, neglecting terms with σ'' and the factor $-\sigma'_p \frac{290}{T_{p_m=1}}$, where $T_{p_m=1}$ signifies the temperature at which the vapour pressure is equal to 1 atmosphere—

$$-\frac{Q_{p(290)}}{T_{p_m=1}} = 3{\cdot}5 \, ln T + \gamma$$

as an approximation formula. Since the values of γ, according to the table, lie in the neighbourhood of 13·7, it follows that $-\frac{Q_{p(290)}}{T_{p_m=1}}$ equals—

29·7	for T =	100
33·6	,, =	300
35·3	,, =	500
37·7	,, =	1000

According to Le Chatelier and Forcrand, as has been pointed out earlier in this work, the values found are about 32, with variations of ±7.

Thus the quotient $-\frac{Q_{p(290)}}{T_{p_m=1}}$ does not appear to be constant, but has different values according to the temperatures for which $p_m = 1$, and also according to the γ value of the substance. The same applies to those analogous quotients from the heats of vaporization and absolute boiling-points at atmospheric pressure, the constancy of which has been maintained by Desprez, Pictet, and Trouton.

The application of Nernst's hypothesis to the gas reactions leads finally to the following numerical expressions, which satisfy the measurements carried out on the gas equilibria concerned :—

$$(a) \quad H_2 + \tfrac{1}{2}O_2 \rightleftarrows H_2O + A$$

$$A = 57{,}300 - 1{\cdot}75 T \, ln T - 0{\cdot}0003 T^2 - RT \, ln \frac{p_{H_2O}}{p_{H_2} \times p_{O_2}^{\frac{1}{2}}} + 0{\cdot}46 T$$

$$Q = 57{,}300 + 1{\cdot}75 T + 0{\cdot}0003 T^2$$

$$(b)\quad CO + \tfrac{1}{2}O_2 \rightleftarrows CO_2 + A$$

$$A = 67{,}650 - 1{\cdot}75T\,ln\,T + 0{\cdot}0015T^2 - RT\,ln\,\frac{p_{CO_2}}{p_{CO} \times p_{O_2}^{\frac{1}{2}}} - 8{\cdot}2T$$

$$Q = 67{,}650 + 1{\cdot}75T - 0{\cdot}0015T^2$$

$$(c)\quad \tfrac{1}{2}N_2 + \tfrac{1}{2}O_2 \rightleftarrows NO + A$$

$$A = 21{,}600 - 0{\cdot}0002T^2 - RT\,ln\,\frac{p_{NO}}{p_{O_2}^{\frac{1}{2}} \times p_{N_2}^{\frac{1}{2}}} + 4{\cdot}6T$$

$$Q = 21{,}600 + 0{\cdot}0002T^2$$

$$(d)\quad \tfrac{1}{2}H_2 + \tfrac{1}{2}Cl_2 \rightleftarrows HCl + A$$

$$A = 22{,}000 - RT\,ln\,\frac{p_{HCl}}{p_{H_2}^{\frac{1}{2}} \times p_{Cl_2}^{\frac{1}{2}}} + 1{\cdot}8T$$

$$Q = 22{,}000$$

It is here assumed that the heat of formation of hydrochloric acid gas does not sensibly alter with the temperature. This assumption is certainly not in harmony with the observations on the specific heats of the gases. Nernst, however, assumes that the value for the specific heat of chlorine is influenced by deviations of this gas from the gas laws. The thermodynamically indeterminate constants of these equations follow for those four cases from the former table as follows[1] :—

$(a)\quad \gamma_{H_2O} - \gamma_{H_2} - \tfrac{1}{2}\gamma_{O_2} = 16{\cdot}9 - 10 - 6{\cdot}4 = 0{\cdot}5$

$(b)\quad \gamma_{CO_2} - \gamma_{CO} - \tfrac{1}{2}\gamma_{O_2} = 14{\cdot}6 - 16{\cdot}5 - 6{\cdot}4 = -8{\cdot}3$

$(c)\quad \gamma_{NO} - \tfrac{1}{2}\gamma_{O_2} - \tfrac{1}{2}\gamma_{N_2} =$ about $16{\cdot}9 - 6{\cdot}4 - 5{\cdot}95 =$ about $4{\cdot}55$

$(d)\quad \gamma_{HCl} - \tfrac{1}{2}\gamma_{Cl_2} - \tfrac{1}{2}\gamma_{H_2} = 13{\cdot}7 - 6{\cdot}85 - 5 = 1{\cdot}85$

The values of the thermodynamically indeterminate constant satisfy in all four cases the measurements which have been carried out on the gas equilibria concerned with perfect exactness. Since the γ values have been obtained from the values of a in an approximate manner (by multiplication with 5), and the values of a themselves are, as Nernst emphasizes, uncertain to some extent, considerably different values of the thermodynamically indeterminate constant would not necessarily contradict Nernst's calculations.

As a final example Nernst gives—

$$(e)\quad \tfrac{1}{2}N_2 + \tfrac{3}{2}H_2 \rightleftarrows NH_3 + A$$

$$A = 12{,}000 - 3{\cdot}5T\,ln\,T - RT\,ln\,\frac{p_{NH_3}}{p_{N_2}^{\frac{1}{2}} \times p_{H_2}^{\frac{3}{2}}} - 11{\cdot}9$$

$$Q = 12{,}000 + 3{\cdot}5T$$

[1] Small differences in the foregoing and following values arise only from rounding off in the calculations of γ from C, as given in the above table.

This expression only claims approximate validity on account of the omission of σ''. It gives for the dissociation of ammonia at 893° abs. the value which has been experimentally found at 1293° abs. It would be very interesting to determine whether the thermodynamical indeterminate constants, which in this way result from the γ values, satisfy the observations on other gas reactions.

It is certain that Nernst's system is the most interesting and comprehensive in this field.

FOURTH LECTURE

EXAMPLES OF REACTIONS WHICH PROCEED WITHOUT A CHANGE IN THE NUMBER OF MOLECULES

General observation.

WE will now undertake to apply the relationships we have just deduced to a few examples. We must, however, preface our remarks by an observation of a general nature.

All the reactions to be considered are reversible. We may therefore write them as though proceeding in both directions. As, for instance—

$$H_2 + Cl_2 = 2HCl \quad . \quad . \quad . \quad . \quad . \quad . \quad (1)$$

$$\text{or} \quad 2HCl = H_2 + Cl_2 \quad . \quad . \quad . \quad . \quad . \quad (2)$$

Whether we get a positive or negative value for the heat of reaction depends on which direction we choose. Our definition of the equilibrium constant also depends upon this same choice. We can, for instance, in the case just cited, consider either—

$$\frac{C^2_{HCl}}{C_{H_2} \times C_{Cl_2}} \quad \text{or} \quad \frac{C_{H_2} \times C_{Cl_2}}{C^2_{HCl}}$$

as the equilibrium constant. The former corresponds to equation (1), the latter to equation (2). Our choice here is entirely free, and we find no fixed usage in the literature. I shall therefore choose, once for all, the *formation* reaction as a starting-point—that is, in the case just considered, the reaction represented by equation (1).

On this scheme the *compounds* always appear in the numerator of the equilibrium constant, and the *components* in the denominator. The heats of reaction are thus always heats of formation, and not heats of decomposition. A further simple consequence of this is that large values of the equilibrium constant always indicate very complete combination, small values far-reaching decomposition.

We have a further choice in the number of molecules we take. Thus, the equation for the formation of hydrochloric acid may be written either as—

$$H_2 + Cl_2 = 2HCl$$

$$\text{or} \quad \tfrac{1}{2}H_2 + \tfrac{1}{2}Cl_2 = HCl$$

In the former case the heat of formation is that of two mols of hydrochloric acid, in the latter of but a single one. The equilibrium constant in the first case would be—

$$\frac{C^2_{HCl}}{C_{H_2} \times C_{Cl_2}}$$

in the latter—

$$\frac{C_{HCl}}{C^{\frac{1}{2}}_{H_2} \times C^{\frac{1}{2}}_{Cl_2}}$$

Both are equally admissible.[1] I prefer the second, because the equilibrium constants thereby acquire convenient values. I believe that this advantage should not be underestimated. In many instances, such as in the formation of sulphur trioxide, one can see at a glance many useful relationships, provided the equilibrium constants have low and convenient values.[2]

Since we are only considering cases where the reaction proceeds without a change in the number of molecules, it is immaterial whether we use concentrations, per cents. by volume, or partial pressures in our calculations. Further, it is more probable here than elsewhere that the thermodynamically indeterminate constant which we discussed in our Second and Third Lectures would be zero.

The differences of the specific heats of the appearing and disappearing gases are here the same at constant volume as at

[1] In the first instance the equilibrium constant is simply the quotient of the velocity constants of the formation and decomposition reactions; in the second, it is the square root of this quotient. The pedagogic advantage secured in the first instance for the comparative treatment of velocity and equilibrium constants does not seem to me very important. As far as the computation is concerned, the roots can be very conveniently handled by means of the tables given in the five-place logarithm tables of F. Gauss (Halle, 1905).

[2] In this connection see the statements in Lunge, "Sodaindustrie," 3rd edit., vol. i. p. 950, where the equilibrium constants for the sulphur trioxide process at 515° and 465° are given in numbers which lie between 10 and 100 millions.

constant pressure. The quantities σ'_p and σ'_v are therefore identical, and the indices may be omitted.

Classification of the cases concerned.

We will consider three cases under this head—

(1) The formation of nitric oxide from the elements.

(2) The formation of hydrochloric, hydrobromic, and hydriodic acid gas from the elements.

(3) The water-gas reaction.

The first and especially the third cases are of marked technical interest. We must, however, postpone any consideration of them from this point of view until the final lecture.

As far as specific heats are concerned, we may divide the gases taking part in these reactions into two classes. Oxygen, nitrogen, and nitric oxide have nearly identical specific heats at ordinary temperatures, and they all increase but slowly with rising temperature. The triatomic gases, carbon dioxide, and water-vapour, on the other hand, have specific heats differing considerably from each other, and even more from that of the diatomic gases. Not only is this true at ordinary temperatures, but their specific heats increase much more rapidly with the temperature than do the specific heats of the above diatomic gases. The hydrogen halides are very like the other diatomic gases in this respect, though the halogens themselves belong rather to the class of carbon dioxide and steam.

Case I. Formation of NO.

Nernst[1] has determined accurately for two temperatures, and approximately for two others, the equilibrium conditions governing the combustion of air to nitric oxide. He gives the following composition of the gaseous mixtures at the four temperatures:—

T	% N_2	% O_2	% NO	$\frac{p_{NO}}{p_{N_2}^{\frac{1}{2}} \times p_{O_2}^{\frac{1}{2}}}$	$\log^{10} \frac{p_{NO}}{p_{N_2}^{\frac{1}{2}} \times p_{O_2}^{\frac{1}{2}}}$	$R \ln \frac{p_{NO}}{p_{N_2}^{\frac{1}{2}} \times p_{O_2}^{\frac{1}{2}}}$
[1811	78·92	20·72	0·37]	0·0091	−2·039	−9·296
2033	78·78	20·58	0·64	0·0159	−1·799	−8·203
2195	78·61	20·42	0·97	0·0242	−1·616	−7·369
[3200	76·6	18·4	5·0]	0·1331	−0·876	−3·995

The ratio, $\frac{p_{NO}}{p_{N_2}^{\frac{1}{2}} \times p_{O_2}^{\frac{1}{2}}}$, representing the equilibrium constant

[1] "Göttinger Nachrichten" (1904), p. 261.

K, the Briggsian logarithm of this quantity, and finally the product $Rln\frac{p_{NO}}{p^{\frac{1}{2}}_{N_2} \times p^{\frac{1}{2}}_{O_2}}$, which recurs so often in these computations, are added to the above table. The more approximate values are bracketed.[1]

We may make the assumption that the specific heats of N_2, O_2, and NO are the same, for it is a general rule that the specific heats of the permanent diatomic gases are nearly identical. The heat of reaction at all temperatures would then be the same as Berthelot found at ordinary temperatures, viz.—

$$N + O = NO - 21,600 \text{ cal.}$$

Thomsen's value for this is 21,500 cal. Our next assumption, then, is that our equation (Second Lecture, p. 46) takes the simplest form—

$$A = Q - RTln\frac{p_{NO}}{p^{\frac{1}{2}}_{N_2} \times p^{\frac{1}{2}}_{O_2}} \quad . \quad . \quad . \quad . \quad (3)$$

where A becomes equal to zero at equilibrium. To test this, we will place the value of $\frac{Q}{T}$ beside those of $Rln\frac{p_{NO}}{p^{\frac{1}{2}}_{N_2} \times p^{\frac{1}{2}}_{O_2}}$.

1	2 $\frac{Q}{T}$	3 $Rln\frac{p_{NO}}{p^{\frac{1}{2}}_{N_2} \times p^{\frac{1}{2}}_{O_2}}$	4 Difference between 3 and 2.
[1811	−11·877	−9·296	+2·581]
2033	−10·625	−8·203	+2·422
2195	− 9·841	−7·369	+2·472
[3200	− 6·750	−3·995	+2·755]

[1] Recently this table has been completed by Nernst and his co-workers Jellinek and Finckh (*Zeitschr. f. anorg. Chem.*, 45 (1905), 116; 49 (1906), 212; 49 (1906), 229). The additional values are—

T	% N_2	% O_2	% NO
1877	78·89	20·69	0·42
2023	—	—	0·52 to 0·80
2580	78·08	19·88	2·05
2675	77·98	19·78	2·23

Compare also Häusser's experiments on the same subject, "Verhandl. d. Vereins zur Beförd. des Gewerbefleisses in Preussen" (1906), 37.

We see that there is no agreement. We must add a number $+2{\cdot}45T$ to equation (3) to satisfy Nernst's measurements.

To see whether, after all, we must not assume a difference of specific heats, let us calculate by means of formula (25) (p. 64) the heat of reaction between 2033 and 2195° (absolute), from the observed values of the equilibrium constant. We get—

$$Q_m\left(\frac{1}{T'} - \frac{1}{T''}\right) = R\ln K' - R\ln K''$$

$$Q_{(2114°)} = -23{,}000 \text{ cal.}$$

This differs but slightly from the value of the heat of reaction observed at ordinary temperatures.[1] Still, it indicates that the heat of reaction tends toward a greater negative value with rising temperature. In accordance with Kirchhoff's law, such an increase must take place if the specific heat of NO is greater than that of a mixture of its components. If we assume that this difference is independent of the temperature and approximately equal to $-0{\cdot}31$, then the thermodynamically indeterminate constant, which we have just estimated as equal to $-2{\cdot}45$, becomes zero. We then find for the heat of reaction at absolute zero—

$$Q_0 = -21{,}500 \text{ cal.}$$

From this we compute the following table:—

1 T	2 $\frac{Q_0}{T}$	3 $R\ln\frac{p_{NO}}{p^{\frac{1}{2}}_{N_2} \times p^{\frac{1}{2}}_{O_2}}$	4 $\ln T$	5 $\sigma'\ln T$	6 Sums of 3 and 5.	7 Difference of 2 and 6.
1811	−11·828	−9·296	7·493	−2·323	−11·619	0·209
2033	−10·581	−8·203	7·609	−2·359	−10·562	0·020
2195	− 9·800	−7·369	7·853	−2·434	− 9·803	0·003
3200	− 6·722	−3·995	8·062	−2·499	− 6·494	0·228

[1] The reliability of the value for $Q_{(2114°)}$ depends, not only on the accuracy of the equilibrium constant, but also very markedly on that of the temperature measurements. The temperature was determined photometrically, as will be explained in the last lecture. The assumption therein made that the total brightness increases as the 13th or 14th power of the temperature is of such an approximate nature that our temperature measurements can easily be wrong by several degrees. An error of a degree at this high temperature changes Q_m by more than a hundred calories.

From this table we see that the assumption of a difference in specific heats of—

$$\tfrac{1}{2}c_{N_2} + \tfrac{1}{2}c_{O_2} + 0{\cdot}31 = c_{NO} \text{ or } \sigma' = -0{\cdot}31$$

brings the expression—

$$A = Q_0 - \sigma'TlnT - RTln\frac{p_{NO}}{p^{\frac{1}{2}}_{O_2} \times p^{\frac{1}{2}}_{N_2}}$$

into agreement with Nernst's observations.

This surmise that σ' is negative, to which the values of the heat reaction give rise, is further confirmed by the observations of Regnault, who determined the specific heat of all three gases at room temperature. He found a difference of the same sign and of corresponding magnitude. We take the data in regard to this from Berthelot's compilation.[1]

$$\begin{aligned} \tfrac{1}{2}c_{p(O_2)} &= 3{\cdot}48 \\ \tfrac{1}{2}c_{p(N_2)} &= 3{\cdot}42 \\ & \overline{6{\cdot}90} \end{aligned} \qquad c_{p(NO)} = 6{\cdot}96$$

The difference amounts to $-0{\cdot}06$ cal., instead of the $-0{\cdot}31$ cals. assumed. The present state of our experimental knowledge of the specific heats of gases does not permit us to say more. We are especially in ignorance as to how the difference changes with the temperature. It therefore seems best to retain for practical use later the expression [2]—

$$A = Q - RTln\frac{p_{NO}}{p^{\frac{1}{2}}_{O_2} \times p^{\frac{1}{2}}_{N_2}} + 2{\cdot}45T$$

Proceeding now to the second case, let us first consider the reaction— **Case IIa. Hydrochloric acid formation.**

$$Cl_2 + H_2 = 2HCl$$

[1] Berthelot, "Thermochimie, Lois et Données," 2 vols. (Paris, 1897). This work, together with Oswald's "Thermochemie" ("Lehrbuch der Allg. Chem.," vol. ii. part i., 1893) contains the chief compilation of thermochemical data. A table of the specific heats of gases is given in Berthelot, *l.c.*, vol. i. p. 57. In the Sixth Lecture, the specific heat measurements of Regnault are given in tabular form. The comparison, which is given above in the text, has already been made by Regnault himself (*Mem. Inst. de France*, 26 (1862), 303, 311, 322).

[2] Nernst also (*l.c.*) takes Q as independent of the temperature. In his paper, however, on calculation of chemical equilibria from thermochemical measurements (see p. 100), he assumes a small difference σ'', while σ' is taken as zero.

Dolezalek[1] has determined the definitive data of this reaction at 30° by measuring the electromotive force of a chlorine electrode in hydrochloric acid solution of known partial pressure, against a hydrogen electrode in the same solution. The action of the cell involves the disappearance of hydrogen and chlorine, and the formation of hydrochloric acid. It makes no difference whether we consider the hydrochloric acid formed to be a gas or to be in solution, so long as gas and solution are assumed to be in equilibrium. The gas pressures of the chlorine and hydrogen were not accurately determined in Dolazalek's experiments. Still we shall not be far wrong if we assume that they were both equal to atmospheric pressure minus the partial pressures of the water and of the hydrochloric acid, for both gases were passed in at atmospheric pressure. Now, the vapour pressure of pure water at 30° is equal to 31·5 mm. Hg, or 0·0414 atmosphere. It is, of course, less over the hydrochloric acid solutions; but since a small deviation of 0·01 atmosphere is of no numerical importance in the following calculation, we shall call it 0·02 atmosphere over all the solutions. We obtain the partial pressure of the chlorine and the equal partial pressure of the hydrogen by subtracting Dolezalek's observed partial pressures of hydrochloric acid from 0·98 atmosphere. This gives us all the necessary data for evaluating the expression—

$$\frac{p_{HCl}}{p_{Cl_2}^{\frac{1}{2}} \times p_{H_2}^{\frac{1}{2}}}$$

We combine in the following table the normality of the hydrochloric acid, its partial pressure, the value of the expression—

$$\frac{p_{HCl}}{p_{H_2}^{\frac{1}{2}} \times p_{Cl_2}^{\frac{1}{2}}}$$

just discussed, and finally the value of—

$$RT ln \frac{p_{HCl}}{p_{Cl_2}^{\frac{1}{2}} \times p_{H_2}^{\frac{1}{2}}}$$

expressed in calories for T = 303° absolute, based on the results of Dolezalek. We also add the electromotive force of the cell as found by Dolezalek, and the reaction energies (free energies) computed from it by multiplication with 23,110 (the number

[1] *Z. f. Phys. Chem.*, 26 (1898), p. 334.

23,110 is the product of the number of coulombs, 96,540, which, according to Faraday's law, must pass through the cell in the formation of a mol of hydrochloric acid, into the factor which converts the product of volts and coulombs, that is joules, into gram calories. This factor is 0·2394).[1]

1 Normality of the HCl.	2 p_{HCl} atmos.	3 $\frac{p_{HCl}}{p^{\frac{1}{2}}_{Cl_2} \cdot p^{\frac{1}{2}}_{H_2}}$	4 $RTln\frac{p_{HCl}}{p^{\frac{1}{2}}_{Cl_2} \cdot p^{\frac{1}{2}}_{H_2}}$	5 E in volts.	6 A=E · 23,110 cal.	7 $A+RTln\frac{p_{HCl}}{p^{\frac{1}{2}}_{Cl_2} \cdot p^{\frac{1}{2}}_{H_2}}$
4·98	$0·316 . 10^{-3}$	$0·31 . 10^{-6}$	−4849	1·190	27,501	22,652
6·43	$0·906 . 10^{-3}$	$0·89 . 10^{-3}$	−4215	1·147	26,507	22,292
11·20	0·175	0·217	− 912	1·005	23,226	22,314
11·62	0·249	0·341	− 647	0·999	23,087	22,440
12·14	0·412	0·725	− 192	0·981	22,671	22,479
12·25	0·443	0·825	− 116	0·974	22,509	22,393

Because of an incompleteness in the electrical measurements of Dolezalek, the values given in the sixth column, separately considered, appear uncertain by some 118 cals. As a first trial let us form the expression—

$$A = Q - RTln\frac{p_{HCl}}{p^{\frac{1}{2}}_{Cl_2} \times p^{\frac{1}{2}}_{H_2}}$$

and so compare the calorimetrically determined values with the sums of—

$$A + RTln\frac{p_{HCl}}{p^{\frac{1}{2}}_{Cl_2} \times p^{\frac{1}{2}}_{H_2}}$$

The sums are given in the last column of the above table. Their mean is 22,428 cals, while Berthelot and Thomsen both found 22,000 cals. There is thus a slight discrepancy between the two. Whether we should attach any importance to it is very questionable. It would require, at the low temperature of 303° abs., that we assume a constant of 1·34 if we would express the observations by the formula—

$$A = 22{,}000 - RTln\frac{p_{HCl}}{p^{\frac{1}{2}}_{Cl_2} \times p^{\frac{1}{2}}_{H_2}} + 1{\cdot}34T \quad . \quad . \quad (4)$$

without reference to the specific heats. The observations of Dolezalek would be satisfied by this formula.

[1] The last decimal place is uncertain. The physikalisch-technische Reichsaustalt recently puts it at 0·239.

Dissociation of hydrochloric acid at higher temperatures.

Measurements of the equilibrium at higher temperatures which we could use are wanting. Mallard[1] and Le Chatelier, to be sure, exploded numerous mixtures of hydrogen and chlorine in closed vessels, and based certain conclusions on the pressures there developed. These observations, however, afford us no certain knowledge of the dissociation. Henry St. Claire-Deville was able, by means of a most ingenious experimental contrivance, the so-called "cold-hot" tube, to show qualitatively a trace of dissociation at 1300°. This apparatus[2] consisted of a heated porcelain tube with a brass tube passing through its centre. Water flows through the brass tube. The gases are sent through the annular space between the two tubes. Deville, by this simple arrangement, could prove qualitatively the dissociation, not only of carbonic acid and ammonia, but also of hydrochloric acid, carbon monoxide,[3] and sulphurous acid, the temperatures being hardly above 1300° C. Victor Meyer and Langer[4] showed the same thing qualitatively at 1700° by measuring the density of HCl in a platinum vessel. Platinum is very porous to hydrogen at these high temperatures, and consequently the gas leaked through the walls of the vessel, leaving chlorine behind, partly as such, and partly as chloride. How far the dissociation went cannot be determined from these experiments. Our formula gives a very slight dissociation at 2000° abs. or 1727° C., that is, at very nearly the same temperature as used by Victor Meyer and Langer. If we put A = zero for the condition of equilibrium, and T = 2000, we get—

$$Rln\frac{p_{HCl}}{p_{Cl_2}^{\frac{1}{2}} \times p_{H_2}^{\frac{1}{2}}} = \frac{22,000}{2000} + 1{\cdot}34 = 12{\cdot}34$$

Hence for every 100 volumes of unchanged hydrochloric acid there is 0·2 volume of chlorine and an equal amount of hydrogen. (This equilibrium, like all gas reactions in which the number of appearing and disappearing molecules is the same, is independent of the pressure, since the volume factors cancel from numerator and denominator.) Victor Meyer and Langer estimated the

[1] *Annales des Mines* (1883), sect. iv. part D.

[2] *Ann. Chim. Pharm.*, 135 (1865), 94.

[3] *Ibid.*, 134 (1865), 124.

[4] Victor Meyer and Langer, "Pyrochem. Untersuchungen" (Braunschweig, 1885), p. 67.

dissociation at 1700° as much greater than this. Yet their observation can be explained equally well without this assumption, provided the velocity with which the dissociation equilibrium adjusts itself be great, as is indeed very probably the case. A more rigorous treatment of the question is impossible with our present insufficient knowledge of the specific heats of these gases. If we use Strecker's and Regnault's values as given in Wüllner's compilation[1] for room temperatures, we find—

$$\begin{array}{r} \tfrac{1}{2}c_{H_2} = 3{\cdot}41 \\ \tfrac{1}{2}c_{Cl_2} = 4{\cdot}08 \\ \hline (\tfrac{1}{2}c_{H_2} + \tfrac{1}{2}c_{Cl_2} = 7{\cdot}49) \end{array} \qquad c_{HCl} = 7{\cdot}08$$

All the values refer to constant pressure. According to these measurements, the true specific heats of the disappearing substances exceeds that of the substance formed by 0·41 at room temperature. Yet Regnault's direct determinations of the specific heat of chlorine differ considerably from the indirect determinations of Strecker. Still, it seems unquestionable that the difference is positive at ordinary temperatures, and assumes a still larger positive value with rising temperatures, for Berthelot, Mallard, and Le Chatelier agree that the specific heat of chlorine at high temperatures greatly exceeds that of hydrochloric acid, which always remains similar to that of hydrogen.

In order to at least approximately inform ourselves as to what significance this difference of specific heats has, and to see how well Dolezalek's measurements harmonize with the assumption that the thermodynamically indeterminate constant becomes zero when we take account of a difference of specific heats, let us recur to the expressions developed in the Second and Third Lectures (p. 51, equation (28*a*), and p. 60, equation (9). For such a comparison it is often convenient to substitute $Q_T - \sigma'_v T - \sigma'' T^2$ for Q_0 in the formula—

$$A = Q_0 - \sigma'_v T lnT - \sigma'' T^2 - RT\Sigma\nu' lnc' + \text{const. } T$$

$\sigma'_v T$ and $\sigma'_v T lnT$ can then be collected in one term, and we obtain in this case, for example—

$$A + \sigma'_v T(1 + lnT) + 2\sigma'' T^2 + RT ln \frac{p_{HCl}}{p_{Cl_2}^{\frac{1}{2}} \times p_{H_2}^{\frac{1}{2}}} = Q_T + \text{const. } T. \quad (5)$$

[1] Wüllner, "Experimentalphysik," 5th edit., vol. iii. p. 554 (1896).

If we give a small negative value, say $-0{\cdot}25$, to the difference σ'_v at absolute zero, and a moderate positive value, say $+0{\cdot}001$, to the difference σ'' of its increment with the temperature, we obtain, in the first place, a difference between the *true* specific heats at ordinary temperatures agreeing both in sign and in magnitude with the experimental results. In the second place, we find, with the help of the calorimetrically measured heat of reaction and the values calculated above from Dolezalek's experiments, that the thermodynamically indeterminate constant (in 5) becomes very nearly equal to zero. Finally, the dissociation in the case of equilibrium at 2000° abs., calculated on these assumptions, is found slightly greater than we deduce from formula (4). Against these assumptions regarding σ' and σ'' may be urged the fact that Strecker, measuring the velocity of sound in chlorine, which is dependent on the ratio of the true specific heats $\left(\frac{c_p}{c_v}\right)$, could find no certain change in this ratio between 17° and 343°. Yet Strecker's values are of varying accuracy, and his conclusion that the specific heat of chlorine is the same between 17° and 343° is too incompatible with the large temperature coefficient of the specific heat at high temperatures found by Mallard and Le Chatelier, to allow us to attach any importance to an objection based upon it. We can here, as in the former case, merely say that the facts are by no means opposed to the assumption of a zero value for the thermodynamically indeterminate constant. Such an assumption is not, however, *proved* by the observations, as we have already mentioned (Second Lecture, p. 46). For further use in the next lecture we shall retain, for simplicity's sake, equation (4), especially since only temperatures of a few hundred degrees are there involved.

Case IIb. The formation of hydrobromic acid.

Measurements bearing on the energy of formation of hydrobromic acid have been made by Bodenstein and Geiger.[1] The relationships here are identical with those in the case of hydrochloric acid, and their discussion may consequently be somewhat abbreviated. Bodenstein and Geiger have measured the electromotive force of the cell $Br_2 - BrH - H_2$ at 30°, using solutions of hydrobromic acid containing respectively 0·9291, 0·9250, and 0·7849 g. HBr in one c.c., and to which small quantities of bromine

[1] *Z. f. phys. Chemie*, 49, p. 70 (1904).

had been added (on one side). The vapour pressure of the bromine and the hydrobromic acid were determined in the same solutions and at the same temperature. It was also ascertained that the vapour pressure of the pure hydrobromic acid over its pure aqueous solution was not materially different from that over the solution containing bromine. Further, the pressure of the hydrogen gas passed in was measured. The value of this pressure in mm. Hg, the observed values of the e.m.f. of the cell in volts, and the energy of the reaction computed therefrom, for the formation of one mol of HBr gas (e.m.f. × 23,110) expressed in gram-calories, are given together in the table below. According to our theoretical considerations, the energy of the reaction—

$$\tfrac{1}{2}Br_2 + \tfrac{1}{2}H_2 = HBr$$

taking account of the difference of specific heats, would be—

$$A = Q_0 - \sigma'_v T ln T - \sigma'' T^2 - RT ln \frac{p_{BrH}}{p^{\frac{1}{2}}_{Br_2} \times p^{\frac{1}{2}}_{H_2}} + \text{const. } T \quad (6)$$

If we substitute here as before (see p. 111)—

$$Q_T = Q_0 + \sigma'_v T + \sigma'' T^2$$

for the sake of convenience, we get—

$$A = Q_T - \sigma'_v T(1 + ln T) - 2\sigma'' T^2 - RT ln \frac{p_{BrH}}{p^{\frac{1}{2}}_{Br_2} \times p^{\frac{1}{2}}_{H_2}} + \text{const. } T$$

If we insert for T in this equation the absolute temperature 303°, used by Bodenstein and Geiger, and 12,200 cal. for the heat of reaction, this number lying midway between the value found by Berthelot (12,300) and that which Ostwald calculated from the measurements of Thomsen (12,100), it follows that—

$$A = 12{,}200 - 6{\cdot}714 \times 303 \times \sigma'_v - 2 \times (303)^2 \times \sigma''$$
$$- 1382 \log_{10} \frac{p_{BrH}}{p^{\frac{1}{2}}_{Br_2} \times p^{\frac{1}{2}}_{H_2}} + 303 \times \text{const.}$$

The value of the logarithmic member of this equation is given in the next to the last column of the following table; that of the expression—

$$A + 1382 \log_{10} \frac{p_{BrH}}{p^{\frac{1}{2}}_{Br_2} \times p^{\frac{1}{2}}_{H_2}}$$

in the last column. The mean of these values is very nearly 12,800 cal.

	p_{HBr} mm. Hg.	p_{Br_2} mm. Hg.	p_{H_2} mm. Hg.	EMK Volt.	A cal.	$RTln\frac{p_{BrH}}{p^{\frac{1}{2}}_{BrH} \times p^{\frac{1}{2}}_{H_2}}$	$A + RTln\frac{p_{BrH}}{p^{\frac{1}{2}}_{Br_2} \times p^{\frac{1}{2}}_{H_2}}$
(1)	12·0	0·682	742·5	0·573	13,242	− 373	12,869
(2)	2·25	1·509	753·6	0·625	14,444	−1625	12,819
(3)	1·19	1·448	760·6	0·636	14,698	−1997	12,701

We neglect the difference of the specific heats.

If we arrange the equation as follows:—

$$A - Q_T + RTln\frac{p_{BrH}}{p^{\frac{1}{2}}_{Br_2} \times p^{\frac{1}{2}}_{H_2}} = -\sigma'_v T(lnT + 1) - 2\sigma'' T^2 + \text{const. } T$$

and take 12,200 cal. for Q_T, we can see from the table that the value of the left-hand member amounts to some 600 cal. If we insert this value, dividing through by 303 (calling $\frac{600}{303} = 2$) and transposing, we get—

$$\text{const.} - 2 = 6{\cdot}714\sigma'_v + 606 \times \sigma''$$

If, first, we put σ'_v and σ'' equal to zero, the const. becomes 2, and we get, as in the case of hydrochloric acid, the approximation—

$$A = 12{,}200 - RTln\frac{p_{BrH}}{p^{\frac{1}{2}}_{Br_2} \times p^{\frac{1}{2}}_{H_2}} + 2T \quad . \quad . \quad (7)$$

The assumption that the heat of reaction is the same at all temperatures finds expression in this equation.

Influence of the specific heats.

If, on the other hand, we examine the available data concerning the specific heats, we find that at ordinary temperatures the true specific heat of

$$\tfrac{1}{2}Br_2 + \tfrac{1}{2}H_2$$

exceeds that of 1HBr by nearly a unit. The specific heats of hydrogen and hydrobromic acid are per mol approximately equal. We may assume without hesitation that their increase with the temperature would also be equal. On the other hand, the specific heat of bromine is much greater, and undoubtedly its increase with the temperatures is also more rapid. We may assume here, as in the case of HCl, a negative value ($-0{\cdot}5$) for σ'_v, and a positive value ($+0\ 0027$) for σ'', and thereby cause the thermodynamically indeterminate constant to disappear, and at the same time make the difference between the true specific heats at ordinary temperatures correspond to that experimentally determined. Yet by thus fixing the value of σ' and σ'' we are forced to assume a more rapid change of the specific heats of bromine with the temperature than is at all compatible with the acoustic measurements which Strecker has made upon this gas. Since we possess no other evidence regarding the change of specific heats with the temperature, we must reserve judgment.

Dissociation of hydrobromic acid at higher temperatures.

The dissociation of hydrobromic acid vapour can be easily calculated from equation (7) if, for instance, we put T = 1000 abs. (727° C.); it then follows for the equilibrium, where A = 0, that—

$$\log_{10}\frac{p_{BrH}}{p_{Br_2}^{\frac{1}{2}} \times p_{H_2}^{\frac{1}{2}}} = 3{\cdot}114$$

$$\text{and} \quad \frac{p_{BrH}}{p_{Br_2}^{\frac{1}{2}} \times p_{H_2}^{\frac{1}{2}}} = 1300$$

Now, at equilibrium, for every one part of pure hydrobromic acid supposed to have been originally present, there is $1 - x$ part still unchanged, while $\frac{x}{2}$ part is present as bromine, and the same amount as hydrogen. We find then that—

$$\frac{p_{BrH}}{p_{Br_2}^{\frac{1}{2}} \times p_{H_2}^{\frac{1}{2}}} = \frac{(1-x)}{\left(\frac{x}{2}\right)^{\frac{1}{2}} \times \left(\frac{x}{2}\right)^{\frac{1}{2}}} = \frac{2(1-x)}{x} = 1300$$

$$\text{or} \quad \frac{1}{x} = 651, \text{or } x = 0{\cdot}15 \times 10^{-2}$$

Expressed in per cents., the degree of dissociation $x = 0{\cdot}15$ per cent. at 727°.

Bodenstein[1] states that Reichenbach found the degree of dissociation at 650°–750° to lie between 0·3 per cent. and 0·9 per cent. He calls attention to the fact that the measurements probably gave too high results. The assumption of a negative value for σ' and a positive value for σ'', leads to a somewhat higher value for the degree of dissociation.

Case IIc. The formation of hydriodic acid.

We will complete this discussion by considering the equilibrium conditions of hydriodic acid. Here we have equilibrium measurements made by Bodenstein, who subjected this matter

[1] *Z. f. phys. Chemie*, 49 (1904), 61.

to a thorough and searching study.[1] We will now show in the usual way the bearing of the values obtained, upon the formation of hydriodic acid, according to the equation—

$$\tfrac{1}{2}I_2 + \tfrac{1}{2}H_2 = HI$$

The "degree of dissociation" x of the hydriodic acid was the quantity directly determined. From 1 mol hydriodic acid, $1-x$ mol remains undecomposed at equilibrium, and $\frac{x}{2}$ mol each, of hydrogen and iodine are formed. The equilibrium constant can therefore be written[2]—

$$K = \frac{2(1-x)}{x} \quad . \quad . \quad . \quad . \quad . \quad . \quad . \quad (2)$$

Bodenstein's decomposition values were—

° C. ...	508	487·2	443	427	410	393	374	356
x ...	0·2408	0·2340	0·2198	0·2157	0·2100	0·2058	0·2010	0·1946
° C. ...	328	302	283					
x ...	0·1885	0·1815	0·1787					

From these values Bodenstein computed a table of equilibrium constants which we will alter only by substituting values recomputed according to our definition of the equilibrium constant. The equilibrium constants labelled "found" are derived from the adjoining decomposition values, which in turn are computed by interpolation from the observed values of x given above. The interpolation formula deduced and used by Bodenstein is—

$$x = 0{\cdot}13762 + 0{\cdot}00007221t + 0{\cdot}00000025764t^2$$

The values of the heat of reaction also labelled "found" are

[1] *Zeitschr. f. physik. Chem.*, 29 (1899), 295, where earlier papers are also given.

[2] Bodenstein takes as the equilibrium constant—

$$K_B = \frac{x^2}{4(1-x)^2}$$

We therefore get our constant from his by making—

$$K = \frac{1}{K_B^{\frac{1}{2}}}$$

Bodenstein gives the values as $ln K_B$. Our constant, in the terms of this, therefore is—

$$ln K = -\tfrac{1}{2} ln K_{(B)}$$

calculated, interval by interval, from the equation of Van't Hoff—

$$R \ln \frac{K''}{K'} = Q_m \left(\frac{1}{T''} - \frac{1}{T'} \right)$$

From these values of Q_m, the heat of reaction at any temperature may be computed by means of the interpolation formula—

$$Q_T = 89{\cdot}575 - 1{\cdot}575T + 0{\cdot}00549T^2$$

This formula corresponds to our expression—

$$Q_T = Q_0 + \sigma'_v T - \sigma' T^2$$

Finally, with its help, we get the equation for the energy of the reaction—

$$A = 89{\cdot}575 + 1{\cdot}575T \ln T - 0{\cdot}00549T^2 - RT \ln \frac{p_{IH}}{p_{I_2}^{\frac{1}{2}} \times p_{H_2}^{\frac{1}{2}}} + \text{const. } T$$

Bodenstein determined the constant by calculating the value $\ln$K for the condition of equilibrium (A = 0) and comparing it with the value of the $\ln$K "found." This gave −2·67 as the thermodynamically indeterminate constant.[1]

t	T	*x*	K	*ln*K found.	*ln*K calculated.	Q_T found.	Q_T calculated.
520	793	0·24483	6·2	1·8195	1·8197	2222	2222
500	773	0·23813	6·4	1·8562	1·8562	2084	2084
480	753	0·23164	6·6	1·8922	1·8924	1951	1950
460	733	0·22535	6·9	1·9279	1·9280	1821	1821
440	713	0·21927	7·1	1·9631	1·9631	1696	1696
420	693	0·21339	7·4	1·9978	1·9979	1575	1575
400	673	0·20772	7·6	2·0319	2·0319	1459	1461
380	653	0·20226	7·9	2·0654	2·0657	1343	1347
360	633	0·19700	8·2	2·0983	2·0982	1237	1239
340	613	0·19195	8·4	2·1306	2·1306	1135	1136
320	593	0·18711	8·7	2·1621	2·1622	1038	1038
300	573	0·18247	8·9	2·1929	2·1929	942	943
280	553	0·17803	9·2	2·2229	2·2231		

The equations of Bodenstein can be used to calculate the **Heat of reaction.**

[1] Bodenstein's equation for the equilibrium runs—

$$\ln K_s = -\frac{90{\cdot}48}{T} - 1{\cdot}5959 \ln T + 0{\cdot}0055454T + 2{\cdot}6981$$

If we replace (see above note) $\ln K_s$ by $-2 \ln K$, and multiply through by RT we obtain the above equation.

heat and the energy of the reaction at ordinary temperature. The value of Q_T, according to Bodenstein's equation given above, amounts for ordinary temperature to only +96 cal. This value is appreciably smaller than we should expect from the thermochemical data. Berthelot gives −6400 cal. for the formation of gaseous hydriodic acid from solid iodine and gaseous hydrogen, while from Thomsen's data we get −6100 cal. for the same quantity. The heat of sublimation of solid iodine can be calculated from measurements of its vapour pressure at different temperatures. Such measurements have recently been made by Baxter, Hickey, and Holmes,[1] and they show that 15,100 cal. are used up in converting one mol of solid iodine into vapour at ordinary temperature. Therefore the heat of formation of one mol HI from the gaseous elements is equal to 7400 − 6400, *i.e.* 1000 cal. at ordinary temperature. Using Thomsen's data, we get 7400 − 6100, *i.e.* 1300 cal.

Reaction energy at higher temperature.

The equation of the reaction energy can be tested by determining the e.m.f. of a galvanic cell, which consists of an iodine and a hydrogen electrode in a solution of hydriodic acid. The partial pressures of iodine vapour and of hydriodic acid gas, which are in equilibrium with the solutions at the electrodes, and the pressure of the hydrogen gas must be known. Recently the author, in collaboration with Gottlob, has carried out preliminary measurements on such cells at 43·7°. According to Bodenstein's equation of the reaction energy, the e.m.f. at this temperature should be—

$$E = 0{\cdot}06757 - 0{\cdot}0625 \log \frac{p_{IH}}{p_{I_2}^{\frac{1}{2}} \times p_{H_2}^{\frac{1}{2}}} (E = \text{volts})$$

Five cells were tested, the concentration of the hydriodic acid varying between 3·5 normal and 6·44 normal, the concentration of iodine at the positive electrode being from 0·55 normal to 1·076 normal. Hydrogen was passed in at the negative electrode under atmospheric pressure. The vapour pressures of hydriodic acids with and without addition of iodine were determined by the so-called dynamic method. All values of the e.m.f. were found to be larger than calculated from the

[1] *Journ. Am. Chem. Soc.*, Jan. 1906. Former determinations have been made by Ramsay and Young (*Jour. London Chem. Soc.*, 49, p. 453 (1886), and by V. Richter (*Berl. Ber.*, 19 (1886), p. 1060).

above equation. The mean difference amounted to 0·052 volt, corresponding to a difference of 1200 cal. in the reaction energy.

These differences between calculated and experimental values for the reaction energy as well as the reaction heat suggest the question whether the iodine in Bodenstein's experiments behaved as an ideal gas. It may be remembered that, according to Guye and Radice,[1] the critical temperature of iodine is some degrees above 500°. Now the specific heats of gases at temperatures appreciably below their critical temperature show that the pressures are not as independent as is presupposed in the deduction of our formulæ. It seems possible that Bodenstein's experiments, especially in the neighbourhood of 300°, were influenced by this factor.

It is of interest to compare the various determinations of the specific heats of the gases concerned. According to Bodenstein— **The specific heats.**

$$\sigma'_v + \sigma''T = -1·575 + 0·00549T$$

This value represents the difference between the mean specific heats (from 0° abs. to T°) of $\frac{1}{2}I_2 + \frac{1}{2}H_2$ and HI. Strecker[2] has tested the ratio $\frac{C_p}{C_v}$ of the specific heats of hydriodic acid by measuring the velocity of sound in it. The true specific heat at constant pressure, calculated from his measurements, is 7·04, without any perceptible change with the temperature between 21° and 100°. We may certainly take the Regnault value of 6·82 per mol as the true specific heat of hydrogen at constant pressure at ordinary temperatures. Both values can, indeed, be employed up to 300° without any perceptible error. Testing the iodine, Strecker found a ratio of specific heats in the neighbourhood of 300° (220° – 375°) which gave a specific heat at constant pressure of 8·53, again quite independent of the temperature. According to these measurements, the difference of the specific heats of $\frac{1}{2}I_2 + \frac{1}{2}H_2$ and HI at 300° (573° abs.) is +0·62. Bodenstein's formula, on the other hand, would lead us to expect the much higher value of +4·73 for this difference. The discrepancy is remarkable, though we have to consider

[1] See Landolt-Börnstein, "Physikalisch-Chemische Tabellen," 3rd edit. (1905), p. 184.

[2] Wuellner, "Experimentalphysik," 5th edit. (1895), vol. iii. p. 554; Strecker, *Wied. Ann.*, 13, p. 41 (1881).

that Strecker himself has emphasized the difficulty and inaccuracy of acoustic measurements on I_2 at high temperatures.

The dissociation of I_2 into atoms.

Concerning Bodenstein's values at high temperature (500°), there is another point to be considered. The reaction studied by Bodenstein changes its nature with rising temperature. The iodine, as well as the hydriodic acid, begins to dissociate. The two reactions—

$$2I \rightleftarrows I_2$$
$$I_2 + H_2 \rightleftarrows 2HI$$

therefore infringe one upon the other. The heat of formation of molecular iodine from atomic iodine is certainly very large. Planck[1] has calculated from several measurements by Meier and Crafts,[2] on the assumption that the specific heat of iodine at constant volume is not changed by dissociation into atoms, that it amounts to 29,000 cal. (at constant volume). He has further calculated from two of Bodenstein's numbers, that the heat of formation (which he assumes as independent of the temperature) of a mol of hydriodic acid from molecular iodine and hydrogen is 1300 cal. We may use these results for a rough estimation of the relative importance of the two reactions, for, according to them, the decomposition of a mol of hydriodic acid into $\frac{1}{2}$ mol of iodine and hydrogen uses up only $\frac{1}{12}$ $\left(i.e. \frac{1300}{1300 + 29,000}\right)$ of the total heat of the two reactions, while the further conversion of the $\frac{1}{2}$ mol of molecular iodine into atomic iodine uses up the remaining $\frac{11}{12}$ of this total heat of reaction. The decomposition of hydriodic acid therefore entails a greater heat absorption the higher we raise the temperature, because thereby more iodine atoms, together with iodine molecules, are produced. The heat of reaction Q_T given by Bodenstein's measurements is nothing other than this heat consumption with changed sign. We can, therefore, no longer consider it as simply the heat of formation of hydriodic acid from molecular iodine and hydrogen, but must rather recognize in it the sum of two heats of reaction, that of the reaction—

$$I + \tfrac{1}{2}H_2 = HI$$

[1] "Thermodynamik," 2nd edit. (1905), p. 216.

[2] *Berl. Ber.* (1880), 851, 1316; *Compt. Rend.*, 90, 360, 92, 39. Compare V. Meyer, *Berl. Ber.* (1880), 394, 1010, 1103; (1881) 1453; and Naumann, the same (1880), 1150.

and—

$$\tfrac{1}{2}I_2 + \tfrac{1}{2}H_2 = HI$$

Doubtless the share taken by the first reaction is very small, but vanishingly small it certainly is not, and at 520° it represents a measureable though small fraction of Q_T. The dissociation of iodine into atoms under atmospheric pressure reaches 25 per cent. at 1043°, 50·2 per cent. at 1275°, and 73·1 per cent. at 1468°, according to Crafts and Meier. At lower temperatures the result of density measurements are not very sharp, yet even at 680° the decrease of the density is noticeable.

The heat increment—

$$1{\cdot}575 + 2 \times 0{\cdot}00549T$$ [1]

of the reaction does not therefore represent merely the difference of the specific heats of the reacting substances, but is rather determined by the fact that every increase of the temperature increases the relative frequency of the process involving the large heat change $I + \tfrac{1}{2}H = HI$, and decreases somewhat the frequency of the process with the small heat change—

$$\tfrac{1}{2}I_2 + \tfrac{1}{2}H_2 = HI$$

The third, and technically the most important of the class of reactions we are considering, is the water-gas reaction— **Case III. The water-gas reaction.**

$$CO_2 + H_2 \rightleftarrows H_2O + CO$$

The heat of reaction has here been accurately determined at ordinary temperature. It represents the difference between the heats of reaction of the following reactions:—

$$\begin{array}{l} H_2 + \tfrac{1}{2}O_2 = H_2O + 58{,}000 \text{ gram-cals.} \\ CO + \tfrac{1}{2}O_2 = CO_2 + 68{,}000 \quad \text{,,} \\ \hline CO_2 + H_2 = CO + H_2O - 10{,}000 \text{ gram-cals.} \end{array}$$

These values are based on the critical discussion of the thermochemical data relating to the formation of water and carbonic acid, given in Ostwald's large text-book. Berthelot's results lead to the value of $-10{,}100$ cals., differing but 1 per cent. from the above.

The determination of the equilibrium of this reaction has **History of the water-gas reaction.**

[1] The factor 2 is here present because the heat increment of the reaction is equal to the difference of the true specific heats (see p. 29). This has the factor 2 in front of the second number.

a history.[1] It began with the experiments of Bunsen. He exploded a mixture of carbon monoxide and hydrogen with a quantity of oxygen insufficient for complete combustion, and studied the distribution ratio of the oxygen between the carbon monoxide and the hydrogen. Bunsen thought that, under these conditions, carbonic acid and water were always formed in simple ratios. E. von Meyer shared, at first, the same view, influenced by experiments of a similar nature. Then Horstmann called attention to the incorrectness of their interpretations, and, on the basis of results of his own, developed a correct theoretical treatment of the problem. This affirmed that the reaction proceeded at every temperature to the equilibrium represented by the expression—

$$\frac{C_{H_2O} \times C_{CO}}{C_{CO_2} \times C_{H_2}} = K$$

The specific heats of the gases present in the water-gas equilibrium.

Bunsen and E. von Meyer accepted this view after renewed examination of the problem. Horstmann could not attempt a calculation of the change of the equilibrium constant with the temperatures, because at that time the specific heats of the gases at high temperatures were not known. After research in this direction had furnished new material, Hoitsema[2] attempted to give a numerical treatment of Horstmann's observations.

It will help us in understanding these calculations and the developments based on them, if we first review what is known about the specific heats of the gases concerned. The methods of their determination will be only lightly touched upon here, because they will receive special treatment in a later lecture. Regnault first showed, in his comprehensive research on the specific heats of gases (1862), that the specific heats of hydrogen and carbon monoxide were nearly the same both at ordinary and at slightly elevated temperatures. He found 6·86 cals. per mol as the specific heat of carbon monoxide, and 6·87 per mol as that of hydrogen.[3] All later investigations of the behaviour

[1] See Haber and Richardt, "Ueber das Wassergasgleichgewicht in der Bunsen-flamme und die chemische Bestimmung der Flammentemperatur," *Zeitschr. f. anorg. Chem.*, 38 (1904), 5.

[2] *Zeitschr. f. physik. Chem.*, 25 (1898), 686.

[3] As regards the value of the Regnault calorie, see Wüllner, "Experimentalphysik," 5th edit. (Leipzig, 1896), vol. ii. p. 523; also Dieterici, Drude's *Ann.*, 16 (1905), 593.

of these gases at higher temperatures agree in affirming that their specific heats change in an identical way with rising temperature, so that their difference may be considered as approximately zero at all temperatures. Consequently, in the calculation of the heat of reaction and of the reaction energy of the water-gas reaction, only the difference in specific heats of carbon dioxide and water-vapour counts, provided we calculate the difference in the specific heats of the appearing and disappearing substances in the usual way. All investigators have not only found a difference in the specific heats of these gases, but have also found that this difference varies with the temperature.[1] Regnault first determined these quantities at low temperature.

Specific heat of water-vapour at low temperature.

He determined the specific heat of water-vapour at constant pressure by an indirect method. That is, he first conducted steam at ordinary pressure, heated to 120°, into a calorimeter, and measured the heat of condensation. He next performed the identical operation with steam heated to 220° C. The difference of the two quantities of heat evolved was evidently equal to the heat given off by the steam in cooling from 220° to 120°. It amounted in Regnault's experiments to 7 per cent. (in round numbers) of the total heat measured. This difference represents the product of the specific heat of steam between 120° and 220°, multiplied into the temperature difference (100°). Regnault found 0·48 gram-cal. as the mean specific heat per gram of water-vapour at atmospheric pressure from three experiments of this kind. This corresponds to a specific heat of 8·64 gram-cal. per mol. Though these measurements were made with the greatest care, we should not fail to recognize that they cannot guarantee any great accuracy of results. Let us consider

[1] Fliegner ("Vierteljahrsschrift der naturforschenden Gesellschaft," in Zürich" (1899), p. 192) has raised certain objections to this conclusion which have found wide circulation in the literature of the engineering profession. He discusses the experiments of Mallard and Le Chatelier (*Ann. des Mines*, 4 (1883), 379) and those of Berthelot and Veille (*Ann. chim. phys.*, 4 (1885), 13), and seeks to prove that all gases so far investigated exhibit a constant specific heat. According to his views, the former measurement of Mallard and Le Chatelier gave false results because the influence of the walls was insufficiently considered. These doubts and the view based upon them that the specific heat at 2000° remains constant are disproved on the one hand by the experiment of Langen, and on the other by the results of Holborn, Austin, and Henning.

the last series of Regnault's experiments, for instance, which gave an evolution of heat of 690·18 cal., when the steam was heated to 216·03°, and 645·44 cal. when heated to 122·75°, or a difference of 44·74 cal. (the liquid water formed is assumed to be cooled to 0°). This corresponds to a temperature difference of the vapour of 93·28°. From this we get 0·47963 cals. as the specific heat of the vapour per gram. Now, the temperature change of the calorimeter itself in the single experiments amounts to some 20°. The 44·74 cal. corresponds to a temperature change of about 0·7° in the calorimeter. But the corrections (based on observation) which must be applied to the calorimeteric results because of conduction and radiation before and after the introduction of water-vapour, amount to from 0·5 to 1° in single cases, or are approximately as great as the quantity investigated. These unfavourable conditions are operative in all the series of experiments. But even if we admit the accuracy of Regnault's determinations, we cannot deny that the values found by him can have but slight significance. Water-vapour under one atmosphere pressure and between 120° and 229° is by no means a perfect gas. It is still far beneath its critical temperature of 360°. We would expect, therefore, that its specific heat would exhibit peculiarities. Such peculiarities did indeed come to light in an investigation carried out by Lorenz.[1] Here superheated steam was conducted through a calorimeter at such a temperature that it did not cool to the point of saturation and condensation, but after a slight cooling made its exit, still in the condition of vapour. The following tables contain Lorenz's values. The temperatures at which the steam entered the calorimeter are given under A, the temperatures at which it left under B. Under M are given the means of these two temperatures, under p the vapour-pressure in Kg. per square centimetre, and, finally, under c_p the true specific heats in small calories per mol of water-vapour under the given pressures (p) and at the given mean temperatures (M)—

Experiment of Lorenz.

[1] *Zeitschrift des Vereins deutscher Ingenieure*, 1904, pp. 1189 and 698. Compare also the observations of Bach and of Weyrauch, which he also calls attention to (*Zeitschr. des Vereins deutscher Ingenieure*, 1876, p. 77, and 1904, p. 24). Similar experiments carried out by A. H. Peake (*Proc. Royal Soc., London*, 76 (1905), 185) did not give quite satisfactory results, as the author himself observes.

A ...	224·1	228·9	243·5	240·8	291·1	313·8	343·3
B ...	191·0	182·0	161·6	204·0	232·9	244·8	244·7
M ...	207·6	205·4	207·6	222·4	262·0	279·3	290·0
p ...	6·829	4·311	1·932	8·97	8·98	6·82	1·89
c_p ...	11·20	11·49	9·56	12·56	11·52	10·22	9·29

A ...	326·6	367·5	361·3	381·5
B ...	262·4	260·1	281·7	296·3
M ...	295·0	313·8	321·0	338·9
p ...	4·05	8·985	6·945	3·936
c_p ...	8·71	10·26	9·99	8·60

Lorenz deduced the formula—

$$c_p = 0{\cdot}43 + 3{\cdot}6 \times 10^7 \frac{p}{T^3} (p = \frac{\text{kg.}}{\text{sq. cm.}}, T = \text{abs. Temp.})$$

as representing his experimental results for the true specific heat per gram of water-vapour. Regnault's values for 453° abs. do not fit this formula at all well. Its applicability presumes, of course, that the temperature should not be far outside the temperature studied (205° – 339°). Over this interval it shows that the more the pressure sinks, and consequently the nearer the water-vapour behaves like an ideal gas, so much more does the specific heat per mol at constant pressure approach the value 7·74 cal. Experiments of Jones,[1] to which Lorenz (*l.c.*) refers, afford a partial confirmation of Lorenz's results. An exhaustive discussion of the specific heats of water-vapour has been published by Thiesen[2] in connection with the measurements of Regnault, and his own numerous determinations of the velocity of sound in water-vapour. We learn from his discussion that the specific heat of aqueous vapour under small constant pressures falls off very considerably between 0° and 80°. It reaches a minimum of 7·34 per mol at 80°, and then rises again, reaching the value 7·94 at 480°.[3]

Thiesen's calculation.

[1] Jones' values are given the preference by Carpenter. See Lorenz's paper for a comparison of the two series of results, as well as for a discussion of the formula and the thermodynamic relationships connected with it (see also Planck, "Thermodynamik," 2nd edit. (1904), p. 123).

[2] Thiesen, Drude's *Ann.*, 9 (1902), 88. See also Tumlicz (*Wiener Acad. Ber.*, 1897, Abt. II*a*. 654, and 1899, II*a*. 1395); further, Macfarlane Gray (*Phil Mag.*, [V.] 13, 337 (1882)).

[3] We may take into consideration Winckelmann's (*Wied. Ann.*, 4 (1878), 7) determinations of the thermal conductivity of aqueous vapour, which give indirect evidence on the basis of the kinetic gas theory bearing on the change

Specific heat of water-vapour between 110° and 800°.

Recently Holborn and Henning[1] have very much extended our knowledge of the specific heat of water-vapour at the pressure of one atmosphere by calorimetric determinations carried out between 110° and 800°. The results of their experiments give the relation between the specific heats of water-vapour and air, so that any uncertainty regarding the change of the specific heat of air with rising temperature influences the value for water-vapour.

Holborn and Henning give two expressions for the mean specific heat of water-vapour at the constant pressure of one atmosphere:

$$(a) \quad c_{p(H_2O)} = 8{\cdot}03 + 0{\cdot}00078t$$

or

$$c_{p(H_2O)} = 7{\cdot}60 + 0{\cdot}00078T$$

and

$$(b) \quad c_{p(H_2O)} = 7{\cdot}94 + 0{\cdot}00115t$$

or

$$c_{p(H_2O)} = 7{\cdot}31 + 0{\cdot}00115T$$

In the first case the increase of the mean specific heat of air at the constant pressure of one atmosphere is taken as $3{\cdot}0 \times 10^{-4}t$, in the second case twice this amount.

Specific heat of CO_2 at low temperature.

The specific heat of carbon dioxide at low temperature has been investigated by Regnault between −30° and 10°, +10° and 100°, and between 10° and 210°. Regnault found 0·18427, 0·20246, and 0·21692 gram-cal. per gram. Later experiments, in which he varied the pressure of the gas between 1 and 10 atmospheres, indicated that the mean specific heat between 185° and ordinary temperature is quite independent of the pressure. The Regnault experiments at ordinary pressures have been repeated by

of the specific heat with the temperature. According to these determinations, which Mallard and Le Chatelier have availed themselves of, the true specific heat of water-vapour should be some 12·7 per cent. greater at 100° than at 0°, when the pressure of the water-vapour amounts to but a few millimetres. Other indirect determinations have been carried out by A. H. Peake (*l.c.*) and by A. Griessmann (*Forschungsarbeiten auf dem Gebiet des Ingenieurwesens*, Heft 13 (1904)). They throttled superheated steam. This method is analogous to the procedure of Joule and Thomson (see p. 18). In the case of a *non-ideal* gas a fall of temperature takes place, which has a thermodynamic relation to the specific heat.

[1] Drude's *Ann. d. Physik* (IV.), 18 (1905), 739.

Wiedemann[1] with somewhat different experimental arrangements. According to Wüllner (*l.c.*), Regnault's and Wiedemann's results may be expressed as follows:—

	Regnault.		Wiedemann.
$c_{p(CO_2)m}$ mean specific heat, 0 – $t°$ C.,	$8{\cdot}28 + 0{\cdot}00596t$	...	$8{\cdot}56 + 0{\cdot}005038t$
$c_{p(CO_2)t}$ true " " at $t°$ C.,	$8{\cdot}28 + 0{\cdot}01192t$	...	$8{\cdot}57 + 0{\cdot}010076t$

These numbers have been well checked by determinations of the ratio of the specific heats made by Röntgen,[2] Müller,[3] and Wüllner,[4] according to various methods. It seems, consequently, beyond question that the specific heat of carbon dioxide rises rapidly with the temperature under atmospheric pressure in the interval of temperature studied. On the other hand, the results of Regnault are somewhat inaccurate, since Lussana[5] has shown with certainty that under heavy pressures the specific heat is a function of the pressure between the temperatures of 13·2° and 114·9°. At first sight there is something surprising in the fact that this specific heat should increase with rising temperature under constant atmospheric pressure, while that of water-vapour, according to Lorenz's observation, decreases under somewhat higher pressures. The phenomenon becomes more comprehensible when we notice that the same fall in the specific heat has also been observed by Lussana in carbon dioxide itself, when the pressure is greater than 54 atmospheres. Under lower pressure there is, according to him, first a decrease, and then later an increase in its specific heat.

The specific heat of carbon dioxide under the constant pressure of one atmosphere has been determined calorimetrically by Holborn and Austin.[6] They carried out their investigations between 20° and 800°, and derived the expression— Specific heat of CO_2 between 20° and 800°.

$$c_{p(CO_2)} = 8{\cdot}923 + 0{\cdot}003045t - 0{\cdot}000000735t^2$$

or

$$c_{p(CO_2)} = 7{\cdot}097 + 0{\cdot}003045T - 0{\cdot}000000735T^2$$

A linear expression which can be deduced from Langen's

[1] Wiedemann, *Pogg. Ann.*, 157 (1876), 1.
[2] *Pogg. Ann.*, 148 (1873), 580.
[3] *Wied. Ann.*, 18, 94.
[4] *Wied. Ann.*, 4 (1878), 321.
[5] *Fortschritte der Physik für* 1896, p. 345, and 1897, p. 331.
[6] *Sitzungsber. der Kgl. Preuss. Akad.*, 1905, p. 175.

determinations at higher temperatures is in fairly good agreement with the results of Holborn and Austin, and will be used later on instead of these expressions.

Specific heat at high temperatures. As far as relationships at high temperature are concerned, Mallard and Le Chatelier were the first to study the specific heats at constant volume. They exploded gases in closed spaces and observed the pressures generated.[1] From these, assuming the gas law ($pv = RT$), the mean specific heat between the ordinary and the explosion temperature could be calculated basing their conclusion on them. We will later recur to these experiments. It will here suffice to state that Mallard and Le Chatelier recommended the experimental formula—

$$c_v = 4 \cdot 33(T \times 10^{-3})^{0 \cdot 367}$$

for the specific heat of carbon dioxide at constant volume. Later, they resorted to a simpler linear formula.

For water-vapour they obtained the following values for the mean specific heats between 0° and $t°$:—

$t°$	3360	3380	3320
c_v	16·5	16·2	17·1

They deduced the formula—

$$c_{v(0°, t° C.)} = 5 \cdot 61 + 3 \cdot 28 \times 10^{-3} t.$$

to express these values. Only Regnault's value for ordinary temperature was known at that time, and could be compared with their observations at about 3350°, so that there was support for nothing more than a linear function. Wiedemann, as above mentioned, made measurements of thermal conductivity, and the deduction therefrom also agreed with this formula, though this, to be sure, was but scant proof.

[1] Berthelot and Veille (*Ann. Chim. Phys.*, 4 (1885), 13) instituted similar experiments, but, instead of the static pressure, measured the acceleration which a movable piston experienced under the influence of the explosion. The procedure was original with Veille. Such high temperatures were used throughout the measurements that dissociation set in. They therefore got "apparent" specific heats, that is, values which were affected by the heat of dissociation. The calculation of the real specific heat from these "apparent" values can hardly be considered very reliable. Le Chatelier made no further use of these values when, a few years after their publication, he was studying the dissociation of carbon dioxide from a theoretical standpoint.

To test their formula, Mallard and Le Chatelier made use of three determinations of the specific heat of mixtures of carbon dioxide and water-vapour. They assumed that the above-mentioned formula for the specific heat of water-vapour was correct, and calculated with its help the specific heat of carbon dioxide from the pressure measurements made on the mixture of carbon dioxide and water-vapour, and compared these with the values given by their exponential formula. The data are—

	$c_v(H_2O)$ $(0^\circ, t^\circ)$	$c_v(CO_2)$ $(0^\circ, t^\circ)$	$c_v(CO_2)$ from exponential formula.
2100°	12·3	14·4	13·8
1540°	10·6	11·6	12·6
1280°	9·7	11·3	11·8

The agreement is, as we see, a good one. The deviations amount to from 5 to 8 per cent. In the determinations of Mallard and Le Chatelier, it is assumed that at the moment of explosion the combustion is complete. If this is not the case, but instead, great dissociation takes place, the values obtained would be useless, because the heats of combustion would not be as great as assumed. The specific heat would appear too high. We have an illustrative case in nitrogen tetroxide at low temperatures. This gas (N_2O_4) dissociates, on warming, into $2NO_2$, and the heat-change which accompanies this gives rise to entirely erroneous values in specific heat measurements.[1] The inference that dissociation has taken place can, however, be tested here by the results of the explosion method. The pressure change after the explosion follows a different course when dissociation occurs than when it does not. Mallard and Le Chatelier's experiments lead to the strange result that hydrogen and oxygen unite completely even at a temperature of 3500°. A closer examination shows that a complete combination at this temperature is inconceivable in the equilibrium condition. The equilibrium must then have been overshot, or, better expressed, circumvented. We can imagine that the formation of water does not take place by the path *a*, in the following diagram, but by the path *b*, *c*, and *d*.

[1] Berthelot and Ogier, *Ann. Chim. Phys.* (5), 30 (1882), 382.

$$\begin{array}{ccc} 2H_2 + O_2 & \xrightarrow{b} & H_2O_2 + H_2 \\ \downarrow a & & \downarrow c \\ 2H_2O & \xleftarrow{d} & 2H_2O \\ xH_2 + yO_2 & & \end{array}$$

If the process takes place exceedingly fast, it is conceivable that it only has time to pass through the steps *b*, *c*, but not *d*, so that only undissociated water is formed, while had the heating lasted longer, free hydrogen and oxygen could not help being formed, just as though the process had taken the direct course *a*. This unusual condition of affairs will perhaps be somewhat elucidated by an illustration. We may picture a brook flowing down the side of a valley into a pool. The water cannot, of course, flow from the pool up the opposite side of the valley. The only way it can reach a point on this opposite side is to first flow around the head of the valley. The sole requirement is that everywhere on this circuitous route there be a certain minimum down grade. The water loses a part of its available energy of position in flowing around the head of the valley. It loses the rest of this energy as it flows down the slope into the pool. Whether the direct or the indirect path is followed depends here, just as it does in a chemical reaction, on the relative *resistances* of each path, or, as one would say in the chemical case, on the *reaction velocity*. The experiments of M. Traube and C. Engler[1] confirm the assumption that hydrogen peroxide, and not water, is the primary product of the union of hydrogen and oxygen.

Conditions are quite different in the case of carbon dioxide, a very perceptible dissociation being observed here, even below 2000°. Mallard and Le Chatelier estimate that at 2000°, under a pressure of 6 atmospheres, it amounts to something less than 5 per cent. In order quite to exclude any effects due to dissociation, these investigators employed another method of calculation, taking as a basis the experiments of Sarreau and Veille with the crusher manometer as a starting-point. This manometer is a sort of explosion bomb. The pressure developed in the explosions, amounting to thousands of atmospheres, compresses a copper cylinder fitted into an opening in the wall of the bomb.[2] By means of a compression machine, a hydrostatic pressure is

[1] *Berl. Ber.*, 33 (1900), 1110; Nernst, *Z. phys. Chemie*, 46 (1903), 720.

[2] Described and illustrated in the Sixth Lecture.

found which will deform similar cylinders to an equal amount. In this way the gas pressure developed by the explosion is measured. From the nature and amount of the explosion, we know the heat evolved, as well as the mass and nature of the gaseous product of the explosion. We then know that a known quantity of heat can cause these gases to exert a measured pressure within a certain volume. We are therefore able, with the help of the gas laws, to calculate the specific heats of the gases at constant volume. The method has this advantage that the enormous pressures generated prevent any dissociation taking place which would render the reaction incomplete at high temperatures. In this manner, Mallard and Le Chatelier calculated the mean specific heat per mol at constant volume to be—

$4{\cdot}76 + 0{\cdot}00122t$ for the permanent gases
$5{\cdot}78 + 0{\cdot}00286t$ for water-vapour
$6{\cdot}5 \ + 0{\cdot}00387t$ for carbon dioxide

Langen has recently measured the specific heats of these gases by Mallard and Le Chatelier's original method, and has found—

$4{\cdot}8 + 6{\cdot}0 \times t \times 10^{-4}$ for the permanent gases
$5{\cdot}9 + 2{\cdot}15 \times t \times 10^{-3}$ for water-vapour
$6{\cdot}7 + 2{\cdot}6 \times t \times 10^{-3}$ for carbon dioxide [1]

His numbers all signify mean specific heats at constant volume between 0° and t°. They particularly apply to the temperature interval of 1300° – 1700°, and, in this region, correspond well with the observed pressures. Even up to a temperature of 2000° they are not in error by more than 3 per cent. Langen's method of calculating the results of his explosion experiments has been criticized by Schreber.[2] Langen got three linear equations for the mean specific heats of known mixtures of carbon dioxide, water-vapour, and the permanent gases between 0° and 1300°, 0° and 1500°, and 0° and 1700°. Since the specific heats of the three components: (1) permanent gas, (2) water-vapour, (3) carbon dioxide, are the three unknown quantities, it is possible to calculate them directly from the three equations. Schreber found in this way—

[1] "Mitteilungen über Forschungsarbeiten auf dem Gebiete des Ingenieurwesen," vol. viii. (Berlin, 1903).

[2] *Dinglers polytechn. Journ.*, 318 (1903), 433.

Permanent gas	$4{\cdot}879 + 0{\cdot}00053t$
Water-vapour	$7{\cdot}456 + 0{\cdot}001165t$
Carbon dioxide	$7{\cdot}771 + 0{\cdot}00189t$

It may be remarked that the increase of the mean specific heat of water-vapour, according to Schreber's calculation, is almost the same as calculated by Holborn and Henning from their calorimetric determinations, the increase of the mean specific heat of air being taking as—

$$6{\cdot}0 \times 10^{-4}t$$

Langen's own computation starts from the assumption that the mean specific heat of the permanent gases is correctly given by the formula—

$$c_v = 4{\cdot}8 + 0{\cdot}0006t$$

which Mallard and Le Chatelier derived from their explosion experiments, and on this basis got the above values. It is clear from Schreber's criticism that the mean specific heat of water-vapour between 0° and 1300° or 1700°, as determined by Langen, is relatively uncertain. The values of the mean specific heats of the permanent gases are checked by Steven's measurements[1] of the ratio (κ) of the two specific heats of air at high temperatures made by the use of Quincke's acoustic thermometer. His value of $\kappa = 1{\cdot}34 \pm 0{\cdot}01$ harmonizes well with the values which Mallard and Le Chatelier, as well as Langen and Schreber, have calculated; while its agreement with the values of Mallard and Le Chatelier based on the experiments with the crusher manometer is faulty. There are no further results to be mentioned in this connection.[2] We will now attempt to summarize the various specific heat formulæ already given, and also append a few additional ones.

For carbon dioxide we have the following formulæ:—

(*a*) $c_v = 4{\cdot}33(\mathrm{T}10^{-3})^{0{\cdot}367}$ according to Mallard and Le Chatelier (explosion method). Basis: explosive pressure in gas explosions at 2000°, and Regnault's numbers.

(*b*) $c_v = 6{\cdot}5 + 2{\cdot}6t10^{-3}$ according to Langen. Basis: explosive pressures at 1300°, 1500°, 1700°. The calorimetric determinations of Holborn and Austin between 0° and 800° agree with this expression.

[1] For the method and for a criticism of the result by Kalähne, see the Sixth Lecture.

[2] Haüssere's results (*Forschungsarbeiten auf dem Gebiet des Ingenieurswesen*, Heft 25 (Berlin, 1905)) are very uncertain.

(*c*) $c_v = 7{\cdot}771 + 0{\cdot}00189t$, according to Schreber. Basis: Langen's above-mentioned experiments.

(*d*) $c_v = 6{\cdot}5 + 0{\cdot}00387$ according to Mallard and Le Chatelier. Basis: experiment with the crusher manometer.

(*e*) It should finally be added that Le Chatelier[1] later concluded to formulate all specific heats on the supposition that the specific heats of gases under constant pressure converge toward 6·5 per mol at the absolute zero. The values which he therefore considered as the most probable are—

Permanent gases	$6{\cdot}5 + 0{\cdot}0006T$
Water-vapour	$6{\cdot}5 + 0{\cdot}0029T$
Carbon dioxide	$6{\cdot}5 + 0{\cdot}0037T$

For water-vapour we have the following formulæ:—

(*a*) $c_v = 5{\cdot}61 + 3{\cdot}28t10^{-3}$ according to Mallard and Le Chatelier. Basis: explosion observations at about 3350°, Regnault's numbers for 120° – 220°, checked by explosion experiments with carbon monoxide, hydrogen, and oxygen based on formula (*a*) for CO_2.

(*b*) $c_v = 5{\cdot}78 + 2{\cdot}86t10^{-3}$ according to Mallard and Le Chatelier. Basis: observations with the crusher manometer.

(*c*) $c_v = 7{\cdot}456 + 0{\cdot}001165t$, according to Schreber. Basis: Langen's experiments.

(*d*) $c_v = 5{\cdot}9 + 2{\cdot}15t10^{-3}$, according to Langen. Basis: explosions at 1300°, 1500°, and 1700°.

In our consideration of the water-gas reaction we are interested in the difference between the specific heats of carbon dioxide and water-vapour, and we shall use these formulæ to compute the value of this difference for a number of temperatures. Further, instead of temperatures Celsius, we shall substitute absolute temperatures.[2]

[1] *Compt. Rend.*, 104 (1887), 1780; and "Cours de chimie industrielle." These are the values which Herr von Jüptner (*Z. anorg. Chem.*, 38 (1904), 63) used.

[2] The following example illustrates the method of calculation. Langen states that—

$$C_{vCO_2}\text{ (between } 0^\circ \text{ and } t^\circ) = 6{\cdot}7 + 0{\cdot}0026t$$

It follows then that—

$$C_{vCO_2}\text{ (true at } t^\circ) = 6{\cdot}7 + 0{\cdot}0052t$$

and that—

$$C_{vCO_2}\text{ (true at } T^\circ) = 6{\cdot}7 + 0{\cdot}0052(T - 273) = 5{\cdot}28 + 0{\cdot}0052T$$

and then—

$$C_{vCO_2}\text{ (between } 0^\circ \text{ and } T^\circ) = 5{\cdot}28 + 0{\cdot}0026T$$

We will first choose Langen's values for constant volume.

$$c_{vCO_2} = 5{\cdot}280 + 2{\cdot}6 \times 10^{-3}T$$
$$c_{vH_2O} = 4{\cdot}726 + 2{\cdot}15 \times 10^{-3}T$$
$$c_{vCO_2} - c_{vH_2O} = 0{\cdot}554 + 0{\cdot}45 \times 10^{-3}T$$

$c_{vCO_2} - c_{vH_2O}$ then becomes, in the interval between 0° abs. and T°, equal to—

° C.	...	1227	1427	1627	1827
T	...	1500	1700	1900	2100
Cal.	...	+1·23	+1·34	+1·40	+1·50

Secondly, we will choose Schreber's data for constant volume—

$$c_{vCO_2} = 6{\cdot}739 + 0{\cdot}00189T$$
$$c_{vH_2O} = 6{\cdot}82 \ + 0{\cdot}001165T$$
$$c_{vCO_2} - c_{vH_2O} = -0{\cdot}081 + 0{\cdot}000725T$$

$c_{vCO_2} - c_{vH_2O}$ then becomes, in the interval between 0° abs. and T°, equal to—

° C.	...	1227	1427	1627	1827
T	...	1500	1700	1900	2100
Cal.	...	+1·00	+1·15	+1·30	+1·44

Thirdly, we choose the values of Mallard and Le Chatelier based on manometer measurements—

$$c_{vCO_2} = 4{\cdot}39 + 0{\cdot}00387T$$
$$c_{vH_2O} = 4{\cdot}22 + 0{\cdot}00286T$$
$$c_{vCO_2} - c_{vH_2O} = 0{\cdot}1685 + 0{\cdot}00101T$$

$c_{vCO_2} - c_{vH_2O}$ in the same interval then becomes—

° C.	...	1227	1427	1627	1827
T	...	1500	1700	1900	2100
Cal.	...	+1·684	+1·886	+2·088	+2·289

Hoitsema's consideration regarding the water-gas reaction.

We will now revert to Hoitsema's treatment of the water-gas reaction. Hoitsema selects the formulæ which Mallard and Le Chatelier computed from their experiments with the crusher manometer to represent the change of the specific heats of the gases with the temperatures, and assumes them to be correct. With their help he calculates the difference between the specific heats of the gases appearing and disappearing in the water-gas reaction as—

$$\begin{array}{r} 6{\cdot}5 \ \ + 0{\cdot}00387t \\ 5{\cdot}78 + 0{\cdot}00286t \\ \hline 0{\cdot}72 + 0{\cdot}00101t \end{array}$$

On this basis he arrives at the expression—

$$Q_t = -10{,}111 + 0{\cdot}72t + 0{\cdot}00101t^2$$

using Berthelot's value for the heat of reaction at 18°, and on the absolute temperature scale—

$$Q_T = -10{,}232 + 0{\cdot}1685T + 0{\cdot}00101T^2$$

Hoitsema showed that the heat of reaction must therefore become zero at 2825°, which is easily seen to be true when we substitute the value (2825 + 273) for T in the equation just given. Hoitsema then recalled that the equilibrium constant must show a maximum at the place where the heat of reaction passes through zero. This follows from our former equation (see p. 64)—

$$\frac{dlnK}{dT} = -\frac{Q}{RT^2}$$

For, according to this equation, an increase $dlnK$ will accompany an increase of temperature dT, when Q has a negative value, but afterwards, when Q has acquired the opposite sign at a higher temperature, $dlnK$ must become negative, and consequently the equilibrium constant must become smaller. Now, Hoitsema found that the ratio of the compositions of the gases in Horstmann's experiments on the explosions of carbon monoxide and hydrogen with insufficient oxygen shows a maximum—

$$\frac{C_{H_2O} \times C_{CO}}{C_{H_2} \times C_{CO_2}} = 6{\cdot}25$$

He calculated, with the help of the above-mentioned specific heats, and with the help of the known heats of combustion of carbon monoxide and hydrogen, what the temperature of the gases in Horstmann's experiments must have been, when the equilibrium constant showed this maximum. This temperature came out at 2700°, which is evidently very near to the value 2825°.

Luggin's discussion of the water-gas equilibrium.

We have treated this calculation so exhaustively here because it has played an important *rôle* in the subsequent study of the question. It would seem at first sight as though its results were sufficiently unequivocal to definitely settle the location of the water-gas equilibrium at all temperatures. Thus, as Luggin[1] has shown, if we turn to our formula—

[1] *Journ. f. Gasbel. u. Wasserversorg.* (1898), 713.

$$A = Q_0 - \sigma' T \ln T - \sigma'' T^2 - RT \ln \frac{C_{H_2O} \times C_{CO}}{C_{H_2} \times C_{CO_2}} + \text{const. } T$$

after an easy transformation we find, for the case of equilibrium, where $A = 0$ and $\frac{C_{H_2O} \times C_{CO}}{C_{H_2} \times C_{CO_2}} = K$

$$0 = \frac{Q_0}{RT} - \frac{\sigma'}{R} \ln T - \frac{\sigma''}{R} T - \ln K + k'$$

The constant k' has the value $\frac{\text{const.}}{R}$. If, finally, we insert Briggsian logarithms—

$$\log^{10} K = \frac{Q_0}{2{\cdot}3 \times R \times T} - \frac{\sigma'}{R} \log^{10} T - \frac{\sigma''}{2{\cdot}3R} T + k''$$

Here $k'' = \frac{\text{const.}}{2{\cdot}3R}$. Inserting numerical values, we get—

$$\log^{10} K = -\frac{10{,}232}{2{\cdot}3 \times R \times T} - \frac{0{\cdot}1685}{R} \log^{10} T - \frac{0{\cdot}00101}{2{\cdot}3 \times R} T + k''$$

We can directly calculate the constant k'', if it is certain that the equilibrium constant K for the $T = 2825 + 273$ has the value 6·25. Carrying out the calculation on this basis, and inserting the value of k'' obtained (putting $R = 2$)—

$$\log^{10} K = -\frac{2232}{T} - 0{\cdot}08463 \log^{10} T - 0{\cdot}0002203T + 2{\cdot}4943$$

If we now recur to our main formula, we may write it as—

$$A = -10{,}232 - 0{\cdot}1685 T \ln T - 0{\cdot}00101 T^2 - RT \ln \frac{C_{H_2O} \times C_{CO}}{C_{H_2} \times C_{CO_2}} + 10{\cdot}725T$$

Yet if we critically examine the basis of this formula, we are met by a series of grave objections. The correctness of the whole matter is evidently dependent, in the first degree, upon how close to the truth are Mallard and Le Chatelier's values for the specific heats. Indeed, this dependence is of a double character. If, for instance, the specific heats of carbon dioxide and water-vapour were each slightly altered, the percentage effect of this change on the small difference of the two quantities (0·1685 + 0·00101T) would be large. But a small difference in the value of the second term of this expression (that is, 0·00101T) would be sufficient, at these temperatures, where Q becomes equal to zero, to give entirely different numbers. If we assume the values of Langen as true for all temperatures, the difference of the specific heats of carbon dioxide and water-vapour is 0·554 + 0 00045T. We get, then, by a similar development—

$$Q_T = -10{,}298 + 0{\cdot}554T + 0{\cdot}00045T^2$$

and Q does not became equal to 0 till at about 4000°, or at a temperature some 1000° higher than before.

In the second place, the values of the specific heats determine what temperature one computes for the gas mixture at the moment of explosion from the heats of combustion of carbon monoxide and hydrogen. Here the sum, and not the difference, of the specific heats comes into consideration. An inaccuracy in the difference by no means involves a like inaccuracy in the sum, and *vice versâ*. The agreement of the maximum temperatures, which Hoitsema deduced on the one hand from Horstmann's gas analysis, and on the other hand from the heat of reaction, is therefore just as uncertain as the temperature at which the heat of reaction becomes equal to zero.

The other basis for Hoitsema's value is the number 6·25 for the maximum value, which the ratio—

$$\frac{C_{CO} \times C_{H_2O}}{C_{CO_2} \times C_{H_2}} = K$$

reaches in the explosion experiments. The calculation of Hoitsema can only be right, provided the ratio, after it has adjusted itself at nearly 3000°, in the moment of the explosion, does not alter during the cooling. It is, however, improbable that this cooling is rapid enough to retain unaltered the composition of the gases reached at these high temperatures.

Hoitsema himself noted that Macnab and Ristori had made experiments with the gases produced in the detonation of explosives. These are chiefly carbon monoxide, carbon dioxide, nitrogen, hydrogen, and water-vapour. Their experiments did not agree with Horstmann's results, and could only be brought into accord with them by the assumption that in Macnab and Ristori's experiments the gases reacted further as they cooled off. We should also point out in this connection that repetition of Horstmann's experiments by Bötch,[1] and

[1] *Lieb. Ann.*, 210 (1881), 207.

especially by Dixon,[1] have given results which cannot be reconciled at all with those of Horstmann, and awaken the suspicion that the walls of the vessels affect the results of the explosion process to such a degree as to render such experiments valueless for calculations of the equilibrium.

Experiments of Harries. In view of these considerations, it must at first seem very surprising that Luggin's formula, based on Hoitsema's computation—

$$\log^{10} K = -\frac{2232}{T} - 0{\cdot}08463 \log^{10} T - 0{\cdot}0002203T + 2{\cdot}4943$$

should fit so well those observations of Harries, which Luggin was able to use in testing it. Harries had, at H. Bunte's suggestion, passed water-vapour over glowing coals and determined the composition of the mixture of hydrogen, carbon dioxide, carbon monoxide, and water-vapour formed. Harries' results, as computed by Luggin, were as follows:—

$t°$ C.	H_2	CO	CO_2	H_2O	Gas current. Litres per second.	K obs.	K cal. Luggin.
674	8·41	0·63	3·84	87·12	0·9	1·70	0·49
758	22·28	2·67	9·23	65·82	1·8	0·85	0·70
838	28·68	6·04	11·29	54·09	3·66	1·01	0·98
838	32·77	7·96	12·11	47·15	3·28	0·94	0·98
861	36·48	11·01	13·33	39·18	5·3	0·89	1·07
954	44·43	32·70	5·66	17·21	6·3	2·25	1·41
1010	47·30	48·20	1·45	3·02	6·15	2·12	1·65
1060	48·84	46·31	1·25	3·68	9·8	2·78	1·88
1125	50·73	48·34	0·60	0·303	11·3	0·48	2·11

From the given compositions of the gas mixture, the ratios—

$$\frac{C_{H_2O} \times C_{CO}}{C_{CO_2} \times C_{H_2}} = K$$

are first calculated for the various temperatures. Beside them are placed the values of K as calculated for the various temperatures from Luggin's formula. Though these experiments were merely undertaken to determine the effect of water-vapour on glowing coal, and were executed under experimental conditions which gave no assurance of the constancy of the temperature,

[1] *Phil. Trans. Roy. Soc.*, 175 (1884), 618.

yet, with a few exceptions, there is a surprising agreement between the values of—

$$\frac{C_{H_2O} \times C_{CO}}{C_{CO_2} \times C_{H_2}}$$

and the values of K, calculated from Luggin's formula.

Boudouard then instituted experiments to fix more accurately the equilibrium constant by leading carbon dioxide and hydrogen into a hot evacuated vessel, which contained platinum as a contact substance. After a long heating, the gas was sucked out, and its composition determined. The results are entirely useless, for the composition of the gas before and after the experiment does not satisfy the stoichiometric requirements of the reaction. When, for instance, Boudouard started with a mixture containing 51 per cent. CO_2 and 49 per cent. H_2, there must have been 51 per cent. (CO_2 + CO) and 49 per cent. (H_2 + H_2O) after the experiment. Instead of this he found, for instance, 60·2 per cent. CO + CO_2 and 39·8 per cent. (H_2 + H_2O). Hahn has adduced yet other legitimate objections to Boudouard's experiments. He himself[1] then subjected the problem to a careful study. Experiments of Boudouard

On the one hand, Hahn led mixtures of carbon dioxide and hydrogen over platinum as a contact substance, and, on the other, mixtures of carbon monoxide and water-vapour. The values which he obtained in the first case for the equilibrium are given under (*a*), those in the second case under (*b*). The computation was carried out by Hahn in two ways—first using Luggin's expression exactly as given, and then again using it after the thermodynamically indeterminate constant had been slightly altered (2·4943 to 2·5084). Experiments of Hahn

t	T	*a*	*b*	Mean.	Calculated with 2·4943.	2·5084.
686	959	0·534	—	—	0·505	0·522
786	1059	0·872	0·808	0·840	0·790	0·851
886	1159	1·208	1·186	1·197	1·133	1·170
986	1209	1·596	1·545	1·571	1·520	(1·570)
1005	1278	1·62	—	—	1·597	1·650
1086	1359	—	1·96	—	1·938	2·002
1205	1478	2 126	—	—	2·457	2·538
1405	1678	2·49	—	—	3·320	3·433

[1] *Z. f. phys. Chemie*, 44, 510 and 48, 735.

In the interval between 686° and 1086°, Hahn's experiments agree surprisingly well with the Hoitsema-Luggin curve. The values for 1205° and 1405° are divergent. Hahn later attempted another computation in order to explain these divergences. He used the Le Chatelier supposition that the specific heats of all gases under constant pressure equal 6·5 at the absolute zero point. Like Le Chatelier, he further makes use of the assumption that the specific heats, at any temperature, can be approximately represented by the linear formula $6{\cdot}5 + x\mathrm{T}$, but chooses another value for x than did Le Chatelier. He takes $6{\cdot}5 + 0{\cdot}0042\mathrm{T}$ as the molecular specific heat of carbon dioxide under constant pressure between 0° and T°, and $6{\cdot}5 + 0{\cdot}0024\mathrm{T}$ as that of hydrogen. Le Chatelier himself took $6{\cdot}5 + 0{\cdot}0037\mathrm{T}$ for carbon dioxide and $6{\cdot}5 + 0{\cdot}00029\mathrm{T}$ for water-vapour—as we saw before. Hahn's apparently insignificant change in Le Chatelier's values causes the difference between the specific heats of carbon dioxide and water-vapour to come out entirely different,[1] 0·0018T being obtained instead of 0·0008T. Hahn did not attempt to justify his procedure by a comparison of his values of the specific heats with observation at high temperatures. He was satisfied to find that the observations which he had made at 1205° and 1405° upon the equilibrium appeared less divergent when he made this assumption regarding the specific heats, and consequently used the formula—

$$\log \mathrm{K} = -\frac{2226}{\mathrm{T}} - 0{\cdot}0003909\mathrm{T} + 2{\cdot}4506$$

based on it for his calculation. He obtained in this way—

Cal. K	...	0·57	0·86	1·19	1·55	1·62	1·91	2·33	2·94
Found K	...	0·53	0·84	1·20	1·57	(1·62)	1·96	2·13	2·49
t° C.	...	686	786	886	986	1005	1086	1205	1405

[1] This becomes especially apparent when we compute the value of $c_{p(CO_2)} - c_{p(H_2O)}$, as we have already done above (p. 134). According to Le Chatelier, its value is 0·0008T; according to Hahn, 0·0018T. If we place as before—

t	... ...	1227	1427	1627	1827
T	... ...	1500	1700	1900	2100

we get—

Le Chatelier	...	+1·2	+1·36	+1·52	+1·68
Hahn ...	...	+2·7	+3·06	+3·42	+3·78

While Le Chatelier's values uniformly accord with those of Langen, Schreber, and Mallard and Le Chatelier, Hahn's values are very different, and are quite unconfirmed by any observation.

Hahn's results at 1205° and 1405°.

Hahn's assumption evidently does not bring observation and calculation into agreement. One is therefore led to inquire whether these values of Hahn cannot be otherwise explained. Haber, Richardt, and Allner, who continued the study of the question, investigated how quickly the composition of the gas mixture in the water-gas equilibrium changed with the temperature, when it was allowed to cool without touching the solid walls. They found that when the temperature fell at a moderate rate, the adjustment of the equilibrium took place with sufficient rapidity above 1600°, but not below. From Hahn's own observation we see that platinum lowers this temperature of rapid adjustment to about 700°. It is certainly permissible to assume that the rate of adjustment on porcelain, quartz, and similar solid bodies would have an intermediate value, for gas reactions are, in general, accelerated by contact with hot, solid walls. It consequently seems very probable that the composition of the gas mixture in Hahn's experiments changed at temperatures above 1100° after it had left the platinum mass serving as a contact substance, and while it cooled off in contact with the quartz walls of the apparatus. Hahn did not fail to consider this possibility, but by observing that he obtained the same results with both slow and fast streams of gas, he felt assured that no such changes in composition took place. This proof is, however, not conclusive, as Nernst has shown. The temperature-fall in a gas streaming from a hot tube does not keep pace with the changing velocity of the gas's motion, but depends greatly on the thermal conductivity of the gas and the other thermal properties of the system. The only guarantee against a displacement of the equilibrium in cooling, would have been to have the gas pass directly from the glowing platinum into a cooled tube. A further uncertainty arises from the fact that Hahn used quartz vessels for his experiments, and this has been shown by Villard,[1] by Jacquerod and Perrot,[2] and by Berthelot[3] to be permeable to hydrogen and other gases above 1100.° A statement of Berthelot makes this very plain. He says that 1 c.c. of hydrogen measured at atmospheric pressure was sealed up in a quartz tube of 5 c.c.

[1] *Compt. Rend.*, 130 (1900), 1752.
[2] *Ibid.*, 139 (1904), 789.
[3] *Ibid.*, 140 (1905), 821.

capacity, and the tube heated for an hour to 1300°. After the heating, it was found that 44 per cent. of the hydrogen used had disappeared, while 14 per cent. of nitrogen had entered. In other experiments carbon heated to 1300°—1325° in evacuated quartz tubes, in the air, was attacked with the formation of carbon monoxide, showing that oxygen had diffused through the quartz into the tube. Indeed, we possess no material of which we may construct a vessel impermeable to the components of water-gas at temperatures above 1200°. Platinum, like quartz, is permeable to hydrogen, while porcelain, according to results of Le Chatelier and Boudouard,[1] allows perceptible quantities of hydrogen to pass through at 1200°, and large quantities (2 mg. from a porcelain vessel of 60–70 c.c. contents) at 1320°, according to measurements by Crafts. Besides this, porcelain, because of a small amount of iron oxide which it usually contains, acts chemically on the hydrogen at these high temperatures.

But we have direct evidence against Hahn's results above 1086°. Haber, Richardt, and Allner have determined the water-gas equilibrium in open flames, and have found values between 1250° and 1500°, which agree throughout with Hahn's original calculation—

$$\log K = -\frac{2232}{T} - 0{\cdot}08463 \log^{10} T - 0{\cdot}0002203T + 2{\cdot}5084$$

so that this must pass as a reliable determination of the location of the water-gas equilibrium between about 680° and about 1480°, or over an interval of 800°.

In view of these results, it might be useful to give a brief summary of the values we should get for the equilibrium constant, and the change of this constant with the temperature, when we use other data for the specific heats than those of Mallard and Le Chatelier (from the crusher manometer method). We will therefore calculate the values of K—

(1) Using the heats of reaction according to Berthelot, and the specific heat of carbon dioxide and water-vapour given by Langen, we assume the value of K at 986° = 1·57 as correct.

(2) Using the heat of reaction according to Berthelot, and the specific heat of carbon dioxide and water-vapour which

[1] "Températures élevées" (Paris, 1900), p. 47.

Schreber computed from Langen's observations, we assume here, as before, that the value of K = 1·57 at 986° is correct.

From (1) we get the expression—

$$\log K = -\frac{2245}{T} - 0{\cdot}2783 \log T - 0{\cdot}0000981 \times T + 2{\cdot}9653$$

From (2)—

$$\log K = -\frac{2213}{T} + 0{\cdot}04061 \log T - 0{\cdot}000158T + 2{\cdot}0266$$

Beside this, are arranged (3) the numbers observed by Hahn, (4) the values derived from Hahn's earlier formulæ, and finally (5) numbers computed by means of another entirely different formula, which will be discussed further on. This formula is—

$$\log K = -\frac{2116}{T} + 0{\cdot}783 \log T - 0{\cdot}00043 \times T$$

t	T	1	2	3	4	5
686	959	0·50	0·49	0·534	0·52	0·52
786	1059	0·79	0·78	0·840 ± 0·032	0·85	0·82
886	1159	1·15	1·12	1·197 ± 0·011	1·17	1·19
986	1259	(1·57)	(1·57)	1·571 ± 0·026	(1·57)	1·60
1005	1278	—	—	1·62	1·65	—
1086	1359	2·04	2·05	1·956	2·00	2·04
1205	1478	2·67	2·66	[2·126]	2·54	2 60
1405	1678	3·69	3·75	[2·49]	3·43	3·48
1500	1773	—	4·27	—	—	3 87
1600	1873	4·70	4 81	—	4·24	4·24

We see, in the first place, that the numbers under (1) (up to 1405° at least) do not materially differ from those under (4) and (2). The numbers under (1), however, are based on Langen's specific heats, which, as we saw before, are wholly irreconcilable with Hoitsema's maximum. It is evident, then, that the observations of Harries, Hahn, Haber, Richardt, and Allner do not confirm the existence of Hoitsema's maximum. Considered alone, however, it has no significance, and may be dismissed from our consideration.

If, now, we survey the entire question, we see that the use of all previously mentioned values of the specific heats in our formula— **The thermodynamically indeterminate constant**

$$A = Q_0 - \sigma' T ln T - \sigma'' T^2 - RT ln \frac{C_{CO} \times C_{H_2O}}{C_{CO_2} \times C_{H_2}} + \text{const. } T$$

leads us to assume a large value for the thermodynamically indeterminate constant. Luggin calculates it, on the basis of Mallard and Le Chatelier's values, to be +10·725. Hahn's observations change this number to 11,438. With Langen's specific heats it becomes 13,521. Recalculation according to the values of the specific heat computed by Schreber makes it 9·24. We should expect zero instead of these values. A formula which would give such a value is—

$$A = -9650 + 1{\cdot}55T\,ln\,T - 0{\cdot}00195T^2 - RT\,ln\frac{C_{CO} \times C_{H_2O}}{C_{CO_2} \times C_{H_2}}$$

A simple transformation of this formula gives the above-mentioned expression, by means of which the numbers in the fifth column of the above table were calculated. They agree excellently with the observed values. The quantity $c_{CO_2} - c_{H_2O}$ becomes on this basis $-1{\cdot}55 + 0{\cdot}00195T$. This difference of the mean specific heats between 0° abs. and T°, computed as before for a series of temperatures, gives the numbers—

t° C.	1227	1427	1627	1827
T	1500	1700	1900	2100
Gram-cal. ...	+1·37	+1·76	+2·15	+2·54

These numbers, both as regards magnitude and sign, are very close to those which we previously found from the results of Mallard and Le Chatelier, Langen and Schreber. Since it is precisely this interval (between 1300° and 1700°) which is best studied, the agreements of our assumption with the other formulæ gives it added support. According to our formula, the sign of the difference changes at 522°. This does not agree with the results of Holborn and Austin and of Holborn and Henning. Calculation based upon their observations shows that the differences between the mean specific heats of carbon dioxide and water-vapour between 0° abs. and T° are—

t° C.	527	727	927
T	800	1000	1200
Gram-cal. ...	+0·54	+0·63	+0·69

The differences as calculated from our formula are greater at temperatures above, and less at temperatures below 850°. Yet in reality there is no contradiction between the two, for the measurements of Holborn and his associates did not extend higher than 850°, and our formula does not claim to be applicable at any temperature below 650°. But the discrepancy is small between 650° and 850°, and, indeed, we shall see in the next lecture that we, using our formula, obtain correct values for the energy of formation of water-vapour from its elements at temperatures a great deal lower than 650°.

There is still a word to be said regarding the heat of reaction $Q_0 = -9650$, which we have assumed. Since we maintain the accuracy of our formula only for high temperatures, we are no longer justified in using it to calculate the value of Q_0. A correct procedure would be to first calculate the value of Q for say 700° C. from its value at ordinary temperatures (−10,000 to −10,100 cal.), using the formulæ of Holborn and his co-workers. From this we could calculate Q_0, using our expression $-1{\cdot}55 + 0{\cdot}00195T$. But the heat of reaction changes so slowly with the temperature that for the present we may call Q_0 approximately equal to −9650 cal. When the equilibrium constant of the water-gas reaction and the specific heats of the gases taking part in it are better known, this can be easily corrected. Meanwhile the equation—

$$A = -9650 + 1{\cdot}55T\,lnT - 0{\cdot}00195T^2 - RT\,ln\frac{C_{CO} \times C_{H_2O}}{C_{CO_2} \times C_{H_2}}$$

may be recommended for calculating the location of the equilibrium for high temperatures (above 650°).

Another fact worthy of notice appears from Hahn's observations. Hahn found that the equilibrium constant becomes equal to unity at 830°. It follows from this that carbon monoxide and hydrogen are equally strong reducing agents, and carbon dioxide and water-vapour equally strong oxidizing agents, at this temperature. At lower temperatures carbon monoxide, at higher temperatures hydrogen, is the stronger reducing agent.

Concluding remarks.

Reviewing now the cases discussed in this lecture, we see that it is possible to consider the thermodynamically indeterminate constant as equal to zero without in any way conflicting with the facts. Our conclusion is not, however, a rigorous one, for we have everywhere found great uncertainty regarding the specific heats. Now, only the logarithms of the ratios of the concentrations enter into our equation, while the specific heats appear as multiples of their differences. That is, in the expression for $Rln K$ we have σ' multiplied by the factor lnT, and σ'' multiplied by the still much larger factor T. Two observations may be made in this connection.

If we consider adherence to the equation $pv = RT$ as a first criterion of an ideal gas reaction, and the equality of the specific heats and their constancy with the temperature as a second, then we should expect that the ordinary gas reaction would satisfy the first much better than the second criterion. For the *logarithm* of a *ratio* is but slightly affected by small changes in the separate quantities, while the *multiple* of a *difference* is very sensitive to any such changes. It follows from this that the study of chemical equilibria at high temperatures offers much better conditions for the accurate determination of the specific heats of gases at these temperatures than does either the acoustic or the calorimetric method. This fact is also of importance in comparing the results of Bodenstein's study of the hydriodic acid equilibrium with the acoustic measurements of Strecker. To be sure, we only get the difference of the specific heats from such equilibria measurements. But the study of dissociations where a single gas is in equilibrium with solids allows us, as was shown at the end of the last lecture, to find the specific heats of many gases provided we know the specific heats of the solid substances and the dissociation pressures. Thus carbon dioxide could be studied in the dissociating carbonates, hydrogen in hydrides, nitrogen in nitrides, and oxygen in oxides.[1]

[1] The phenomena of "aging" shown by many substances at ordinary temperatures (see, for instance, Haber and van Oordt, *Z. f. anorg. Chem.*, 38 (1904), 378) might very considerably complicate these dissociations at high temperatures. This view has been especially emphasized by Wöhler (*Z. f. Electrochemie* (1905), 836), who investigated the dissociation of the palladium oxides. See also in this connection Joulin's experiments with the carbonate of manganese, *Ann. Chim. Phys.*, (4) 30 (1873), 277.

FIFTH LECTURE

SOME EXAMPLES OF REACTIONS INVOLVING A CHANGE IN THE NUMBER OF MOLECULES

THE second class of reactions, that is, reactions where $\Sigma\nu'$ is not zero, but where instead the number of molecules changes during the reaction, also requires a brief preliminary discussion. Since pressure is here the governing factor, we shall use partial pressures in our equations in accordance with our previous discussion (p. 52). This usage is not the general one. Following Le Chatelier and Planck, it has been usual to employ so-called "numerical concentrations." These are identical with the partial pressures when the total pressure is one atmosphere. At other pressures they are equal to the quotient of the partial pressure over the total pressure. If we call the numerical concentrations $\bar{c}$, then, taking as an example the formation of carbon dioxide—

$$\bar{c}_{CO_2} = \frac{p_{CO_2}}{P}, \quad \bar{c}_{CO} = \frac{p_{CO}}{P}, \quad \text{and } \bar{c}_{O_2} = \frac{p_{O_2}}{P}$$

and so—

$$\frac{p_{CO_2}}{p_{CO} \times p^{\frac{1}{2}}_{O_2}} = \frac{\bar{c}_{CO_2} \times P}{\bar{c}_{CO} \times P \times (\bar{c}_{O_2} \times P)^{\frac{1}{2}}} = \frac{\bar{c}_{CO_2}}{\bar{c}_{CO} \times \bar{c}^{\frac{1}{2}}_{O_2}} \times \frac{1}{\sqrt{P}}$$

where P signifies total pressure.

If now we write the expression for the energy of the formation of carbon dioxide, using numerical concentrations,[1] we get—

$$A = Q_0 - \sigma'_p T ln T - \sigma'' T^2 - RT ln \frac{\bar{c}_{CO_2}}{\bar{c}_{CO} \times \bar{c}^{\frac{1}{2}}_{O_2}} \times \frac{1}{\sqrt{P}} + \text{const. } T$$

[1] Planck's definition of equilibrium says ("Thermodynamik," 2nd edit. (1905), para. 241)—

$$c_1^{-\nu'} \times c_2^{-\nu''} \ldots = a\left(\frac{T}{p}\right)^{\nu_1 + \nu_{11} + \ldots} \times e^{-\frac{b}{T}} \times T^c$$

or—

$$A = Q_0 - \sigma'_p T ln T - \sigma'' T^2 + RT ln P^{\frac{1}{2}} - RT ln \frac{\bar{c}_{CO_2}}{c_{CO} \times c_{O_2}^{\frac{1}{2}}} + \text{const. } T$$

The result simply is, as we see, that the effect of pressure finds a particular expression in the term $RTlnP^{\frac{1}{2}}$, while in our usual notation—

$$A = Q_0 - \sigma'_p T ln T - \sigma'' T^2 - RT ln \frac{p_{CO_2}}{p_{CO} \times p_{O_2}^{\frac{1}{2}}} + \text{const. } T$$

nothing is specially emphasized. No use will be made of this rearrangement in what follows.

After a short preliminary discussion concerning compounds of oxygen and nitrogen, we shall consider in detail five reactions, namely—

If we take the logarithms of this expression, we obtain our formula—

$$\Sigma\nu' ln \bar{c}' = lna + \Sigma\nu' ln T - \Sigma\nu' ln p - \frac{b}{T} + c ln T$$

Here p represents the total pressure, and may be combined with the numerical concentrations, when we obtain—

$$0 = lna + \Sigma\nu' ln T - \Sigma\nu' ln p' - \frac{b}{T} + c ln T$$

Here p' represents partial pressure, and $\Sigma\nu' lnp'$ the equilibrium constant K_p. Planck defines $(-b)$ as $\frac{Q_0}{R}$ and $(-c)$ as $\left(\frac{\sigma'_v}{R}\right)$ in our notation. Substituting, we get—

$$0 = \frac{Q_0}{RT} - \frac{\sigma'_v}{R} ln T + \Sigma\nu' ln T - \Sigma\nu' ln p' + lna$$

But—

$$\sigma'_p = \sigma'_v - R\Sigma\nu'$$

Substituting this in the above equation, and multiplying by RT at the same time, it follows that—

$$0 = Q_0 - \sigma'_p T ln T - RT\Sigma\nu' ln p' + (Rlna)T$$

The equation then has exactly the same meaning as ours. A is given the particular value 0 applying to the equilibrium condition, and σ'' is considered as negligible. Then, too, in place of our constant, there is an expression $Rlna$, where a is also a constant chosen chiefly for convenience in calculations. Planck in general omits to consider the differences in specific heats at constant volume, and thus assumes that the heat of reaction at constant volume does not change with the temperature. The "dictate of experience that an element has the same specific heat in its various compounds" is considered by Planck himself as merely a first approximation (*l.c.*, para. 51). For the history of this principle see p. 66.

1. The formation of carbon dioxide from carbon monoxide and oxygen.

2. The formation of water from its elements.

3. Deacon's process for the manufacture of hydrochloric acid.

4. The formation of sulphur trioxide from sulphur dioxide and oxygen.

5. The formation of ammonia from its elements.

Nitrogen dioxide and nitrogen tetroxide. The formation of nitrogen tetroxide from nitrogen dioxide is a frequent text-book example of an equilibrium with an unequal number of reacting molecules. It has been studied by Playfair and Wanklyn,[1] by R. Müller,[2] by Deville and Troost,[3] by Ed. and Lad. Natanson,[4] by Nauman,[5] and by von Salet.[6] Gibbs,[7] Boltzmann,[8] van't Hoff,[9] Swart,[10] and Schreber,[11] have treated the case mathematically.[12]

This intensive treatment has been of great importance in the historical development of the theory of gaseous chemical reactions. We must, therefore, at least mention it, although our treatment must be very brief, as it does not immediately concern us at present. We shall make use of the data for the equilibrium constant which Schreber[13] has critically computed from the experimental results of the Natanson brothers.

1 *Ann. Chem. Pharm.*, 122, 245.

2 *Ibid.*, 122, 1.

3 *Compt. Rend.*, 64, 237 (1867); comp. also *Berl. Ber.* (1878), 2045.

4 *Wied. Ann.*, 24, 454 (1885), and 27 (1886), 606.

5 *Ann. Chem. Pharm.* (1868); suppl. vi. 205.

6 *Compt. Rend.*, 67, 488.

7 "Thermodynamische Studien," translated into German by Ostwald, p. 210 (Leipzig, 1892).

8 *Wied. Ann.*, 22, 72 (1884).

9 "Studien zur chem. Dynamik," von van't Hoff-Cohen (Leipzig, 1896), p. 156.

10 *Z. f. phys. Chemie*, 7, 120 (1891).

11 *Ibid.*, 24 (1897), 651.

12 Comp. also Nernst, "Theoret. Chem.," 4th edit., p. 437.

13 Schreber makes use of another definition of the equilibrium constant than we have chosen. His refers (1) not to formation, but to decomposition; (2) not to ½ mol N_2O_4, but to one mol; (3) his unit of pressure is not an atmosphere, but a millimetre of mercury. Our equilibrium constant, therefore, bears the following relation to his:—

$$K_p = \sqrt{\frac{760}{K_{p\,(Schr.)}}}$$

We see, in the first place, that to the reaction—

$$NO_2 \rightleftarrows \tfrac{1}{2}N_2O_4$$

corresponds the equilibrium constant—

$$K_p = \frac{p^{\frac{1}{2}}_{N_2O_4}}{p_{NO_2}}$$

Equilibrium constants for the formation of nitrogen tetroxide.

This can be calculated from measurements of the density of the gaseous mixtures, since this increases as the formation of nitrogen tetroxide progresses.[1]

The following table was so obtained:—

° C.	T	K_p	$R \ln K_p$
0·0	273·0	8·06	4·132
18·3	291·3	3·71	2·598
49·9	322·9	1·116	0·218
73·6	346·6	0·544	−1·207
99·8	372·8	0·273	−2·568

We can easily calculate the degree of dissociation at a given total pressure P of the mixture from the equilibrium constant K_p, if we remember that the total pressure is simply the sum of the partial pressures—

$$P = p_{NO_2} + p_{N_2O_4}$$

We can further see that if x represents the degree of dissociation, $1 - x$ is the fraction of the mols of tetroxide originally present still unchanged, and $2x$ the number of mols of dioxide which has been formed. The total pressure is therefore exerted by $(1 - x) + 2x$, that is $1 + x$ mols. The partial pressure P_{NO_2} is exerted by $2x$ mols of NO_2, and the partial pressure $P_{N_2O_4}$ is exerted by $(1 - x)N_2O_4$. But since the partial pressure bears the same ratio to the total pressure as the mols of gas producing it do to the total number of mols, it follows that—

$$\frac{p_{NO_2}}{P} = \frac{2x}{1 + x} \text{ and } \frac{p_{N_2O_4}}{P} = \frac{1 - x}{1 + x}$$

[1] For the relation of partial pressure and density in this case, see Nernst, "Theor. Chemie," 4th edit., p. 437.

The equilibrium constant is then represented by—

$$\frac{p^{\frac{1}{2}}_{N_2O_4}}{p_{NO_2}} = Kp = \frac{\left(\frac{1-x}{1+x}\right)^{\frac{1}{2}}}{\frac{2x}{1+x}} \times \frac{1}{P^{\frac{1}{2}}} = \frac{\sqrt{1-x^2}}{2x} \times \frac{1}{P^{\frac{1}{2}}}$$

So nitrogen peroxide is, for instance, half dissociated ($x = 0{\cdot}5$) under atmospheric pressure when the equilibrium constant has the value 0·865, that is, according to the table, at about 64° C. If we based our table on other measurements of density, we should get slightly different numbers. Thus, according to Deville and Troost (*l.c.*) the half-dissociated stage is reached somewhat sooner, since according to them 52·84 per cent. ($x = 0{\cdot}5284$) is dissociated at 62°. The equilibrium constant decreases rapidly with rising temperature. The degree of dissociation thus approximates nearer and nearer to unity. But the formula shows that it can never reach this value, since otherwise (at finite pressure) the equilibrium constant would acquire the impossible value of zero ($ln K_p = -\infty$). Yet the dissociation has progressed so far at 150° that we may consider it as practically complete. Only under tremendously high pressures can tetroxide exist above 150° in quantities worth mentioning. **Degree of dissociation of nitrogen tetroxide.**

Let us now proceed to the application of our general formula to this case. Omitting the term $\sigma''T^2$, which always has small values at low temperatures, and of whose existence the measurements give us no hint, we get—

$$A = Q_0 - \sigma'_p T ln T - RT ln \frac{p^{\frac{1}{2}}_{N_2O_4}}{p_{NO_2}} + \text{const. } T$$

At the equilibrium A becomes equal to zero, and the equilibrium constant is to be substituted in the logarithmic term for the ratio for the partial pressures. We then obtain—

$$R ln K_p = \frac{Q_0}{T} - \sigma'_p ln T + \text{const.}$$

As a first approximation we may assume that the specific heats at constant volume of NO_2 and $\frac{1}{2}N_2O_4$ are identical. At constant pressure the former would be greater by R and the latter by $\frac{1}{2}$R than at constant volume. The difference σ'_p would therefore have the value $\frac{R}{2}$, or about 1. Schreber computed the former at Planck's suggestion, making this assumption. He found in this way that $Q_0 = 6566$ cal. The heat of formation of $\frac{1}{2}$ mol N_2O_4 from 1 mol NO_2 at constant pressure would therefore be 6860 cals. at ordinary temperatures. Actual calorimetric measurement gave the smaller number 6450 cal.

If we insert in our formula the value 6566 for Q_0 and $\frac{R}{2}$ for σ'_p, we get a very large value for the constant, namely $-14·3$. The assumption that σ'_v equals zero, which is responsible for this large value, is, however, an improbable one. Experience has shown us that the specific heats of condensed gases at constant volume are generally smaller than the sum of the specific heats of their components. This is evident from the following examples taken from a compilation in Berthelot's Thermochimie (all the values refer to ordinary or slightly elevated temperatures):—

	True specific heats of the components.		of the compound.	
	Const. pressure.	Const. vol.	Const. press.	Const. vol.
$CO + \frac{1}{2}O_2$ ...	$\underbrace{6·83 + 3·41}_{10·24}$	7·26	8·59	6·61
$H_2 + \frac{1}{2}O_2$ ...	$\underbrace{6·82 + 3·41}_{10·23}$	7·25	8·65	6·67
$\frac{1}{2}N_2 + \frac{3}{2}H_2$ ...	$\underbrace{3·42 + 10·23}_{13·65}$	9·69	8·51	6·53
C_2H_2 + BrH ...	$\underbrace{9·7 + 6·64}_{16·34}$	12·37	12·1	10·3

It is therefore easy to believe that the specific heat of nitrogen dioxide at constant volume perceptibly exceeds that of $\frac{1}{2}$ mol of nitrogen tetroxide.

Difference of the specific heats in the case of nitrogen tetroxide. According to the Natanson brothers, we may take 1·31 as the ratio of the specific heats of NO_2 at constant pressure and constant volume. It follows from this that the specific heat at constant volume amounts to some 6·4 per mol NO_2. It is then very like that of carbon dioxide and water-vapour. If we estimate the specific heat of the tetroxide at constant volume as about the same as that of ethylene bromide, that is about 10, then the difference of the specific heats of 1 mol NO_2 and $\frac{1}{2}$ mol N_2O_4 at constant volume should be 1·4, and at constant pressure 2·4. If, then, we put the difference $\sigma'_p = 2{\cdot}4$, it follows from the experimentally determined values of the heat of reaction at ordinary temperatures ($Q_T = 6450$) and constant pressure that $Q_0 = 5730$ cal, and we obtain—

$$Rln K_p = \frac{5730}{T} - 2{\cdot}4 ln T + \text{const.}$$

It is easy to show that the assumption of the small value −3·6 for the constant will now yield values of $RTlnK_p$ concordant enough with the value in the table. We thus obtain—

°C.	$RlnK_p$ according to earlier tables.	$RlnK_p$ calculated.
0°	4·213	4·00
18·3	2·598	2·46
49·9	0·218	0·28
73·6	−1·207	−1·11
99·8	−2·568	−2·44

We could not expect a more precise argument, because in the temperature interval we are considering the gases are very near to their point of condensation. Indeed, nitrogen peroxide liquefies at ca. 25° under atmospheric pressure. The maximum work, as well as the specific heat, agrees only roughly with the fundamental laws in the neighbourhood of the condensation point.[1]

Cases analogous to that of nitrogen tetroxide.

This case is typical from a chemical point of view with a whole series of other reactions. However, they can hardly be called real gas reactions, although they may be treated according to the same principle. I may mention the formation of the "addition product," methylether hydrochloride.

[1] If we attempt to calculate the velocity of sound in this gaseous mixture on the basis of our assumptions regarding specific heats, we get results entirely at variance with the experimental values obtained by the Natanson brothers. Indeed, we should not expect an agreement in the light of the above consideration. Swart (*l.c.*) treats this case without the use of the fundamental gas equations.

$$HCl + CH_3\text{-}O\text{-}CH_3 = CH_3 - \overset{\displaystyle H}{\underset{\displaystyle Cl}{\overset{|}{\underset{|}{O}}}} - CH_3$$

Friedel[1] has studied the dynamics of this reaction experimentally, and Wegscheider[2] theoretically. Phosphorus pentachloride,[3] ammonium chloride, and ammonium carbamate all dissociate readily when heated, and therefore also belong to this class. These cases are generally explained by assuming the breaking of weak, but real, bonds between the phosphorus trichloride and chlorine, the hydrochloric acid and the ammonia, and between the carbamic acid and the ammonia. Where such an assumption does not agree with the current conception of valence, as in the case of acetic acid, which, like nitric oxide, shows a marked tendency to polymerize[4] just above its boiling-point, "molecular compounds" are assumed. Other explanations are sometimes made use of, and oxygen furnishes us an illustration of this. As long as the idea of divalent oxygen was adhered to, the hydrochloride of methylether was called a molecular compound. But now it is considered to be a salt of hydrochloric acid and methylether, since von Baeyer and Villiger[5] have raised the idea of a tetravalent oxygen to the rank of a principle.

"Molecular" compounds.

In the final analysis the question depends fundamentally on the equilibrium constant. If the equilibrium constant of a substance at ordinary temperature is as a rule rather large in its divalent condition, but small in its tetra- and hexavalent condition, then the latter is difficult to prepare and of limited stability. When we do get such a compound it is spoken of as a molecular compound, in order not to conflict with the simple divalent idea. The heat of reaction has a determinative influence on the magnitude of the equilibrium constant at low temperatures, because the terms $\sigma' p ln T$ and $\sigma'' T$ in our formula do not

[1] Friedel, *Bull. soc. chim.*, 24, 160, and 241 (1875).

[2] Wegscheider, *Sitzungsber. Wiener Akad.*, 106 (1899), p. 119.

[3] For a historical and mathematical treatment of this case see Wegscheider, *Wiener Akad.*, 108 (1899), p. 76.

[4] For data and calculations see Nernst, "Theoret. Chem.," 4th edit., p. 480.

[5] *Berl. Ber.*, 34 (1901), II. 2680.

become large till high temperatures are reached. Actually, then, we only consider substances to be molecular compounds when their formation is accompanied by a small evolution of heat. Thus any sharp division into real and molecular compounds is a thoroughly arbitrary one.

In this connection the case of nitrogen tetroxide gives occasion for yet another observation. The dioxide formed by the dissociation of the tetroxide can itself dissociate farther into nitric oxide and oxygen. The equilibrium constant for the formation of tetroxide is very small, even at 150°, and the tetroxide is therefore almost wholly dissociated. The equilibrium constant of the reaction—

$$NO + O \leftrightarrows NO_2$$

on the other hand, is still large at this low temperature, and the dioxide is consequently stable. Richardson's[1] measurements of the vapour densities of nitrogen dioxide furnish a basis for evaluating a formula for this reaction similar to that just used in the case of the formation of dioxide from tetroxide. The dissociation of the dioxide amounts to only 5 per cent. under atmospheric pressure at 184°, while at 600° under atmospheric pressure it is practically complete. **Nitrogen dioxide.**

We recognize in its much higher heat of formation the chief reason for the stability of the dioxide. The heat of formation of 1 mol NO_2 from NO + O amounts indeed to 13,100 cal. at ordinary temperatures.

One might now conclude that nitric oxide, too, would decompose into nitrogen and oxygen at still higher temperatures. But that is not the case; for, as we saw in the preceding lecture, the equilibrium constant of the formation of nitric oxide from the elements is very small at 1800°, and decreases from there on. **Nitric oxide.**

Nitric oxide is therefore stable below 1800°, simply because the velocity of its decomposition is extremely small. If the dissociation of nitric oxide did not have such a specifically small velocity even at the most intense white heat, nitrogen tetroxide when heated would not decompose into nitrogen dioxide, and this

[1] *Journ. Chem. Soc.*, 51 (1887), 397; and Nernst, "Theoret. Chem.," 4th edit., p. 438.

in turn into nitric oxide, but both would dissociate directly into the elements.

Nitrous oxide. Finally, nitrous oxide, like nitric oxide, is certainly stable at high temperatures. Like nitric oxide, too, its formation at ordinary temperatures is accompanied by a large absorption of heat (− 18,500 cal. per mol NO_2 at constant pressure). Yet we know that at 900° it is almost completely dissociated into its elements.[1] Perhaps the temperature in which it can be formed from the elements to any considerable extent lies very much higher, and it would be interesting to determine whether it too is formed when nitric oxide is produced in the electric arc.

Case I. The formation of carbon dioxide. The discussion of those cases which are of more immediate interest to us may well be opened by a consideration of the dissociation of carbon dioxide.

Sixteen years ago Le Chatelier[2] made certain calculations concerning this reaction, in the course of which he derived a table of the degrees of dissociation of carbon dioxide. The table has become widely known in scientific literature because of the convenience with which it can be used. Its experimental basis[3] consists of three sets of observations—Deville's analysis of dissociated gases; Mallard's and Le Chatelier's measurements of the explosion pressures developed when carbon monoxide and oxygen unite in closed vessels; and density determinations of carbon dioxide at high temperatures.

Le Chatelier (*l.c.*) states that H. St. Claire Deville found, with the help of his "cold-hot" tube, that carbon dioxide is 0·002 dissociated at ca. 1300°. Deville's "cold-hot" tube[4] was simply a porcelain tube through which a brass tube filled with flowing water was inserted. The porcelain tube was heated from without, the gas being contained in the annular space between the two tubes. Deville was able by means of this simple and elegant

[1] Victor Meyer and Langer, "Pyrochemische Untersuchungen. Concerning its behaviour at 520° see Berthelot, *Comp. Rend.*, 77 (1873), 1448; *Bull. soc. chim.* (2) 26, 191.

[2] *Ann. des Mines*, (viii.) 13 (1888), 274; *Z. f. phys. Chemie*, 2 (1888), 782; Nernst, "Theoret. Chem.," 4th edit., p. 443.

[3] Trevor and Kartright's discussion of this matter was not available to me in the original (*American Chem. Journ.*, 16, 782). For the use of Le Chatelier's numbers in calculating the oxy-hydrogen cell, see Preuner, *Z. f. phys. Chemie*, 42 (1902), 50.

[4] *Ann. Chem. Pharm.*, 135 (1865), 94.

device to detect qualitatively, not only the dissociation of carbon dioxide and ammonia, but also of carbon monoxide,[1] HCl and SO_2, at temperatures which could hardly have exceeded 1300°. The arrangement is not suitable for quantitative measurements. Deville determined the dissociation quantitatively by a different method. He placed a porcelain tube filled with porcelain chips in a wider porcelain tube, fitted this in turn inside a still wider iron tube, and then covered the latter with a layer of clay. He heated the whole to a temperature which he did not measure, but which he estimated with his eye to be about 1300°. A rapid stream of carbon dioxide was kept passing through the inner tube and into a solution of potassium hydrate: 7·83 litres of carbon dioxide were passed through the tube in an hour. After absorption in the caustic potash a gaseous residue of 20–30 c.c. remained, which contained on the average 30 per cent. O_2, 62·3 per cent. CO, and 7 per cent. N_2. If the same stream of carbon dioxide were passed directly into the caustic potash without traversing the heated tube, 1·4 c.c. of a gaseous residue was obtained in the same time, consisting of 14 per cent. N_2 and 86 per cent. of O_2. We conclude from this that 12 to 20 c.c. out of the 7·83 litres of the carbon dioxide was dissociated. This corresponds to a dissociation of between 0·0015 and 0·0025, or in round numbers of about 0·002. Le Chatelier evidently had this experiment in mind in making the above statement. But Deville himself remarked that perhaps the dissociation came out too small because of a *recombination* of the gases as they cooled. Besides, it is impossible to say whether the temperature in the inside tube was really as high as 1300° or not, for the double porcelain pocket must have been a poor conductor of heat, and the gas stream was rapid. In any case, we may consider 0·002 as a minimum value of the degree of dissociation at 1300°. The equilibrium constant of the reaction—

$$CO + \tfrac{1}{2}O_2 \rightleftarrows CO_2$$

would therefore be at 1300° (1573 abs.)—

$$\frac{p_{CO_2}}{p_{CO} + p^{\frac{1}{2}}_{O_2}} = K < \frac{0{\cdot}998}{0{\cdot}002 + 0{\cdot}001^{\frac{1}{2}}};\ i.e.\ 1{\cdot}58 \times 10^4$$

[1] *Ann. Chem. Pharm.*, 134 (1865), 124.

Deville's experiment with the flame. We have Deville to thank for another much more original, but at the same time much less accurate, measurement of the degree of dissociation of carbon dioxide. Deville produced, as it were, a continuous carbon monoxide explosion by allowing a stream (47 c.c. per second, or 170 litres per hour) consisting of 64·3 per cent. CO, 33·3 per cent. O_2, and 2·3 per cent. N_2 to issue from a burner having an opening 5 sq. mm. in cross-section, and then igniting it. It formed a conical flame 10 mm. high, whose base was the aperture of the burner, and whose surface represented the reaction zone. An aureole extended above this, decreasing in brightness with the height and consisting of the escaping reaction products of the inner cone, mixed with the outer air, and thereby cooled as well as completely burned. Thus the whole flame was some 67 to 70 mm. high. Deville inserted a thin-walled silver tube of about 1 cm. bore into this flame, and sucked water through it. At one place on the tube there was a hole 0·2 mm. in diameter. The tube was so placed that the hole came precisely in the vertical axis of the flame and pointed downwards. The suction prevented water leaking out of the hole, and at the same time sucked gas in. This gas was very suddenly cooled by the water, carried along by it, and collected. The carbon dioxide was then removed by caustic potash, and the residual gas analyzed. The results were as follows. Visual observations of the temperature are appended.

Height above mouth of burner in mm.	Composition of the gaseous product.			Temperature.
	CO	O_2	N_2	
67 [1]	0·2	21·3	78·5	Melting point of silver and higher
54	6·2	28·1	65·7	„ „ gold
44	10·0	20·0	70·0	Platinum almost white hot
35	17·3	24·8	57·9	„ white hot
28	19·4	26·5	54·1	„ a brilliant white
18	29·0	25·1	45·9	„ blinding white
15	40·0	32·9	27·1	„ begins to melt
12	47·0	36·0	17·0	„ melts
10 [2]	55·3	35·3	9·4	„ „ quickly and sublimes
10 [3]	55·1	36·5	8·4	Very hottest point
0	64·4	33·3	2·3	Initial gas

[1] Edge of flame. [2] Near the top of the inner cone.
[3] Top of the inner cone.

Deville considered that at the hottest point the gases were at most not more than two-thirds combined.

These experiments form a beautiful qualitative illustration of the strong dissociation of gases at very high temperatures and their recombination during cooling. But their quantitative value is slight. Deville mixed nitrogen with his initial gases in order to estimate the progress of the carbon dioxide formation from the increase of nitrogen as the tube was placed higher and higher above the mouth of the burner. In principle, the admixture of air above the cone does not cause any disturbance, because we may consider the gases sucked off as a mixture of air and (partially) burnt original gas, and calculate from the known composition of both, the amounts of each component present. Yet actually we are unable to get any useful results from this sort of computation. Even at the hottest point, where no outer air is supposed to have entered, we find that—

From 100 vols. of initial gas mixture containing	64·4 CO	33·3 O_2	2·3 N_2
27·4 vols. of the final gas mixture were formed, containing	15·1 CO	10·0 O_2	2·3 N_2
And there had therefore been consumed	49·3 CO	23·3 O_2	

The amounts of carbon monoxide and oxygen which disappeared must stand in the stoichiometric ratio of 2 : 1, which, as we see, is only approximately the case. In a repetition of the experiment it would seem advisable to let the flame strike into the cavity of a *double-walled* cooling tube in order to prevent the analytical uncertainties arising from the different solubilities of the gaseous constituents in water, which prevented Deville from making any determination of the carbon dioxide. Since the initial gas mixture brings with it all the oxygen needed in the combustion, it would also seem advisable to entirely exclude the external oxygen, as, for instance, Haber, Richardt, and Allner have done. Further, the tube for removing the gases from the flame should have more favourable relative dimensions, so that the little flame should not be too greatly deformed by the thick cooling tube. Still, from the gases which Deville removed from the hottest part of the flame, we conclude that some 48 vols. of CO_2 [mean of 49·3 and 2 (23·1)] are present with 15·1 vols. CO, 10·0 vols. O_2, and 2·3 vols. N_2. If we assume that this composition represents the equilibrium condition,

$$K = \frac{p_{CO_2}}{p_{CO} \times p^{\frac{1}{2}}_{O_2}} = \frac{0·637}{0·199 \times 0·133^{\frac{1}{2}}} = 8·7$$

Le Chatelier's views regarding Deville's experiments. Le Chatelier concluded from Deville's experiments that the degree of dissociation of the carbon dioxide at the hottest point of the flame amounted to—

$$\frac{CO}{CO + CO_2} = 0·4$$

giving an equilibrium constant of 3·67. It is, indeed, very difficult to fix on any exact value of the (assumed) equilibrium in Deville's experiments. That Le Chatelier had no better data at his disposal on which to base his calculations of the dissociation, illustrates the paucity of our knowledge in this field.

Yet the uncertainty regarding the value of the equilibrium constant itself is even less than the uncertainty regarding the temperature at which the equilibrium was established. Deville's statements about the temperature at various heights places us in a most difficult position. In the upper parts of the flame where the temperatures are estimated with some definiteness, we are sure that there is no equilibrium, for the reaction velocity is not sufficiently high.[1] But in the lower parts of the

[1] We know, for instance, that gold melts at 1065° ± 10° according to the critical discussion of all measurements given by Le Chatelier and Boudouard [" Températures élevées," (1900), p. 81]. If, then, at the melting-point of gold there was observed to be present 6·2 pts. of CO, 28·1 pts. O_2, and 65·7 pts. N_2, this indicates that the dissociation is greater at this temperature than Deville found it to be in his previously mentioned experiments at 1300°. If we overlook the oxygen for a moment, and only observe that in the initial gas 64·4 c.c. CO was present to every 2·3 parts of N_2, while at the point under discussion (54 mm. from the mouth of the burner) there is but 0·218 c.c. CO, we see that more than 64·182 c.c. CO_2 cannot be present besides this 0·218 c.c. CO. Actually the amount of carbon dioxide must be much smaller, because, as the high content of the oxygen shows, a large part of the nitrogen comes from the air. Yet 0·218 c.c. CO to 64·182 c.c. CO_2 would be more than would be possible according to Deville's results in his experiments with porcelain tubes, especially in the presence of much oxygen. It is not to be wondered that the equilibrium cannot keep pace with the falling temperature in the upper parts of the flame, when we remember that the flame-gases must have a velocity of about 10 m. per second at the mouth of

flame where the reaction velocity is probably great enough, the statements are not sufficiently precise to permit us to estimate the temperature with any accuracy.

Mallard and Le Chatelier's explosion experiments.

To fill this gap, Le Chatelier made use of some explosion experiments which Mallard and he had carried out with mixtures of carbon monoxide and oxygen. As mentioned in the previous lecture, they exploded these mixtures in closed vessels, and measured the pressures developed by means of a registering manometer. In a series of six experiments they were able to depress the temperature of the explosion to as low as 2000° by the addition of large amounts of carbon dioxide. The addition of carbon dioxide also had the effect, through its mass action, of driving back to a minimum slight dissociation of the carbon dioxide already present. From these experiments, Mallard and Le Chatelier obtained 13·6 as the mean specific heat of carbon dioxide between 0° and 2000°, and by combining this value with Regnault's values at ordinary temperatures, they got the experimental formula for the specific heat of carbon dioxide given in the previous lecture. We will explain the relation between the maximum explosion pressure and the specific heat a little more minutely.

The relation between explosion pressure and explosion temperature.

If at the temperature T_a for every $\frac{2}{3}$ mol of carbon monoxide and every $\frac{1}{3}$ mol of oxygen, z mols of a foreign gas are forced into the bomb, then after the combustion there will be left $\frac{2}{3}$ mol of carbon dioxide for every z mols of the foreign gas. If the pressure of the gas before the explosion is p_a, then after the explosion, and after all the heat generated has been dissipated, the pressure will be—

$$p = p_a \frac{\frac{2}{3} + z}{1 + z}$$

the burner to permit a consumption of 47 c.c. a second. In the flame itself the cross-section of the hot mass is greater, but the volume, too, is much greater because of the higher temperature, and the air carried along with it increases it further. The velocity will therefore be of the same order of magnitude as at the mouth of the burner. Now, since the temperature in Deville's flame evidently decreases hundreds of degrees per cm. rise, the equilibrium constant would have to change enough to correspond to a temperature change of several hundred degrees in a thousandth part of a second. According to the experience of Haber, Richardt, and Allner, this cannot safely be assumed to take place at a temperature beneath 1600°. It may be that the reaction velocity of the formation of CO_2 is much higher than that of the water-gas reaction. But near the melting-point of gold both reactions are certainly slow.

If the maximum pressure P was attained[1] during the explosion, then—

$$P : p = T_e : T_a$$

where T_e signifies the maximum temperature. Thus we find—

$$T_e = \frac{PT_a}{p} = \frac{P}{p_a} T_a \frac{1+z}{\frac{2}{3}+z} \quad . \quad . \quad . \quad . \quad (1)$$

On the other hand, the quantity of heat Q is set free in the formation of $\frac{2}{3}$ mol CO_2 from the $\frac{2}{3}$ mol CO and $\frac{1}{3}$ mol O_2, and this heats the $\frac{2}{3}$ mol CO_2, together with the z mols of the foreign gas, from T_a to T_e. This yields the equation—

$$\tfrac{2}{3}Q = (\tfrac{2}{3}c'_v + zc''_v)(T_e - T_a)$$

c'_v is here the specific heat of carbon dioxide, and c''_v that of the admixed foreign gas. If this foreign gas is simply carbon dioxide, as it was in Mallard and Le Chatelier's experiments, the equation simplifies to—

$$\tfrac{2}{3}Q = (\tfrac{2}{3} + z)c_v(T_e - T_a) \quad . \quad . \quad . \quad . \quad (2)$$

Taken with (1) this equation permits us to evaluate T_e and c_v. The specific heat c_v, deduced in this way, is the mean specific heat per mol CO_2 at constant volume between the temperatures T_a and T_e.

Mallard and Le Chatelier carried out four other experiments at temperatures where the dissociation had become perceptible. In one they added a given relatively small quantity of carbon dioxide as a foreign gas, in a second they added carbon monoxide, and in a third nitrogen. In a fourth decisive experiment they took pure carbon monoxide and oxygen in equivalent amounts, no foreign gas being present except 1·2 vol. water-vapour to every 100 vols. of the explosive mixture. At these temperatures, the calculation is somewhat different. Only the fraction x of one volume of the explosive mixture

[1] More accurately, we cannot call the observed maximum pressure the real maximum pressure without making a correction for that little heat which is radiated to the walls of the vessel during the combustion. Mallard and Le Chatelier estimated this correction in a very roundabout way to be 4 per cent. The highest pressure observed, increased by 4 per cent., then represents the maximum pressure P used in the above questions. Fliegner believed this correction to be an entirely mistaken one (see his criticism previously mentioned, p. 123).

The relation between explosion pressure and degree of dissociation.

now combines at the highest temperature, while $1 - x$ vol. remains dissociated. If we imagine the gas to be cooled down to $T_a°$ without any change in the degree of dissociation, then the pressure p after the explosion is now connected with the initial pressure by the equation—

$$p = p_a \frac{1 + z - \frac{1}{3}x}{1 + z}$$

This follows from the fact that if $\frac{2}{3}x$ mols of carbon monoxide plus $\frac{1}{3}$ mol of oxygen disappear, $\frac{2}{3}$ mol carbon dioxide is formed. The relation of maximum pressure to maximum temperature is then given by—

$$T_e = \frac{PT_a}{p} = \frac{P}{p_a} T_a \frac{1 + z}{1 + z - \frac{1}{3}x} \quad . \quad . \quad . \quad . \quad (1a)$$

and the equation connecting the quantities of heat becomes—

$$\tfrac{2}{3}xQ = \{(1 - x)c_v' + zc_v'' + \tfrac{2}{3}xc_v'''\}(T_e - T_a) \quad . \quad (2a)$$

Here c_v' is the specific heat of the explosive mixture of carbon monoxide and oxygen, c_v'' the specific heat of the foreign gas, and c_v''' the specific heat of carbon dioxide, all between the temperature $T_e°$ and $T_a°$, and at constant volumes. Knowing c_v', c_v'', and c_v''', we can calculate the value of x, the "degree of the combination" of the carbon monoxide and oxygen.[1]

When Mallard and Le Chatelier came to make their calcu-

[1] This, however, is by no means unconditionally true. If we combine equation (1*a*) and (2*a*), then, knowing z, $\frac{P}{p_a}$, T_a, Q, and the specific heats, we obtain a cubic equation for T_e. But this cubic equation does not always have a root which will satisfy (1*a*) and (2*a*). The following is an illustration of this. In Mallard and Le Chatelier's final experiment was z (water-vapour) = 0·012, $\frac{P}{p_a} = 9{\cdot}95$, and $T_a = 273$. If we call Q at ordinary temperatures equal to 68,000, and introduce the specific heats at constant volume between 0° and $t°$, as determined by Mallard and Le Chatelier in their experiment with the crusher manometer,

$$\begin{aligned} c_v \text{ (permanent gas)} &= 4{\cdot}76 + 0{\cdot}00122t \\ c_v \text{ (H}_2\text{O vapour)} &= 5{\cdot}78 + 0{\cdot}00286t \\ c_v \text{ (carbon dioxide)} &= 6{\cdot}5 \ \ + 0{\cdot}00387t \end{aligned}$$

we get the cubic equation for Centigrade temperatures

$$t^3 - 1156{\cdot}7t^2 - 24{,}743{,}970t + 62{,}507{,}297{,}500 = 0$$

One will try in vain to find a value of t which will satisfy this equation.

lations, they were apparently confronted with a serious difficulty. Their experimental formula for the mean between 0° and 2000° specific heat of carbon dioxide, based on the above-mentioned data of Regnault and Wiedemann at ordinary temperatures and on their own "undissociated explosions" at 2000°, was their sole means of calculating the specific heat at these very much higher temperatures. That is, they must extrapolate without having a sufficiently precise basis to work from. They discussed no less than five possible formulæ[1] for the mean specific heat of carbon dioxide between 0° and $t°$ at the constant volume. These were—

$$6{\cdot}3 + 0{\cdot}00564t - 0{\cdot}000000108t^2$$
$$6{\cdot}3 + 0{\cdot}006t - 0{\cdot}000000118t^2$$
$$6{\cdot}26 + 0{\cdot}00367t$$
$$4{\cdot}74 \times (\mathrm{T} \times 10^{-2})^{\frac{1}{3}}$$
$$4{\cdot}33 \times (\mathrm{T} \times 10^{-2})^{0{\cdot}367}$$

They chose the last.

Since the amount of water-vapour present was small, it mattered little what value was selected for its specific heat. The specific heat of the carbon monoxide-oxygen mixture is that of all permanent gases. Mallard and Le Chatelier used the formula $4{\cdot}8 + 0{\cdot}006t$ (mean specific heat at constant volume between 0 and $t°$) to express the results of their varied observations. Extrapolating on this basis, Mallard and Le Chatelier came to the conclusion that the temperature in the explosion experiment, where 1·2 vols of water-vapour was mixed with 100 vols. of $CO + \frac{1}{2}O_2$ was 3130° C., and that the degree of combination was 0·61. The degree of dissociation was therefore 1 − 0·61, or 0·39. We should remember, too, that the maximum pressure was some 10 atmospheres.

Le Chatelier returned to these old numbers when he came to study dissociation phenomena. In the mean time[2] he had discovered the remarkable fact that it was possible to express the specific heats per mol of the most diverse gaseous systems at constant pressure, at least approximately, by the formula—

$$c_p = 6{\cdot}8 + a(t + 273) = 6{\cdot}8 + a\mathrm{T}$$

[1] *Compt. Rend.*, 93 (1881), 1014; *Ann. des Mines*, 4 (1883), p. 524; *ibid.*, pp. 525 and 526.

[2] *Compt. Rend.*, 104 (1887), 1780.

Le Chatelier puts the coefficient a at zero for permanent gases, and at 0·0072 for carbon dioxide. It follows from this assumption that the mean specific heat per mol of carbon dioxide at constant pressure is—

$$c_p = 6{\cdot}8 + 0{\cdot}0036\text{T}$$

where 6·8 is taken as the specific heat of all permanent gases at constant pressure. Le Chatelier appends the remark that one may express the true specific heat of the permanent gases by—

$$c_p = 6{\cdot}5 + 0{\cdot}0008\text{T}$$

and hence that of carbon dioxide by—

$$c_p = 6{\cdot}5 + 0{\cdot}0044\text{T}$$

However, he laid no stress on this remark, but took the specific heat of the permanent gases as constant, and hence the specific heat of carbon dioxide as 6·8 + 0·0036T. He has evidently computed the earlier experiments of Mallard and himself on this basis, for he now remarks that the degree of dissociation comes out as 0·34, and the temperature as 3300° C.

From all the facts, Le Chatelier concludes that the temperature of Deville's flame was 3000° C. It is impossible to tell whether this is merely a rough estimation, made by subtracting 10 per cent. from the explosion temperature of 3300°, or not. Uncertainty of Le Chatelier's estimation.

There are three aspects of the matter to be considered—

(1) Though we admit that the whole method as described is capable of yielding correct values for the temperature and the degree of dissociation, it still does not follow that the degree of dissociation found represents equilibrium conditions. According to Mallard and Le Chatelier, the explosion experiments with oxy-hydrogen mixtures in closed vessels show absolutely no dissociation. Even if we find dissociation to take place in the carbon monoxide-oxygen explosions, we cannot be quite sure that this dissociation corresponds to the formula— (1) The attainment of the equilibrium is not certain.

$$\text{CO} + \text{O} \rightleftarrows \text{CO}_2$$

The formation of carbon dioxide from carbon monoxide and oxygen seems to proceed in several stages, just as the formation of water from the elements does. The fact that a dry mixture of carbon monoxide and oxygen does not explode,[1] and the fact that the velocity of propagation of the explosion wave in a mixture of carbon monoxide and oxygen is strongly affected by the presence of water-vapour, support this view. It is reasonable to suppose that a reaction of this kind taking place in several stages might yield different results in an explosion bomb than it would in a freely burning flame. The pressure and the velocity with which the explosion is propagated are constant in the inner cone of the flame, but variable in the explosion bomb. So long, therefore, as we know nothing further about these circumstances, we cannot be sure what equilibrium temperature we are to assign to the samples of gas which Deville drew off from the hottest part of his flame.

(2) The maximum temperatures are not directly comparable.

(2) If we overlook the above objections, we may still ask whether or not the maximum temperature calculated from the pressure developed in any explosion is directly comparable with the temperature prevailing in a stationary flame. The extreme temperature which we compute from our pressure measurements is the average of the temperatures prevailing in all parts of the explosion bomb, and they will be widely different. In explosions where no dissociation ensues, we may, of course, take this average temperature as a true homogeneous temperature without limitations. But in explosions which give rise to dissociation the case is different, at least where we come to calculate the maximum temperature and the degree of dissociation in the above-mentioned way. For the degree of dissociation does not change in so simple a way with the temperature, nor in so slight a measure as do the specific heats. So this uncertainty is bound to creep in when we try to estimate the temperature of Deville's flame from explosion experiments in a bomb.

(3) The cooling action of the cold tube is difficult to allow for.

(3) Aside from (1) and (2) there still exists the question whether enough allowance has been made for the effect on the combustion zone of the thick, cold silver tube running right across Deville's flame. It is a well-known and fundamental experiment in the theory of illumination that a cold vessel when introduced into a luminous hydrocarbon flame makes it non-luminous, because it cools it off. If a basin with a flat bottom be filled with water and brought into a Bunsen flame,

[1] Dixon, *Phil. Trans.*, 175 (1884), 630; M. Traube, *Ber. d. d. Chem. Ges.* 15 (1882), 666; Dixon, *Journ. Chem. Soc.*, 49 (1886), 95.

we can actually see the cooling effect, for the flame never touches the bottom of the basin, the gases being so strongly cooled that they cannot unite.[1]

This cooling effect would be immaterial in our case if we could assume that the reaction in the hottest zone progressed instantaneously to equilibrium. But no matter how great a velocity of chemical combination we assume at that point, it can never be so great that a massive and cold body will not have time to absorb some heat from the gases during the reaction, and so lower the temperature of the combustion. We must admit that this cooling effect amounts to less in this particular case than it would in a flame where the dissociation was less (CO and air, for instance), because the recombination of the gases on cooling yields fresh heat, and so acts like a brake on the falling temperature. But this objection, nevertheless, increases the uncertainty regarding the temperature of Deville's flame.

Not long after his computations of dissociation, Le Chatelier obtained new values for the specific heats at constant volume from his experiments with the crusher manometer. These seemed better suited as a basis for calculating the dissociation of carbon dioxide than previous ones.[2] In the experiments mentioned above, dissociation became appreciable at 2000°, but the high pressures in the crusher manometer prevented any dissociation taking place even at very high temperatures, and this allows us to extend our calculation of the specific heats over a much longer range. However, Le Chatelier does not seem to have undertaken a recalculation using these new values. After he had made a final revision of the specific heats of gases in the light of his theory that at constant pressure the specific heats of gases should all converge towards 6·5 at absolute zero, and had expressed the specific heats of carbon dioxide and the permanent gases as

Le Chatelier's later statements in regard to specific heats.

$$c_{p(CO_2)} = 6{\cdot}5 + 0{\cdot}0037T$$
$$c_{p(\text{perm. gases})} = 6{\cdot}5 + 0{\cdot}0006T$$

[1] Haber, "Habilitationsschrift" (Munich, 1896), published by Oldenbourg, 1896, sec. 3. "Ueber die Verbrennung an gekühlten Flächen."

[2] We must indeed admit that the accuracy of these measurements all depends on how well we can separate the static pressures we are looking for from the effect of sudden impacts. For the theory, and a more detailed explanation, see the Sixth Lecture.

he apparently never returned to the question of how much carbon dioxide is dissociated at high temperatures.

Observations of V. Meyer and Langer.

We have mentioned yet a third method of measuring the dissociation of carbon dioxide, namely the determination of its density at high temperatures. Dissociation increases the volume, and therefore decreases the apparent density. It is not easy to carry out accurate density determinations at high temperatures. It is much easier to observe the dissociation by some such arrangement as Deville used at 1300°. We are not, therefore, surprised that neither Bõttcher,[1] working at 1400°, nor Crafts at a somewhat higher temperature, were able to detect any dissociation of carbon dioxide. Victor Meyer and Langer (*l.c.*) then repeated the density determination at 1690° in platinum vessels. They, too, found nearly the normal density. They express surprise that Deville had found an appreciable dissociation at 1300°, and mention that Victor Meyer and Züblin[2] had confirmed the experiment of Deville, and considered it contradictory that the dissociation should be perceptible in a porcelain tube at a temperature as low as 1300°, but still could not be clearly detected in density determinations of carbon dioxide. They believed the explanation to lie in an observation of Menschutkin and Konowalow,[3] according to which certain organic vapours are more dissociated in the presence of asbestos and rough glass surfaces than in their absence. This explanation rests upon a misunderstanding. The rough solid substances do indeed hasten the *process* of dissociation, but they do not alter the *degree* of dissociation. The only way in which we could imagine them to have any effect in the present case would be to assume that by their aid equilibrium can be reached at 1300° in Deville's rapid current of gas, while without their aid it is not established in a platinum vessel heated to 1690° for a much longer time. Yet this is extremely improbable. The natural explanation is rather to be found in the formula representing the reaction energy of the formation of carbon dioxide—

$$\mathrm{A} = \mathrm{Q}_0 - \sigma'_p \mathrm{T} ln \mathrm{T} - \sigma'' \mathrm{T}^2 - \mathrm{RT} ln \frac{p_{\mathrm{CO}_2}}{p_{\mathrm{CO}} \times p^{\frac{1}{2}}_{\mathrm{O}_2}} + \text{const. T}$$

[1] Dissertation, Dresden, 1900; "Ueber die Dissoziationstemperaturen der Kohlensäure und des Schwefelsäureanhydrides."

[2] I have been unable to find any more extensive mention of the matter.

[3] *Berl. Ber.*, 17, 1361.

which, for equilibrium when A = 0, becomes—

$$R ln K_p = \frac{Q_0}{T} - \sigma'_p ln T - \sigma'' T + \text{const.}$$

If we substitute numerical values derived from Deville's experiments, we find that the density determinations at atmospheric pressure should show no evidence of dissociation. The temperature of the flame was 1300° (1573° abs.) and we calculated K_p to be $1 \cdot 58 \times 10^4$. Substituting the values for the specific heats at constant pressure which Langen obtained from his observations of explosion pressures (recalculated for constant pressure), and taking the heat of reaction at ordinary temperatures and at constant pressure to be—

$$CO + \frac{1}{2}O_2 = CO_2 + 68{,}000 \text{ cal.}$$

we obtain for the difference between the specific heats of factors and products—

$$\begin{aligned} CO &= 6 \cdot 45 + 0 \cdot 0006T \\ \tfrac{1}{2}O_2 &= 3 \cdot 23 + 0 \cdot 0003T \\ \hline & 9 \cdot 68 + 0 \cdot 0009T \\ CO_2 &= 7 \cdot 26 + 0 \cdot 0026T \\ \hline & 2 \cdot 42 - 0 \cdot 0017T \end{aligned}$$

and hence for Q_0 the value 67,440. Inserting these values, we get with the help of Deville's observation—

$$4 \cdot 56 \log^{10} 15{,}800 = \frac{67{,}440}{1573} - 2 \cdot 42 ln 1573 + 0 \cdot 0017 \times 1573 + \text{const.} \quad . \quad . \quad . \quad . \quad . \quad . \quad . \quad . \quad (3)$$

from which we find that the constant has the value $-8 \cdot 59$. Taking this result and then calculating the value of K_p when $t = 1690°$, or $T = 1963°$ abs., we get—

$$4 \cdot 56 \log^{10} K_p = \frac{67{,}440}{1963} - 2 \cdot 42 ln 1963 + 0 \cdot 0017 \times 1963 - 8 \cdot 59$$

or $K_p = 231$. It follows from this that the carbon dioxide ought to be about 3 per cent. dissociated at atmospheric pressure. But V. Meyer and Langer's experiments deviate among themselves by as much as 2 to 3 per cent., and are hence unsuited to show dissociation of this amount. One must also keep in mind that observations of dissociation in platinum vessels at these temperatures may well give rise to fictitious results, for platinum at a white heat is not indifferent towards oxygen, and it might easily remove a trace of oxygen which was formed, and so conceal any change of density due to dissociation.

Nernst's experiments.

Nernst[1] has measured the density of carbon dioxide at still higher temperatures. He worked at 1973° in an iridium vessel in the presence of air. The dissociation calculated by the above formula should be 13 per cent., but the presence of the air drives it back. Nernst's results show that the dissociation is certainly not very large. Nothing more definite can be deduced from them.

Le Chatelier's calculations of the dissociation of carbon dioxide.

In the above discussion of the data on which a calculation of the dissociation of carbon dioxide at different temperatures may be based, we introduced and evalued a formula different from that which Le Chatelier had used. He started from the conception that the mean specific heats at constant pressure had the following values:—

$$\text{Permanent gases} = 6{\cdot}8$$
$$\text{Carbon dioxide} = 6{\cdot}8 + 0{\cdot}0036\text{T}$$

This leads to a difference between the mean specific heats of factors and products of—

$$3{\cdot}4 - 0{\cdot}0036\text{T}$$

and taking the heat of reaction at ordinary temperatures Q = 68,000 cal., it follows that the heat of reaction at absolute zero would be—

$$Q_0 = 67{,}300\ ;$$

and hence—

$$\text{R}ln\text{K}_p = \frac{67{,}300}{\text{T}} - 3{\cdot}4ln\text{T} + 0{\cdot}0036\text{T} + \text{const.} \quad (3)$$

Le Chatelier now brings in his assumption that the degree of dissociation in Deville's experiments was 0·4 at equilibrium and under atmospheric pressure, and that the temperature was 3000°. We have already seen that this is equivalent to fixing 3·67 as the value of the equilibrium constant. He then gets—

$$4{\cdot}56 \log 3{\cdot}67 = \frac{67{,}000}{3273} - 3{\cdot}4 \times 2{\cdot}3 \log 3273 + 0{\cdot}0036 \times 3273 + \text{const.} \quad (4)$$

[1] *Z. f. Elektrochemie*, 9 (1903), p. 625.

from which it follows that

$$\text{constant} = -2 \cdot 28.$$

Taking this value of the constant to calculate the equilibrium at 1300° (1573° abs.), we obtain—

$$4 \cdot 56 \log K_p = \frac{67,300}{1573} - 3 \cdot 4 ln\, 1573 + 0 \cdot 0036 \times 1573 - 2 \cdot 28$$

and hence—

$$K_p = 4 \cdot 37 \times 10^4$$

Before, we saw that, according to Deville's observation—

$$K_p < 1 \cdot 58 \times 10^4$$

This discrepancy between the equilibrium constants involves no inconsiderable disagreement in the degrees of dissociation calculated from them. At such slight dissociation, where the partial pressure of the carbon dioxide is very nearly equal to the total pressure P of the gas mixture, we may write without appreciable error—

$$\frac{p_{CO_2}}{p_{CO} \times p^{\frac{1}{2}}_{O_2}} = \frac{P}{p_{CO} \times p^{\frac{1}{2}}_{O_2}} = K_p$$

But since, when carbon dioxide dissociates, oxygen and carbon monoxide are formed in the ratio of 1 to 2, this goes over into—

$$\frac{P}{K_p} = p_{CO} \times \sqrt{\frac{p_{CO}}{2}}$$

Finally, under one atmosphere pressure we get—

$$\frac{1}{K_p} = \sqrt{\frac{p_{CO}^3}{2}}$$

It is easy to see that the partial pressure of the carbon monoxide is but half as large as Deville found it, if K_p equals $4 \cdot 37 \times 10^4$.

Making use of Le Chatelier's assumption, we compute the following values of the equilibrium constant and the degree of dissociation, for a number of temperatures and pressures :—

	t	T	K_p	Degree of dissociation.	Total pressure atmos.
1	1300	1573	$4 \cdot 37 \times 10^4$	$1 \cdot 0 \times 10^{-3}(0 \cdot 3 \times 10^{-3})$	1
2	1500	1773	$4 \cdot 48 \times 10^3$	$0 \cdot 4 \times 10^{-2}(0 \cdot 8 \times 10^{-2})$	1
3	2000	2273	$1 \cdot 07 \times 10^2$	$0 \cdot 3 \times 10^{-1}(0 \cdot 35 \times 10^{-1})$	6
4	3000	3273	3·67	0·4 (0·4)	1
5	3300	3573	2·29	0·28 (0·27)	10

The numbers added in brackets are the values as calculated by Le Chatelier. The discrepancy between our values and his, which is especially marked at low temperatures, is due to a mistake in Le Chatelier's calculation.

Comparison of Le Chatelier's results with actual observation.

The last three values of the above five agree satisfactorily with the estimations or calculations which Le Chatelier made on the basis of his explosion experiments, carried out in collaboration with Mallard. The small value at 1500° agrees with the fact that the density at that temperature is normal. It is impossible to say whether the values for 1300° are in agreement with experiment or not, for Deville's determinations of the degree of dissociation in the porcelain tube at this temperature are too uncertain. We cannot, therefore, come to any decision in the matter till Deville's experiments are repeated more carefully. At present it seems probable that the dissociation at 1300° is greater than it should be according to Le Chatelier's formula.

Our opinion regarding Le Chatelier's calculations differs according to whether we view them from a theoretical or a practical standpoint. One must admit that the whole complex of phenomena above 2000°, which Le Chatelier has treated so ably, is affected with a very considerable uncertainty as regards the true temperature. It is quite possible that the equilibrium condition which Le Chatelier assumes to exist at 3000° may actually belong to a temperature different by some hundreds of degrees. Yet there exist no observations which permit any improvement on Le Chatelier's assumption regarding these extreme temperatures.

If we now look at the technical side of the matter, particularly as to what temperatures can be attained by burning carbon monoxide, and what limits the dissociation of carbon dioxide sets to the use of carbon monoxide for heating purposes, the above uncertainty does not trouble us greatly, and the formula given by Le Chatelier certainly yields an amply satisfactory answer. This is due to the fact that the dissociation below 1700° C. is small anyway under working conditions, and even an uncertainty of 100 per cent. in its evaluation would be quite immaterial. Above 1700°, conditions are a little different. But the demands of accuracy which technicians make become smaller the farther we go above this temperature. Up to 1700° the degree of dissociation at partial pressures of carbon dioxide between 0·1 and 0·2 atmosphere is practically all that interests us technically. This degree of dissociation limits the maximum temperature which we can attain by burning our ordinary heating material with a barely sufficient air supply and at ordinary pressures. The partial pressure of carbon dioxide attainable in flue gases depends stoichiometrically upon the composition of the combustible substance. In the combustion of pure carbon it reaches a pressure of 0·21 atmosphere, since each of the 21 volumes of oxygen contained in 100 volumes of air can be replaced by 1 volume of carbon dioxide.[1] The attainable partial pressure of carbon dioxide is less in the combustion of substances containing hydrogen. For instance, in the combustion of illuminating gas where some 2 c.c. of water-vapour are formed for every 1 c.c. of CO_2, 100 c.c. of air can, at most, contain 10½ c.c. of CO_2 and 21 c.c. of water-vapour to every 79 c.c. of nitrogen. The partial pressure of the carbon dioxide cannot then exceed $\frac{10 \cdot 5}{110 \cdot 5}$, or, in round numbers, 0·1 atmosphere. The dissociation corresponding to this partial pressure is very small below 1700°. Smaller partial pressures, where the dissociation has a greater percentage value, need not be considered, because they are occasioned by the presence of an excess of air, and the greater this excess is, the less important for several reasons does dissociation become. In the first place, the excess of air drives back the dissociation because of the oxygen it contains. In the second place, as long as the flue gas contains but little carbon dioxide the temperature reached must remain low, for the large mass of the diluting gas must be heated by the heat of combustion. The lower the temperature, the smaller the dissociation will be; that is, the *direct* cooling effect exerted by diluting with a quantity of air far exceeds, up to 1700°, the *indirect* heating effect arising from the decreasing dissociation.

It is equally true that the degree of dissociation (up to 1700°) does not come into play in combustion in explosion engines. For there the total pressure is much greater than an atmosphere, and consequently the partial pressure of the CO_2 is higher, and the percentage dissociation smaller. Indeed, Langen could detect no evidence of dissociation up to 1700° in his explosion experiments.

[1] Compare Bunte's "Heizlehre," in Muspratt's "Technischer Chemie." 4th edit. (Brunswick, 1893), p. 314.

Now, all technical processes of combustion occur in the temperature interval below 1700° C., with the exception of the processes taking place in certain zones of the blast furnace. No illuminating flame, except that of acetylene, greatly exceeds this temperature. If we would attain temperatures much higher than this in combustion processes, and make any calculation about them, we are also obliged to take into consideration the fact that loss of heat through convection and radiation increases rapidly with rising temperature. The uncertainties which these factors introduce into the calculations thus become so great that any uncertainty about the degree of dissociation does not make much difference. We can, therefore, get along satisfactorily enough in this whole field by means of Le Chatelier's formula.

Our appraisal of the value of this formula from a theoretical point of view is a very different one. In the first place, the formula assumes σ'', the difference between the increments of the specific heats of 1 mol CO plus ½ mol O_2 and 1 mol CO_2, to be 0·0036, and this is not satisfactorily confirmed by the experimental data we possess for temperatures up to 2000°. If it were not for the calculations and observations of Le Chatelier on the explosion of mixtures of carbon monoxide and oxygen, and for the experiments of Deville with his flame, one would certainly prefer formula 3, based on Deville's experiments in porcelain tubes and Langen's explosion experiments, as best expressing the reaction energy of carbon dioxide formation. But we can easily convince ourselves that its use for temperatures much above 2000° is precluded, unless we are willing to assume that the dissociation which Le Chatelier calculates for 3000° really existed some 700° lower. An error of this magnitude seems, however, out of the question. If, instead of Langen's specific heats, we choose those determined by Mallard and Le Chatelier from their observations with the crusher manometer at much higher temperatures, we get

$$A = 67{,}200 - 3{\cdot}28T\,ln\,T + 0{\cdot}00204T^2 - RT\,ln\frac{p_{CO_2}}{p_{CO} \times p^{\frac{1}{2}}_{O_2}} - 2{\cdot}8T \quad (5)$$

Here again Deville's experiments in porcelain tubes are assumed as correct. We can bring this formula, too, into harmony with what little we know about the dissociation up to

2000°, but we then find the equilibrium constant is equal to 3·67 at 2300°, which, again, is a temperature very different from that required by Le Chatelier's formula. Under these circumstances, we will retain Le Chatelier's formula 4, but to its numerical evaluation we will add such figures as result from formulæ 3 and 5. We shall see that the special application which we shall make will not be much affected by this difference.

A further deduction may be made from the formula representing the formation of carbon dioxide, by combining it with our previous formula for the water-gas equilibrium. If we write— Case II. The formation of water.

$$CO + O = CO_2 + A$$
$$\text{and} \quad CO_2 + H_2 = CO + H_2O + A'$$

and add the two, we get—

$$O + H_2 = H_2O + A + A'$$

that is, we get the energy of formation of water A″ by adding the reaction energies of the water-gas reaction and of the formation of carbon dioxide. This gives us—

$$A = \;\; 67{,}300 - 3{\cdot}4\ Tln T + 0{\cdot}0036\ T^2 - RTln \frac{p_{CO_2}}{p_{CO} \times p^{\frac{1}{2}}_{O_2}} - 2{\cdot}28T$$

$$A' = -9{,}650 + 1{\cdot}55 TlnT - 0{\cdot}00195 T^2 - RTln \frac{p_{CO} \times p_{H_2O}}{p_{CO_2} \times p_{H_2}}$$

$$A'' = \;\; 57{,}650 - 1{\cdot}85 TlnT + 0{\cdot}00165 T^2 - RTln \frac{p_{H_2O}}{p_{H_2} \times p^{\frac{1}{2}}_{O_2}} - 2{\cdot}28T$$

This expression assumes a difference between the mean specific heats at constant pressure between 0° and T° of—

$$c_{(H_2 + \frac{1}{2}O_2)} - c_{H_2O} = 1{\cdot}85 - 0{\cdot}00165T$$

This value is somewhat different from that usually given for the specific heats of water-vapour and the permanent gases.[1]

[1] For instance—

I. Mallard and Le Chatelier (old values)—

$1\frac{1}{2}$ mol permanent gases	9·93 + 0·0009T
1 mol H_2O vapour	6·69 + 0·00328T
	3·24 − 0·0023T

II. Mallard and Le Chatelier (crusher manometer)—

$1\frac{1}{2}$ mol permanent gases	9·21 + 0·00183T
1 mol H_2O vapour	6·98 + 0·00287T
	2·23 − 0·00104T

There is but little experimental data at our disposal with which we may test it. There are measurements of the reaction energy of the formation of water made with the oxyhydrogen cell. Victor Meyer and Langer have also observed that water-vapour shows a barely perceptible dissociation at 1200°, while Nernst, working under the same conditions as with carbon dioxide, could detect no certain variation of the density of water-vapour from the normal at 1973°. Putting A'' = 0 and calculating the equilibrium constant for 1473° abs. and 2246° abs., we find—

$$K_{p(1473°)} = 4{\cdot}6 \times 10^6 \ (1{\cdot}8 \times 10^5) \ [22 \times 10^5]^{1}$$
$$K_{p(2246°)} = 654 \ (146) \ [148]$$

Since in both cases observations were made at atmospheric pressure, it follows, from the formula developed for the case of carbon dioxide—

$$\frac{1}{K_p} = \sqrt{\frac{p_{H_2}^{3}}{2}}$$

that the partial pressure of the hydrogen is about 1×10^{-4} to 2×10^{-4} atmospheres at 1473° abs., and about 0·025 atmosphere at 2246° in equilibrium. The first number corresponds to between 1½ to 3 c.c. of hydrogen and oxygen to 10 litres of water-vapour. Victor Meyer and Langer state that on passing

III. Le Chatelier (latest values)—

1½ mol permanent gases	9·75 + 0·0009T
1 mol H_2O vapour	6·5 + 0·0024T
	3·25 − 0·0015T

IV. Langen—

1½ mol permanent gases	9·93 + 0·0009T
1 mol H_2O vapour	7·29 + 0·00215T
	2·64 − 0·00125T

V. Langen (Schreber's recalculation)—

1½ mol permanent gases	9·88 + 0·0008T
1 mol H_2O vapour	8·82 + 0·00116T
	1·06 − 0·00036T

[1] The unbracketed value results directly from our formula. If we use formula (3), p. 170, in deducing the equation for the formation of water-vapour, instead of Le Chatelier's formula (4), p. 170, we get the values enclosed in parentheses. The values included in brackets are derived by use of expression e, p. 133.

a rapid stream of water-vapour for a long time through a platinum tube heated to about 1200° C., they collected several c.c. of an explosive gas. This seems to agree with the requirements of the formula. The second number corresponds to a dissociation of 2·5 per cent., which is within the limits of accuracy of Nernst's vapour-density measurements.[1]

In regard to the oxy-hydrogen cell, we should first point out that our formula for the reaction energy would give for 25° the value— **The oxy-hydrogen cell at ordinary temperatures.**

$$A = 57{,}650 - 3670 - R \times 298 ln \frac{p_{H_2O}}{p_{H_2} \times p^{\frac{1}{2}}_{O_2}}$$

$$= 53{,}980 - R \times 298 ln \frac{p_{H_2O}}{p_{H_2} \times p^{\frac{1}{2}}_{O_2}}$$

Bose[2] has studied the oxy-hydrogen cell at 25°, using the gases under atmospheric pressure. The vapour pressure of his electrolyte can be considered as equal to that of pure water. Then—

$$-R298 \times ln \frac{p_{H_2O}}{p_{H_2} \times p^{\frac{1}{2}}_{O_2}} = +2029 \text{ cal.}$$

and our formula would lead us to expect that A = 56,020 cals. Bose found 52,654 ± 693 cal. The discrepancy is large. While Bose found the electromotive force of the oxy-hydrogen cell to be 1·1392 ± 0·0150 volt (at 760 mm. pressure and 25°), he should have found, according to our formula, 1·212 volt.[3] I would, however, not lay too great stress on this

[1] A. W. v. Hofmann (*Berl. Ber.*, 23 (1890), 2, 3314), reports to the effect that when a very strong current of water-vapour is forced rapidly through a white-hot platinum spiral, one can collect, in a few seconds, enough of a mixture of oxygen and hydrogen to almost break the eudiometer when it explodes. From our formula we should expect a litre of water-vapour to contain 20 c.c. of the oxy-hydrogen mixture at 1723° and at atmospheric pressure. It does not greatly matter which of the three formulæ above mentioned for the reaction energy of formation of CO_2 we use.

[2] *Z. f. phys. Chem.*, 34 (1900), 701, and 38 (1901), 1, where extended references to the literature will be found.

[3] If instead, we use equation (*c*), p. 133, which is based upon Langen's specific heats and Deville's experiments in tubes, we get 1·208 volts. If we use equation (*e*), p. 133, based on Le Chatelier's experiments with the compression manometer, and on the similar experiments of Deville, we get 1·227 volts.

discrepancy of 80 millivolts. It is very possible that Bose's value is too low. It should be pointed out that Smale found 1·073 volts for the oxy-hydrogen cell, and this was generally accepted until Czepinsky, Bose, and Crotogino obtained higher values, those of Bose's being the most carefully determined. A further elevation of the value by some centivolts does not seem improbable when we remember that the oxy-hydrogen cell is not completely understood, at least as far as the oxygen electrode is concerned. There is no proof whatever that a platinized platinum electrode is in equilibrium with the oxygen in an electrolyte saturated with oxygen at atmospheric pressure. According to L. Wöhler,[1] platinum black is not pure platinum, but rather a lower hydroxide of platinum of whose electromotive properties we know nothing. It may well be inherent in the nature of this electrode that in the presence of oxygen it behaves as a not quite saturated solution of oxygen would, and therefore always gives too small an electromotive force for the oxy-hydrogen cell. Abegg and Spencer[2] have shown that it is impossible to get any other value with platinized platinum than that of Bose. But their interesting investigation still leaves open the possibility that the properties of the platinum black affect the value obtained.

Accuracy of the observed value of the oxy-hydrogen cell at ordinary temperatures.

The oxy-hydrogen cell at high temperatures.

On the other hand, we must not fail to recognize that the calculation of the reaction energy at 25° by the help of our formula leads us far from the region in which the observations underlying the formula were made. The fact that the temperature lies far below the critical temperature of water-vapour, and that we are therefore in a region where the specific heats exhibit the irregularities to which we have previously referred, is especially misleading. It therefore seems desirable to test the formula by measurements of the oxy-hydrogen cell at much higher temperatures. Haber and Bruner[3] have done this, using molten caustic soda as an electrolyte. They found—

[1] *Berl. Ber.*, 36 (1903), 3476; and *Z. f. anorg. Chem.*, 40 (1904), 423.

[2] *Z. f. anorg. Chem.*, 44 (1905).

[3] *Z. f. Elektrochemie*, x. (1904), 697.

t^o	T^o	EMK volt.	A″ cal.	A″ calculated by our formula.[1]
312	585	1·24	57,313	$49{,}993 - 2668 \log^{10} \frac{p_{H_2O}}{p_{H_2} \times p_{O_2}^{\frac{1}{2}}}$
412	685	1·15	53,153	$48{,}600 - 3124 \log^{10} \frac{p_{H_2O}}{p_{H_2} \times p_{O_2}^{\frac{1}{2}}}$
532	805	1·03	47,607	$46{,}933 - 3671 \log^{10} \frac{p_{H_2O}}{p_{H_2} \times p_{O_2}^{\frac{1}{2}}}$

The oxygen and hydrogen were at very nearly atmospheric pressure in these experiments, and consequently the denominator of the logarithmic term may be put equal to unity. The vapour tension of water over molten caustic soda is not sufficiently well known to allow us to compare calculation and observation as closely as we would wish. We can only say that our formula seems to give relatively good values when we compare them with the results of these experiments. The intimation is thus supported that the true values of the oxy-hydrogen cell lies even higher than Czepinsky and Bose have found it.[2]

The simple fundamental cases of the dissociation of water and of carbon dioxide have been but slightly investigated, and the expressions for their reaction energy consequently possess less certainty than we would wish. We will, nevertheless, risk Case III. The Deacon chlorine process.

[1] Taking the carbon dioxide equation (*e*), on p. 133, as the basis of our formula, we obtain the values—

45,620 at 312°
43,230 at 412°
40,330 at 532°

[2] To prevent misunderstanding, I should like to call attention to the fact that the correctness of any value obtained for the electromotive force of the oxy-hydrogen cell cannot be tested by seeing whether the quantity A″ fulfils the condition (p. 22) that—

$$A'' - T\frac{dA}{dT} = Q$$

This relation always holds wherever water is formed reversibly. Whether the electrodes are in equilibrium with oxygen and hydrogen at atmospheric pressure remains entirely indeterminate. If the condition of the electrodes corresponds to a lesser gas pressure, A″ comes out smaller, but $T\frac{dA}{dT}$ is just as much larger, and *vice versâ*. Subsequent investigation has confirmed this conclusion. See Appendix to Lecture VII.

pushing the application still further, and use the equation for the formation of water in studying the Deacon process of manufacturing chlorine. This process corresponds to the difference—

$$\begin{array}{r} 2H_2 + O_2 = 2H_2O \\ \text{minus } 2Cl_2 + 2H_2 = 4HCl \\ \hline O_2 + 4HCl = 2H_2O + 2Cl_2 \end{array}$$

We have earlier deduced an expression for the energy of formation of hydrochloric acid. After quadrupling it so that it may correspond to the formation of 4 mols of HCl, we subtract it from the energy of formation of 2 mols of water-vapour—

$$2A'' = 115{,}300 - 3{\cdot}7T\,ln\,T + 0{\cdot}0033T^2 - RT\,ln\frac{p^2_{H_2O}}{p^2_{H_2} \times p_{O_2}} - 4{\cdot}56T$$

$$4A = \quad 88{,}000 \qquad\qquad - RT\,ln\frac{p^4_{HCl}}{p^2_{H_2} \times p^2_{Cl_2}} + 5{\cdot}36T$$

$$A''' = \quad 27{,}300 - 3{\cdot}7T\,ln\,T + 0{\cdot}0033T^2 - RT\,ln\frac{p^2_{H_2O} \times p^2_{Cl_2}}{p_{O_2} \times p^4_{HCl}} - 9{\cdot}92T$$

We may now divide this by 4 so as to make it correspond to the decomposition of one mol of hydrochloric acid in accordance with our general usage, and obtain—

$$A = 6825 - 0{\cdot}925T\,ln\,T + 0{\cdot}000825T^2 - RT\,ln\frac{p^{\frac{1}{2}}_{H_2O} \times p^{\frac{1}{2}}_{Cl_2}}{p^{\frac{1}{4}}_{O_2} \times p_{HCl}} - 2{\cdot}48T$$

The experiments of Lunge and Marmier. We can test this equation for the energy of reaction by the experiments of Lunge and Marmier.[1] To this end, let us examine their experiments a little more closely. Lunge and Marmier led a mixture of hydrochloric acid gas and of oxygen (sometimes of air) over broken bricks which had been soaked with a copper chloride solution, then dried and heated to a temperature of about 450°. The mixture of gases was usually dry, but sometimes it was wet. The reverse experiments with chlorine and water-vapour were not made. The object of the investigation was to determine the degree of decomposition of the hydrochloric acid. If the fraction x of every mol of

[1] *Z. f. angew. Chemie* (1897), 105, and the dissertation of E. Marmier, "Ueber die Darstellung von Chlor nach dem Verfahren von Deacon und Mond" (Zürich, 1897).

hydrochloric acid is decomposed and $\frac{x}{2}$ mols of chlorine produced, then in the final gas mixture—

$$\frac{Cl_2}{HCl} = \frac{x}{2(1-x)}$$

This quotient represents the relative number of chlorine and hydrochloric acid mols, and also the ratio of the per cents. by volume or partial pressures of the two kinds of gases in the gas mixture.

The equilibrium constant of the Deacon process is given by the equation—

$$\frac{p^{\frac{1}{2}}_{H_2O} \times p^{\frac{1}{2}}_{Cl_2}}{p^{\frac{1}{4}}_{O_2} \times p_{HCl}} = K_p$$

In order to introduce the ratio $\frac{Cl_2}{HCl}$ into this expression, we multiply numerator and denominator by $p^{\frac{1}{2}}_{Cl_2}$ and obtain—

$$\frac{p_{Cl_2}}{p_{HCl}} \times \frac{1}{p^{\frac{1}{4}}_{O_2}} \times \frac{p^{\frac{1}{2}}_{H_2O}}{p^{\frac{1}{2}}_{Cl_2}} = K_p$$

or substituting the values just found for the first quotient—

$$\frac{x}{2(1-x)} \times \frac{1}{p^{\frac{1}{4}}_{O_2}} \times \sqrt{\frac{H_2O}{Cl_2}} = K_p \quad . \quad . \quad . \quad (7)$$

The term placed under the radical sign may represent partial pressures, per cents. by volume, or mols per unit of volume, the unit of measurement cancelling. The expression becomes simpler when the initial gas mixture is dry. Then just as much chlorine as water is formed, so that the term $\sqrt{\frac{H_2O}{Cl_2}}$ has the value 1. For this case, then—

$$\frac{x}{2(1-x)} \times \frac{1}{p^{\frac{1}{4}}_{O_2}} = K_p \quad . \quad . \quad . \quad . \quad (7a)$$

This formula tells us the degree of decomposition when equilibrium is attained. It is relatively but little influenced by the excess of oxygen, for only the fourth root of the partial pressure of the oxygen appears in the equation. If, for instance, the partial pressure were in one case 0·9 of an atmosphere and in another only 0·05 of an atmosphere, the fourth root in the one case is 0·974, in the other 0·474. So a change of 18 times in the pressure of the oxygen changes the quotient $\frac{1}{p^{\frac{1}{4}}_{O_2}}$ to a value but twice as great, and the degree of decomposition, which was 0·6 (66 per cent.) in the first place, only changes to about 0·75 (75 per cent.), as can be seen from equation (7), p. 181.

Lunge and Marmier do not state the partial pressure of the oxygen in the gaseous products, but they do tell the composition and pressure of the mixture they started with, and knowing the degree of decomposition, we may calculate the final partial pressure. The following example will illustrate this.

Initial mixture contained 8·5 per cent. HCl, 91·5 per cent. O_2, pressure = 737 mm., x found = 0·83; hydrochloric acid transformed = 0·83 × 8·5 = 7·05 c.c. From this there is found stoichiometrically 3·52 c.c. Cl_2, 3·52 c.c. H_2O, while 1·76 c.c. O_2 were used up. It follows that the gaseous product contained—

Sum.	Cl_2	HCl	O_2	H_2O	Pressure.
98·23 c.c.	3·52 c.c.	1·45 c.c.	89·74 c.c.	3·52 c.c.	Same as initial gas.

Therefore—

$$p_{O_2} = \frac{89 \cdot 74}{98 \cdot 23} \times \frac{737}{760} = 0 \cdot 888 \text{ atm.}$$

$$p^{\frac{1}{4}}_{O_2} = 0 \cdot 968$$

$$\frac{x}{2(1 - x)} = 2 \cdot 44$$

$$\frac{x}{2(1 - x)} \times \frac{1}{p^{\frac{1}{4}}_{O_2}} = K_p = 2 \cdot 51$$

We can express the procedure just illustrated by the formula—

$$p_{O_2} = \frac{O_2 - \frac{x}{4}HCl}{O_2 - \frac{x}{4}HCl + HCl} \times \frac{B}{760}$$

where B is the barometric pressure, O_2 and HCl the percentage content of the initial mixture. The values of p_{O_2} in Table I., p. 185, were calculated in this way.

Lunge and Marmier employed mixtures of gases whose

oxygen content varied from just about the theoretical quantity required to an excess of almost 100 times. In the mixtures containing but a little hydrochloric acid only a relatively small amount had to be transferred in reaching the equilibrium. The smaller the excess of oxygen the greater this quantity need be. Now all the mixtures were conducted over the catalyzer at approximately the same velocity. The result is, as can be easily understood, that at lower temperatures the gas mixtures remain farther from equilibrium the nearer they approach in composition to stoichiometrical quantities. This makes itself apparent in the abnormal falling off of the degree of decomposition with decreasing excess of oxygen. At high temperatures (480°) this phenomenon is no longer prominent, and we may therefore conclude that a reaction velocity has now been attained which is high enough to bring even a stoichiometric mixture up to the equilibrium. Lunge and Marmier ascribed no especial importance to reaction velocity, basing their view on Deacon's statement, that with a sufficiently large catalyzing surface the effect of reaction velocity disappeared. In their above-mentioned experiments, where the degree of decomposition was large, the oxygen was present in great excess, and hence the equilibrium was easily attained, and the catalyzing surface they used was "sufficiently" great; but in the mixtures of gas which are relatively poor in oxygen, this is not true, even at 430°, and the lower the temperature the richer the mixture may be in oxygen without being able, at the velocities employed by Lunge and Marmier, to reach equilibrium and the high degree of decomposition corresponding to it.

The change of the catalyst during the experiment.

The observations of Lunge and Marmier were made with a catalyst, which, according to the statements of the investigators, does not remain unaltered. This phenomenon finds expression in their experimental results in the fact that sometimes more, sometimes less, total chlorine is contained in the gaseous products than was present in the mixture to start with. It seems, however, as if equilibrium in the issuing gases was not disturbed by this. Nor indeed is it theoretically necessary that it should be. Haber and Van Ordt[1] have observed, in the action of hydrogen on the nitride of calcium, that the ammonia equilibrium in the issuing gases is nearly permanent, although

[1] *Z. f. anorg. Chemie*, 44 (1905), 341.

the nitride slowly changes. The same thing occurs in the technical manufacture of water-gas, as will be shown in a discussion contained in the Seventh Lecture. But the change in the catalyzer does trouble us, in that thereby the ratio $\frac{H_2O}{Cl_2}$ differs from 1, even in dry mixtures. If the catalyzer gives off chlorine and takes up oxygen, the term $\sqrt{\frac{H_2O}{Cl_2}}$ occurring in equation 7 will be less than 1 in the escaping gas, and consequently—

$$\frac{x}{2(1-x)} \times \frac{1}{p_{O_2}^{\frac{1}{4}}} > K_p$$

The disturbing effect is of course greatest where the amount of hydrochloric acid which passes over the catalyst during the experiment is smallest, that is, when the excess of oxygen is greatest. For even a small amount of chlorine given off by the catalyzer would change the ratio $\frac{Cl_2}{H_2O}$ greatly when it was mixed with only a small amount of chlorine produced from the hydrochloric acid. Lunge and Marmier did not lay much stress on this point, for they were chiefly interested in the degree of decomposition, and it is but slightly affected at equilibrium by this disturbing influence, for it takes quite a considerable change in $\frac{x}{2(1-x)}$ to produce any change in it. One can easily see this by working through a few examples.

The following table contains data which Lunge and Marmier have collected at 480°. They represent the mean values from parallel experiments numbered in Marmier's dissertation 69-85. The single values lie very close to the mean values. The first value is bracketed, because the difference between the chlorine in the hydrochloric acid taken and the total chlorine contained in the final product was far too great to yield any reasonable value for the equilibrium constant on the assumption that—

$$\sqrt{\frac{H_2O}{Cl_2}} = 1$$

TABLE I.

Pressure. mm. Hg.	Initial gas. % HCl	Initial gas. % O_2	% HCl[1] measured in the single experiments.	x	$p^{\frac{1}{4}}_{O_2}$	$\frac{x}{2(1-x)}$	K_p
726	7·5	92·5	208 and 212	0·87	0·97	(3·34)	(3·2)
726	15·1	84·9	74 „ 119	0·83	0·95	2·44	2·6
721	25·5	74·5	155 „ 104	0·82	0·91	2·28	2·5
723	42·2	57·8	75·6 „ 91·2	0·77	0·85	1·67	2·0
723	49·0	51·0	93 „ 93	0·76	0·82	1·58	1·9
720	57·3	42·7	91 „ 95	0·75	0·76	1·50	2·0
720	68·2	31·8	96 „ 105	0·73	0·68	1·35	2·0
718	80·8	19·2	92 „ 97	0·67	0·496	1·01	2·0

I conclude from this that the equilibrium constant of the reaction underlying the Deacon process equals 2·0 at 480°.

If we calculate the equilibrium constant from our formula for the energy of reaction (p. 180), putting A = 0, we find after a simple transformation that—

$$4{\cdot}56 \log \frac{p^2_{H_2O} \times p^2_{Cl_2}}{p_{O_2} \times p^4_{HCl}} = \frac{27{,}300}{753} - 3{\cdot}7 \times 2{\cdot}3 \log^{10} 753 + 0{\cdot}0033 \times 753 - 9{\cdot}92$$

and from this that—

$$\frac{p^{\frac{1}{2}}_{H_2O} \times p^{\frac{1}{2}}_{Cl_2}}{p^{\frac{1}{4}}_{O_2} \times p_{HCl}} = K_p = 1{\cdot}73\,[2]$$

The agreement is astonishing when we think how long and how beset with uncertainties the path has been by which we have reached these conclusions. We started from the carbon dioxide dissociation, proceeded *via* the water-gas equilibrium and the water-vapour dissociation, and by combining our results with Dolezalek's measurements of the hydrochloric acid cell, finally obtained our formula.

If we now calculate the value of the equilibrium constant again at atmospheric pressure, for the temperature 430°, which Lunge and Marmier considered the most favourable, we find that—

$$\frac{p^{\frac{1}{2}}_{H_2O} \times p^{\frac{1}{2}}_{Cl_2}}{p^{\frac{1}{4}}_{O_2} \times p_{HCl}} = K_{p\,(703\ \text{abs.})} = 2{\cdot}42\,[3]$$

[1] The Cl_2 found is recalculated in terms of HCl and added to the HCl found. The result is expressed in per cents. of the HCl taken.

[2] Had we started from expression (*c*) on p. 133 for the carbonic acid dissociation, we should have found 1·835 for K*p*. This is not very different from the value here obtained.

[3] Had we started from the expression (*c*) on p. 133 for the carbonic acid dissociation, we should have found 2·711.

We see that the location of the equilibrium becomes more favourable for the preparation of chlorine the lower the temperature at which we work. In the Deacon process, therefore, since as complete an oxidation of the hydrochloric acid as possible is desired, we should endeavour to keep the temperature as low as possible. The limit is here set by the reaction velocity. This soon becomes too small. Lunge and Marmier could obtain a very noticeable transformation even at 310°, but technical practice has shown that we must work at temperatures above 400° if we would obtain satisfactory results.

Lunge and Marmier's results at 430°. I reproduce here Table II. containing the experimental results of Lunge and Marmier at 430°, starting with dry gases.

Pressure mm.	% HCl	% O_2	% N_2	Recovered HCl in %	x	$p^{\frac{1}{4}}_{O_2}$	$\frac{x}{2(1-x)}$	K_p
737	8·5	91·6	—	109 and 116	0·83	0·97	2·44	2·51
733	15·7	84·3	—	105, 113, 148	0·84	0·96	2·63	2·74
736·5	16·3	83·7	—	84, 101, 97	0·83	0·95	2·44	2·57
737	21·0	79·0	—	94, 111, 118	0·82	0·93	2·28	2·45
734·6	28·6	71·4	—	102, 103, 104	0·82	0·91	2·28	2·51
729	34·1	65·9	—	106 and 112	0·81	0·88	2·13	2·42
730	40·2	59·8	—	105, 108, 117	0·78	0·86	1·77	2·05
735	51·1	48·9	—	177·3 and 79·7	0·77	0·81	1·68	2·07
734	53·9	46·1	—	105, 99, 101, 102	0·76	0·79	1·60	2·92
734	54·1	45·9	—	102 and 103	0·75	0·78	1·50	1·02
729	67·4	32·6	—	100 and 104	0·58	0·71	0·71	1·00[1]
729	72·5	27·5	—	106 and 143	0·50	0·67	0·50	0·74[1]
725	83·1	16·9		103 and 105	0·39	0·55	0·32	0·58[1]
726	6·6	19·5	73·9	82 and 86	0·79	0·65[2]	1·88	2·88
725	12·7	18·2	69·1	102 and 118	0·75	0·63	1·50	2·38
725	23·3	16·0	60·7	100 and 102	0·74	0·58	1·42	2·45
727	26·0	15·4	58·6	83 and 102	0·71	0·57	1·22	2·14
725	35·1	13·6	51·3	103 and 103	0·65	0·53	0·93	1·75
720	41·4	12·2	46·4	103, 100	0·47	0·52	0·44	0·85[3]
715	51·0	10·2	38·7	91 and 100	0·45	0·46	0·41	0·90[3]

We gather from Table II. that the equilibrium is reached when oxygen is used containing as much as 35 per cent. hydro-

[1] In these experiments the value of x is very uncertain.

[2] The values of p_{O_2} from this number are computed by the formula

$$\frac{O_2 - \frac{x}{4}HCl}{O_2 - \frac{x}{4}HCl + N_2 + HCl} \times \frac{B}{760}$$

which will be readily understood.

[3] In this group of experiments the values of x are very irregular.

chloric acid. If the oxygen is partially replaced with nitrogen, by using air instead of oxygen, then even at 36 per cent. of hydrochloric acid the equilibrium is not completely obtained, and at 35·1 per cent. of HCl the reaction remains a long way from the equilibrium. The favourable influence of an excess of oxygen on the reaction is shown even more plainly by the mixtures containing 50 per cent. of HCl, for of these the one made with pure oxygen still very nearly reaches equilibrium, while that made with air falls far short of it.

Lunge and Marmier expressed their astonishment that in mixtures containing much hydrochloric acid, the degree of decomposition should be greater when the hydrochloric acid is mixed with air than when mixed with oxygen, the ratio $\frac{HCl}{O_2}$ being the same in both cases. Examples of such mixtures would be—

HCl	O_2	N_2	$\frac{HCl}{O_2}$	-
72·5	27·5		$\frac{4}{1·51}$	0·50
35·1	13·6	51·3	$\frac{4}{1·55}$	0·65

The explanation of this phenomenon is evident from our above discussion. The mixture containing air comes very near reaching equilibrium because of its small hydrochloric acid content, while in the mixtures containing only pure oxygen beside the hydrochloric acid the reaction velocity is no longer sufficient to bring about the much larger transformation, in spite of the accelerating effect of the oxygen.

I cannot treat the very valuable experimental material of Lunge and Marmier exhaustively, but must content myself with mentioning two groups of experiments at higher temperatures—

b	% HCl	% O_2	% N_2	Recovered, HCl in %	x	$p^{\frac{1}{4}}_{O_2}$	$\frac{x}{2(1-x)}$	K_p	$t°$C.
718	28·5	14·9	56·6	126 and 113	0·70	0·56	1·17	2·1	510
722	26·6	15·3	58·1	73 and 86	0·66	0·57	0·97	1·7	560

The constant K_p here appears surprisingly large. Yet our formula for the temperature at which the constant becomes equal to unity—

$$2{\cdot}48 = \frac{6825}{T} - 0{\cdot}925 \ln T + 0{\cdot}000825T$$

yields the value T = (about) 850° abs., or t = 577°. But we must recall here that the catalyzer certainly begins to volatilize above 470°. Deacon says it volatilizes even at 428°, and Lunge and Marmier claim that traces of it go over at even lower temperatures. It is, consequently, inevitable that above 500° the catalyzer should distill into the cooler portion of the tube near the outlet. The effect of this would be that the gases would not remain in the unfavourable equilibrium corresponding to the higher temperature, but would react in the cooler portion of the tube toward an equilibrium containing more chlorine.

The chemical "strength" of oxygen and chlorine.

We will close this discussion by a reference to the chemical "strength" of chlorine and oxygen. We have just seen that the equilibrium constant of the Deacon process becomes unity at about 577°. Now, this process depends upon the distribution of hydrogen between chlorine and oxygen, that is, these two substances compete for the hydrogen. We conclude from this that both oxydizing agents are equally strong at 577°. At lower temperatures the oxygen captures more of the hydrogen; at higher temperatures chlorine does this. So oxygen has the greater affinity for hydrogen in the cold (below 577°), chlorine in the heat (above 577°). This, of course, applies only under comparable conditions of concentration, and the question arises, What are comparable concentration conditions? There exists an analogous case where the answer to this question is evident. It is the competition of carbon monoxide and hydrogen for oxygen.

$$\begin{matrix} 2H_2 & + & & \nearrow & 2H_2O \\ & & O_2 & & \\ 2CO & + & & \searrow & 2CO_2 \end{matrix}$$

This is nothing other than the water-gas equilibrium. Its equilibrium constant becomes unity at about 830°. Above this temperature the affinity of oxygen for hydrogen prevails; below it the affinity of oxygen for carbon monoxide. Here the concentrations are comparable when carbon monoxide and hydrogen, on the one hand, and water-vapour and carbon dioxide on the other, have equal partial pressures. We may, therefore, conclude immediately from this that a mixture of water-vapour and hydrogen can exist together with a smaller amount of free oxygen above 830° than can a similar mixture of carbon monoxide and dioxide, assuming, of course, that the pressure is the same in both cases. Therefore, carbon dioxide dissociates more above 830° than does water-vapour. The numerical values which we have found for the dissociation of carbon dioxide and for water-vapour correspond to these conclusions. When we compare chlorine with oxygen the comparable conditions are not so simple, because gaseous chlorine is divalent, and gaseous oxygen is tetravalent. One mol of chlorine unites with but one mol of hydrogen, while one mol of oxygen unites with two mols of hydrogen.

Now, we judge chemical "strength" on the basis of equivalent amounts. That is, we compare—

$$H_2 + O \rightleftarrows H_2O$$

with—

$$H_2 + Cl_2 \rightleftarrows 2HCl$$

In the one case at equilibrium—

$$\frac{p_{H_2O}}{p_{H_2} \times p_{O_2}^{\frac{1}{2}}} = K_{p(H_2O)}$$

in the other—

$$\frac{p^2_{HCl}}{p_{H_2} \times p_{Cl_2}} = K^2_{p(HCl)}$$

Here we must write $K^2_{p(HCl)}$, for we have previously written (p. 108) $K_{p(HCl)}$ as referring to *one* mol HCl. The stronger oxidizing agent will have the higher equilibrium constant. We then use van't Hoff's formula—

$$A = RTlnK_p - RT\Sigma\nu' lnp'$$

to find the comparable conditions in both cases, and obtain—

$$A = RTlnK_{p(H_2O)} - RTln\frac{p_{H_2O}}{p_{H_2} \times p_{O_2}^{\frac{1}{2}}}$$

and

$$A' = RTlnK^2_{p(HCl)} - RTln\frac{p^2_{HCl}}{p_{H_2} \times p_{Cl_2}}$$

If the quotients—

$$ln\frac{p_{H_2O}}{p_{H_2}\times p^{\frac{1}{2}}_{O_2}} \text{ and } ln\frac{p^2_{HCl}}{p_{H_2}\times p_{Cl_2}}$$

are arranged to be equal in value, then each adds just as much, or just as little, to A and A′, and the difference between A and A′ shows directly the difference in chemical strength.

We may similarly determine A and A′ by the aid of galvanic cells. Since the same number of equivalents are transformed, the quantities of electricity generated in each case are equal. Consequently the reaction energies A and A′ stand in the same ratio to one another as the electromotive forces of the two cells. We have Dolezalek's measurements for one cell, and those of Bose for the other. We are already acquainted with both investigations. Bose made observations at but a single tension of aqueous vapour, which we can place at 0·031 atmosphere. Dolezalek, on the other hand, made measurements at various partial pressures of hydrochloric acid, and so we can easily choose comparable concentrations in the light of what has just been said. That is, since in comparable cases—

$$ln\frac{p_{H_2O}}{p_{H_2}\times p^{\frac{1}{2}}_{O_2}} = ln\frac{p^2_{HCl}}{p_{Cl_2}\times p_{H_2}}$$

then also—

$$\frac{p_{H_2O}}{p^2_{ClH}} = \frac{p_{H_2}\times p^{\frac{1}{2}}_{O_2}}{p_{Cl_2}\times p_{H_2}}$$

In the measurements of Bose we may call the pressure of the hydrogen and the oxygen equal to 1 atmosphere without appreciable error. Consequently $p^{\frac{1}{2}}_{O_2}$ equals 1, and the product $p_{H_2}\times p^{\frac{1}{2}}_{O_2}$ also equals 1. In Dolezalek's measurements the same thing does not apply with equal rigour, for the high partial pressure of the hydrochloric acid lowered the partial pressures of the chlorine and oxygen perceptibly when the total pressure was 1 atmosphere. Nevertheless, we may assume for the moment that in this case, too, p_{Cl_2} and p_{H_2}, and hence their product, are equal to 1. This tells us that, to judge the relative chemical strength, we must compare these measurements where—

$$\frac{p_{H_2O}}{p^2_{HCl}} = 1$$

or—

$$p^{\frac{1}{2}}_{H_2O} = p_{HCl}$$

But since the vapour tension of water was 0·031 atmosphere in Bose's experiments, we must use the value Dolezalek found for a partial pressure of hydrochloric acid = $\sqrt{0{\cdot}031}$, or 0·176 atmosphere. For this partial pressure Dolezalek found an electromotive force of 1 volt, in round numbers, while Bose found 1·14 for the oxy-hydrogen gas cell. If we now take the fact just mentioned into consideration—that the partial pressures of the chlorine and hydrogen were less than 1 atmosphere in Dolezalek's experiments—it appears that we must take a cell containing more concentrated hydrochloric acid, and hence one having less electromotive force, if we would obtain comparable conditions. The difference between the oxy-hydrogen cell and the comparable oxy-chlorine cell would thus become still greater. The small difference of 5° between the temperatures of the two cells would affect their electromotive forces too slightly to require consideration here. The result agrees, as was to be expected, with the conclusion we had already arrived at from a study of the Deacon process. In the cold, therefore, chlorine, though more rapidly acting, is the weaker oxidizing agent.

Case IV. The contact process for the manufacture of sulphuric acid.

We will now consider two other gas reactions of technical importance, namely, the formation of sulphur trioxide from sulphur dioxide and oxygen, and of ammonia from the elements.

The formation of the anhydride of sulphuric acid from sulphur dioxide and oxygen—

$$2SO_2 + O_2 \rightleftarrows 2SO_3$$

possesses the very greatest technical importance as the basis of the contact process for the manufacture of sulphuric acid. It has been studied by many investigators.

The heat of reaction, according to Berthelot's statement, is 45,200 g. cal. at ordinary temperatures. The value is calculated as the difference between the heat of formation of gaseous sulphuric acid and sulphurous acid, and, in view of the marked discrepancy between Berthelot's and Thomsen's values for the heat of formation of sulphur dioxide, is not especially certain. It refers to constant pressure.

The mean specific heat of sulphur dioxide at constant pressure has been calorimetrically determined by Regnault to be 9·86 between 10° and 200°. Müller's determination of the ratio between its specific heat at constant pressure and at constant volume is in satisfactory agreement with this, giving 1·256. Regnault found the mean specific heat of oxygen over this same temperature interval to be 6·96. We do not know what the specific heat of sulphur trioxide is, so we cannot compute the change of the heat of reaction with the temperature.

We will again refer the reaction energy to the formation of a single mol of sulphur trioxide. The heat of reaction at ordinary temperatures referred to this quantity is 22,600 cal. The formula for the reaction energy says—

$$A = Q_0 - \sigma'_p T l n T - \sigma'' T^2 - RT l n \frac{p_{SO_3}}{p_{SO_2} \times p_{O_2}^{\frac{1}{2}}} + \text{const. } T$$

At the equilibrium A = 0, and—

$$\frac{p_{SO_3}}{p_{SO_2}} \times \frac{1}{p_{O_2}^{\frac{1}{2}}} = K_p$$

Here the quotient $\frac{p_{SO_3}}{p_{SO_2}}$ is independent of the pressure, since the unit of measure cancels from both numerator and denominator. We might equally well write per cents. by volume or concentrations in its place, so in the future we will omit the measuring factor.

In actual practice the object is to get the highest possible yield of SO_3. At any given temperature this will always be more fully attained the higher the partial pressure of the oxygen in the final product. This is evident when the above expression is written as—

$$\frac{SO_3}{SO_2} = K_p \sqrt{p_{O_2}}$$

Determinations of the equilibrium constant.

The constant K_p becomes greater the lower the temperature. Experience has shown that the formation of sulphur trioxide at high temperatures is considerably reduced by the strong dissociation of the sulphur trioxide into sulphur dioxide and oxygen.

We owe our chief data for determining K_p to Knietsch.[1]

[1] *Berl. Ber.*, 34 (1901), 4069. Since then Knietsch has communicated some further determinations in which sulphur dioxide mixed with air, carbon dioxide, or water-vapour was led over platinized asbestos at constant pressure and temperature, and the unchanged sulphur dioxide determined ("Bericht über den fünften internationalen Kongrefs für angewandte Chemie," vol. i. p. 617 (Berlin, 1904)).

They were obtained by passing a mixture of sulphur dioxide, oxygen, and nitrogen over platinum. He only stated his results in the form of curves, and the partial pressure of oxygen must be indirectly calculated, as in the case of the Deacon process. This has been done by Brode,[1] Bodländer and Köppen,[2] and by J. d'Ans.[3] I shall follow the data given by the last. But for our purposes we must make a slight change in them, because J. d'Ans did not treat the quantity—

$$\frac{SO_3}{SO_2} \times \frac{1}{\sqrt{p_{O_2}}}$$

as a constant, but rather—

$$\frac{SO_2^2}{SO_3^2 p_{O_2}}$$

We will therefore take the square root of the reciprocal of his numbers. The results of the density determinations made by Bodländer and Köppen are appended to the same computation. These two investigators employed a very original manometric method. They used two similar quartz vessels containing rolls of platinum wire netting. They filled one of them with air, and the other with a mixture of sulphur dioxide, oxygen, and nitrogen. Both vessels were now heated to about 600°, and the pressure made one atmosphere in each. The temperature was then lowered to 413°. This occasioned a greater decrease of pressure in the second than in the first. In the first the contraction of the air was due solely to the cooling; but in the second an additional contraction was occasioned by the combination of sulphur dioxide with oxygen. This difference in the decrease of pressure was a measure of the amount of trioxide formed by cooling from the initial temperature to 413°. But since at 413° sulphur dioxide and oxygen unite almost quantitatively, it showed directly how great the dissociation was in the neighbourhood of 600°, and hence showed the location of the equilibrium at this temperature. The use

[1] Given only in Lunge, "Sodaindustrie," 3rd edit., i. 917.

[2] *Z. f. Elektrochemie* (1903), 787.

[3] Dissertation Darmstadt, "Das wasserfreie Ferrosulfat und seine Zersetzung bei höheren Temperaturen," printed by Fiencke in Kiel (1905).

of higher temperatures was precluded by the observation that oxygen was then absorbed by the platinum filings in the quartz vessels. The determinations agree excellently with those of Knietsch.

t°	T°	K_p Knietsch.	K_p Bodländer.	$R \ln K_p$
450	723	187·67	—	10·369
500	773	72·30	—	8·477
515	788	—	65·44	8·290
553	826	—	24·07	6·300
600	873	14·90	—	5·350
610	883	—	10·50	4·658
650	923	—	(4·45)	2·955
700	973	4·84	—	3·124
800	1073	1·81	—	1·172
900	1173	0·57	—	−1·108

There also exist two determinations of the equilibrium constant K_p, made in the laboratories of the Höchst Farbwerke, which, when expressed in our units, give—

t°	T°	K_p
465	738	1175
515	788	536

They deviate very widely, as one can see, from the value in our table. According to the statements of the observers themselves (Lunge, "Sodaindustrie," i. p. 950), the experiments on which they are based are not above criticism, and they may therefore be set aside as less certain than the others. The small values given by Bodländer and Köppen for 650° in the above table fall a little out of line from the other values. Bodländer and Köppen admit the possibility of an experimental disturbance which would have caused it to come out too small. They have attempted to deduce the heat of reaction from their own and Knietsch's observations. For this purpose they have recalculated the constant so that it represents K_c instead of K_p. This

Heat of the reaction according to Bodländer and Köppen.

can be done here by simply multiplying by $\sqrt{0 \cdot 0821T}$.[1] From this they have computed the heat of reaction at constant volume, using van't Hoff's equation (see p. 64)—

$$R l n K_c - R l n K'_c = Q_v\left(\frac{1}{T} - \frac{1}{T'}\right)$$

Bodländer and Köppen found from their measurements that at 550° it equalled 25,500 cal. At constant pressure it would therefore be 25,910 cal. The measurements of Knietsch yield (according to Bodländer and Köppen's computation) the rather variable values—

t	...	550	650	750	850 ° C.
Q_v	...	23,280	19,000	18,620	27,125 cal.

Professor Bodenstein, of Leipzig, has written me personally about a further investigation of this equilibrium which he has just completed. His method consisted in passing a mixture of sulphur dioxide, air, and nitrogen over hot platinized asbestos contained in two quartz vessels. Combination took place in the first vessel until equilibrium was almost reached, so that in the second and main vessel there was no disturbing heat of reaction. The carefully executed determinations gave the following results:— **Bodenstein's investigation.**

t° ...	528	579	627	680	727	789	832	897
T° ...	801	852	900	953	1000	1062	1105	1170
K_p ...	31·3	13·8	5·54	3·24	1·86	0·956	0·627	0·358
$R l n K_p$	6·82	5·20	3·72	2·33	1·23	−0·089	−0·924	−2·035

A calculation of the heat of reaction from interval to interval of the temperature by means of van't Hoff's equation gave much more regular results than did the corresponding calculation of Knietsch's measurements. The numbers lie quite close

[1] This may be found in the following way:—

$$K_p = \frac{SO_3}{SO_2} \cdot \frac{1}{\sqrt{p_{O_2}}}$$

and further

$$p_{O_2} = 0{\cdot}0821 T c_{O_2}$$

where p_{O_2} is expressed in atmospheres and c_{O_2} in mols per litre. Finally

$$K_c = \frac{SO_3}{SO_2} \cdot \frac{1}{\sqrt{c_{O_2}}} = \frac{SO_3}{SO_2} \cdot \frac{1}{\sqrt{\frac{p_{O_2}}{0{\cdot}0821 \times T}}} = K_p \times \sqrt{0{\cdot}0821 \times T}$$

to $Q_v = 21,700$, and show no perceptible tendency to increase or decrease with the temperature. It appears from this that the difference between the specific heats of factors and products, referred to constant volume, must be small. Yet Bodenstein's numbers, although they probably surpass those of Knietsch and of Bodländer and Köppen in accuracy, are hardly accurate enough to furnish us certain knowledge of the specific heats. The expression—

$$Rln K_p = \frac{21,700}{T} - 20{\cdot}4 \quad . \quad . \quad . \quad . \quad (8)$$

may, however, be deduced from Bodenstein's measurements with a degree of approximation sufficiently close for practical purposes, and in what follows we shall use it.

Here, just as in the Deacon process, the technical interest centres chiefly around the "yield," referred in this case to the amount of sulphur dioxide used. If we call this yield x, then—

$$x = \frac{SO_3}{SO_2 + SO_3}$$

where SO_3 and SO_2 represent the fractional parts of the gaseous products which these gases make up. The ratio $\frac{SO_3}{SO_2}$ appearing in our formulæ is then determined by the relation—

$$xSO_2 + xSO_3 = SO_3$$
$$(1 - x)SO_3 = xSO_2$$
$$\frac{SO_3}{SO_2} = \frac{x}{1 - x}$$

We can therefore write—

$$\frac{x}{1 - x} \times \frac{1}{\sqrt{p_{O_2}}} = K_p$$

and since equation (8) tells us the value of K_p with sufficient accuracy over the whole temperature interval, we can calculate the attainable yield. For this purpose it is more convenient to rearrange the last equation to—

$$x = \frac{K_p\sqrt{p_{O_2}}}{1 + K_p\sqrt{p_{O_2}}}$$

We may represent per cents. of the theoretical yield on the basis of $100x$. It can be directly seen from the formula that the yield depends directly on K_p, which in its turn is solely determined by the temperature. Equation (8) may be used without hesitation down to 430°, and according to it, K_p changes from 198 at 430° to 0·36 at 900°. Temperature, therefore, has the very greatest influence. On the other hand, the dilution of the gas in general affects the yield but slightly, because the partial pressure of the oxygen in the gaseous products, which is the only factor which comes into play, exerts an influence which is not proportional to itself, but to its square root.[1]

It is desirable, for practical purposes, to use a formula in which the oxygen and sulphur dioxide contents in the gaseous products appear instead of the partial pressure p_{O_2} of the oxygen. It is easy to formulate such expressions. There is—

$$\begin{array}{l} a = \%\ SO_2 \text{ in initial mixture} \\ b = \%\ O_2 \quad \text{,,} \quad \text{,,} \\ c = \%\ N_2 \quad \text{,,} \quad \text{,,} \\ \hline a + b + c = 100 \end{array}$$

Further, x, as before, shall represent the yield, that is, the ratio of sulphur trioxide formed to that which could have been formed. It is then clear that $0{\cdot}5ax$ of the b parts by volume of oxygen would be used up, and that the volume would decrease from 100 to $100 - 0{\cdot}5ax$. If the total pressure is kept at one atmosphere, the partial pressure of the oxygen in the gaseous product is—

$$p_{O_2} = \frac{b - 0{\cdot}5 \times a \times x}{100 - 0{\cdot}5a \times x}$$

and we obtain—

$$x = \frac{K_p}{K_p + \dfrac{1}{\sqrt{\dfrac{b - 0{\cdot}5ax}{100 - 0{\cdot}5ax}}}}$$

[1] Knietsch, "Bericht über den V. Intern. Kongress," *l.c.*, has found that dilution has an effect in accordance with the theory, at least, when nitrogen or carbon dioxide are used as diluents. On the other hand, the addition of large quantities of water-vapour seems to hamper the reaction.

This expression is not convenient to use, because it is an equation of the third degree with respect to x. It is easy, however, to draw a series of conclusions from it, provided one assumes that in the most favourable case the yield could be 1, that is, the reaction could take place quantitatively. On the other hand, the oxygen content in the initial gas mixture cannot rise higher than 100 per cent. The highest possible value of the radical is therefore 1. Practically all attainable values lie below 1. The smaller the value of—

$$\sqrt{\frac{b - 0{\cdot}5ax}{100 - 0{\cdot}5 \times ax}}$$

the greater will be the value of the denominator of the quotient representing the yield. The greater the value of the radical becomes, the nearer will the denominator approach the most favourable value $K_p + 1$. The greater K_p is, so much less important is it whether the radical equals to 1 or to $\frac{1}{2}$ or $\frac{1}{4}$. The greater b becomes, that is to say, the greater the percentage content of oxygen becomes in the initial gas mixture, the nearer will the radical approach the most favourable value (1) possible. If we have but a trace of sulphur dioxide mixed with pure oxygen in the initial gas mixture, then the radical will not differ appreciably from 1, and x will simply equal $\frac{K_p}{K_p + 1}$. If the temperature lies between 450° and 500°, where the equilibrium constant is about 100, the yield of sulphur trioxide from this trace of sulphur dioxide is $\frac{100}{101}$, or practically quantitative. If the temperature lies some 200° higher, where the equilibrium constant equals about 3, the yield sinks to $x = \frac{3}{4}$, or but 75 per cent., even in this ideal limiting case. If we imagine a trace of sulphur dioxide to be mixed with ordinary air instead of with pure oxygen—

$$\sqrt{\frac{b - 0{\cdot}5ax}{100 - 0{\cdot}5ax}} = \sqrt{\frac{b}{100}} = \sqrt{0{\cdot}209} = 0{\cdot}457$$

and the yield will now be—

$$x = \frac{K_p}{K_p + 2{\cdot}2}$$

Between 450° and 500° the yield will still be $\frac{100}{102{\cdot}2}$, or almost quantitative. But at a temperature 200° degrees higher the trace of sulphur dioxide will not be more than $\frac{30}{52}$, or about 58 per cent. converted. We see, then, that the great difference in nitrogen makes but little difference at lower temperatures, but becomes very important at higher ones.

In order, finally, to make an actual application to technical conditions, we have computed the following table :— **The maximum yield in practice.**

Initial gas mixture.			Temperature °C.	K_p from formula.	Maximum yield in %.
a % SO_2	*b* % O_2	*c* % N_2			
7·0	10·4	82·6	434	181·0	99
7·0	10·4	82·6	550	20·4	85
7·0	10·4	82·6	645	5·14	60

The actual experimental yields may easily exceed the values calculated for the higher temperatures, provided the gas has an opportunity to strike any of the catalyzer as it cools off.

The influence exerted by the composition of the gaseous mixture has been often discussed, because Clemens Winkler, to whom we owe much of our knowledge of this process, came to the erroneous conclusion that the most favourable composition for the initial gas was 2 vols. SO_2 to 1 vol. O_2. He reports regarding the first stages of his procedure in Lunge's "Sodaindustrie." It appears from his description that his first ill success was by no means due to the unfavourable composition of his gaseous mixture. It appears that from 1879 on, in the Muldener smelting works, sulphur dioxide gas mixtures containing 7·0 to 7·5 per cent. SO_2 were passed over platinized asbestos, according to Clemens Winkler's advice, and a 45 per cent. yield of SO_3 obtained, the sulphur dioxide content falling to 4 per cent. "It was discovered in 1889 that the yield of anhydride rose to 85 or 90 per cent. on a single passage over the contact substance, if gas mixtures were used whose oxygen content considerably exceeded that of the sulphur dioxide. Consequently, after that, gas mixtures containing but 6 per cent. of SO_2 were employed." If we assume that there was 7·5 per cent. SO_2, and 3·75 per cent. O_2 in the first case, even then a yield of 85 per cent. ought to **Clemens Winkler's contribution to the sulphur dioxide process.**

have been obtained if the temperature of the contact substance could have been successfully reduced to 470°.[1]

We only need to substitute 7·5 for *a*, 3·75 for *b*, and 0·45 for *x* in our formula to convince ourselves that the constant K_p equals 5·7 in round numbers, and that consequently the temperature used in the Muldener smelters must have been nearly 640°. Conversely, the constant would have needed to be 82·4, and the temperature about 470° C. (perhaps a little lower) if a yield of 85 per cent. were to have been obtained with a gaseous mixture of the same composition. But it is exceeding improbable that the sulphur dioxide gases could have been so poor in oxygen; and even if the later increase of the percentage of air which reduced the SO_2 content to 6 per cent. did double the yield, the mass action of the oxygen certainly had but little to do with this. It is much more probable that the real reason lay in the cooling of the contact substance by the dilution of the gas mixture, and that consequently the equilibrium constant acquired a more favourable value. This view is confirmed by another statement of Clemens Winkler. He says that if the gas mixture which entered the contact chamber containing 7·0 to 7·5 per cent. SO_2, and left it containing but 4 per cent. SO_2, were freed from SO_3 and again passed over the catalyzer, combination went so far that now but 0·2 per cent. SO_2 remained in the escaping gases. In this second operation, then, the yield was $\frac{3·8}{4·0}$ or 95 per cent. How impossible it is to explain this by mass action, and how natural the result is when we remember that in the second operation the diminished heat of reaction of the much poorer gases could not heat the contact substance nearly so hot!

This illustrates, perhaps, how the thermal effects in gas reaction may very often completely mask the influence of mass. Mass action, because of its simplicity, attracts interest and attention, but the thermal effects based on thermodynamic considerations which are less generally known, or rather less appreciated, are neglected.

Other catalysts.

The great advantage of platinum as a contact substance in the preparation of sulphur trioxide lies in the fact that even at

[1] There is no ground for assuming insufficient contact substance, or a too rapid passage of the gas.

a temperature below 500°, it catalyzes the reaction sufficiently to allow us to utilize the very favourable location of the equilibrium at this temperature and yet employ rapid currents of gas. All other catalysts work less satisfactorily. Nothing is easier than to find catalyzers for the formation of SO_3. As is generally the case in gas reactions at high temperatures, almost any indifferent rough surface hastens the reaction. Knietsch has illustrated this effect very prettily by replacing the platinized asbestos with fragments of porcelain. The velocity of formation of sulphur trioxide is much smaller at relatively low temperatures, and the equilibrium is far from reached. But the higher the temperature, the more nearly does the action of the broken porcelain approximate to that of the platinum, and at 850° there is but little difference between the two. Yet at this temperature the location of the equilibrium is so unfavourable ($K_p < 1$) that this catalyzing effect is without technical importance. According to a well-known principle of physical chemistry, first stated by Ostwald, every catalyst accelerates the reaction and the counter-reaction in the same degree, provided it remains itself unaltered. In fact, Knietsch has found that the rate of decomposition of sulphur trioxide by broken porcelain is quite analogous to the rate of formation of sulphur trioxide under the same conditions. On the other hand, when sulphur trioxide was passed at 900° through smooth porcelain tubes under conditions which were otherwise the same, the decomposition was very slight, due to the absence of rough surfaces. We may conclude from this that only small quantities of sulphur trioxide, compared with the theoretical amounts, would be formed in smooth tubes at 900°.

But easy as it is to find substances which will catalyze, it is very difficult to find substances which will catalyze well. Among those which people have hoped would replace platinum, ferric oxide, or more particularly the ashes from the pyrite ovens in the manufacture of sulphuric acid, has acquired the greatest importance. To be sure, its catalytic action compared with that of platinum is shown by Knietsch's curves to be small, and experiments by Lunge and Pollitt[1] have confirmed this. The relation of the equilibrium point to the decomposition potential of ferrous and ferric sulphate which here comes into play, has

[1] *Z. f. angew. Chemie*, 15 (1902), 1105.

been studied by J. d'Ans (*l.c.*) at Keppeler's suggestion. Lunge, in his handbook, states that the relatively high yield of from 60 to 66 per cent. was obtained with these ashes. Other statements indicate a smaller yield. It is still an open question whether equilibrium is reached using this substance as a catalyst.

Case V.—The formation of ammonia. It remains for us to consider the formation of ammonia from the elements. The subject has been carefully investigated by Haber and Le Rossignol[1] after a preliminary treatment by Haber and van Oordt (see p. 259). They determined the quantity—

$$\frac{C_{NH_3}}{C_{N_2}^{\frac{1}{2}} \times C_{H_2}^{\frac{3}{2}}} = K_p$$

using iron, manganese, nickel, chromium as catalyzers.

This reaction is hampered by the peculiar difficulty that even with manganese and iron, so far its best known catalyzers, a quick adjustment of the equilibrium is not reached below a temperature of 750°, where the ammonia is almost completely split into its elements, at least under ordinary pressure.

In the formation of sulphur trioxide it was all important to keep the temperature low where the equilibrium constant has a high value. Here the problem is to find a catalyzer which

[1] *Berl. Ber.*, 40 (1907), 2144. The experiments of Haber and van Oordt being insufficient for a very accurate deduction of the equilibrium, a new investigation was carried out, especially because Nernst privately informed the author that experiments with high pressures gave him comparatively lower figures.

Baur (*Z. anorg. Chemie*, 29 (1902), p. 305, and *Ber. d. d. Chem. Ges.*, 34 (1901), 2395) has made electro-chemical measurements on the electrolysis of ammonia. He concludes from his observations that ammonia is formed at 25° with a reaction energy corresponding to 0·627 volt. Expressed in calorimetric units, this gives 37·470 cal. per mol for the free energy. It is impossible to harmonize this with our measurements, for according to Bauer, ammonia would then have a partial pressure of 0·39 atmosphere when the nitrogen and hydrogen were at atmospheric pressure, or, more precisely, atmospheric pressure minus the partial pressure of the ammonia. I am of the opinion that Bauer's observations do not prove that ammonia is being formed or decomposed reversibly, and consequently do not consider that his measurements can tell us anything certain about the energy of the formation of ammonia. Berthelot's statements regarding the ammonia equilibrium ("Méchanique chimique," ii. (1897), 375) are either wholly meaningless or inapplicable.

will catalyze at these low temperatures, so that they may be used. Iron, the best catalyzer at present known for this reaction, is far inferior even to so poor a catalyzer of the sulphur trioxide reaction as broken porcelain. But we can predict, at least approximately, what results could be attained with a good catalyzer.

We will first write our formula for the reaction energy—

$$A = Q_0 - \sigma'_p T ln T - \sigma'' T^2 - RT ln \frac{p_{NH_3}}{p_{N_2}^{\frac{1}{2}} \times p_{H_2}^{\frac{3}{2}}} + \text{const. } T$$

and observe that 12,000 cal. are evolved in the formation of 1 mol of ammonia at ordinary temperatures and at constant pressures. We further note that, according to Wüllner, the true specific heat of ammonia per mol at constant pressure and at 0° is 8·54; at 100°, 9·07; and at 200°, 9·59; while the mean specific heats of the permanent gases at constant pressure between 0° and T° may be taken with sufficient accuracy as equal to 6·64 + 0·0003T. Putting the mean specific heat of one mol of ammonia at constant pressure between 0° abs. and T° abs.—

$$6{\cdot}1 + 0{\cdot}002T + 2{\cdot}64 \times 10^{-6}T^2$$

the true specific heat will be—

$$6{\cdot}1 + 0{\cdot}004T + 7{\cdot}82 \times 10^{-6}T^2$$

this being at 200° equal to 9·76, in fairly good agreement with Wiedemann's figure. On the other hand, combining this value with the above-given expression for the specific heat of permanent gases, justified by the measurements of Holborn and Henning, we get for the difference in the specific heats of factors and products—

$$c_{p(\text{perm. gases})} = 13{\cdot}28 + 0{\cdot}0006T$$
$$c_{p(NH_3)} = 6{\cdot}1 + 0{\cdot}0002T + 2{\cdot}64 \times 10^{-6}T^2$$

From this we find—

$$Q_0 = 10{,}100 \text{ cal.}$$

and

$$A = 10{,}100 - 7{\cdot}18 T ln T + 1{\cdot}4 \times 10^{-3}T + 1{\cdot}32 \times 10^{-6}T^2 + 21{\cdot}98T - RT ln \frac{p_{NH_3}}{p_{N_2}^{\frac{1}{2}} \times p_{H_2}^{\frac{3}{2}}}$$

The value 21·98 corresponds to the experimental data, as will be seen by computing the equilibrium constants from this formula and comparing them with the experimental results, which were as follows:—

Degrees C.	1000	930	850	800	750	700
10^4K_p found	1·48	2·00	2·79	3·33	4·68	ca. 6·8
10^4K_p calc.	1·45	1·90	2·77	3·60	4·86	6·8

To simplify the calculation the following form of the above equation may be used:—

$$\log_{10} K_p = \frac{2215}{T} - 3{\cdot}626 \log_{10} T + 3{\cdot}07 \times 10^{-4}T + 0{\cdot}29 \times 10^{-6}T^2 + 4{\cdot}82$$

The possible yield.

If we calculate, with the aid of this formula, the composition of mixtures of nitrogen, hydrogen, and ammonia which are in equilibrium at atmospheric pressure, and assume that the nitrogen and hydrogen are present in equilibrium quantities ($N_2:H_2 = 1:3$), we find—

Temperature in °C.	Vol. % H_2	Vol. % N_2	Vol. % NH_3
27	2·76	0·92	96·32
327	73·75	24·58	1·67
627	74·97	24·99	0·039
930	75·0	25·0	0·0065
1000	75·0	25·0	0·0048

It is evident from this table that ammonia could be obtained in good yield at low temperature just as sulphur trioxide could, provided we could only find a catalyst which would act as efficiently on the nitrogen mixture as platinum does on the mixture of sulphur dioxide and oxygen.

The attainable yield of ammonia from a given amount of nitrogen may here be represented with the help of the partial pressures in the gaseous products by—

$$x = \frac{p_{NH_3}}{0{\cdot}5p_{N_2} + p_{NH_3}} \text{ or } \frac{x}{1-x} = \frac{p_{NH_3}}{0{\cdot}5p_{N_2}}$$

Similarly, the yield from a given amount of hydrogen would be—

$$y = \frac{p_{NH_3}}{1{\cdot}5p_{H_2} + p_{NH_3}} \text{ or } \frac{y}{1-y} = \frac{p_{NH_3}}{1{\cdot}5p_{H_2}}$$

The equilibrium condition may be written as—

$$K_p = \frac{p_{NH_3}}{p^{\frac{1}{2}}_{N_2} \times p^{\frac{3}{2}}_{H_2}} = \frac{p_{NH_3}}{0{\cdot}5p_{N_2}} \times \frac{0{\cdot}5p^{\frac{1}{2}}_{N_2}}{p^{\frac{3}{2}}_{H_2}} \text{ or } \frac{p_{NH_3}}{1{\cdot}5p_{H_2}} \times \frac{1{\cdot}5}{p^{\frac{1}{2}}_{N_2} \times p^{\frac{1}{2}}_{H_2}}$$

Substituting the expressions for the yield in this, we get—

$$K_p = \frac{x}{1-x} \times \sqrt{\frac{0{\cdot}707p_{N_2}}{p_{H_2}^3}} = \frac{1{\cdot}5y}{1-y} \times \frac{1}{\sqrt{p_{N_2} \times p_{H_2}}}$$

These expressions are somewhat complicated, but this much is directly evident from them. When we wish to convert nitrogen at all completely into ammonia, we must use a gaseous mixture containing much hydrogen and but little nitrogen. The utilization of the hydrogen is then, of course, about as incomplete as possible. But if we wish to make ammonia in this way we could do as is done at the Muldener Smelters: we could remove the ammonia formed by absorption, and repeat the process after having added a little nitrogen. In this way no hydrogen would be lost.

The reverse attempt to obtain the highest possible yield of ammonia from a given amount of hydrogen never has so good chances of success. If we try to accomplish this by using a mixture containing much nitrogen and but a little hydrogen, then very little ammonia is formed, as the experiments of Haber and van Oordt have shown. Noting that when the content of ammonia is small, the sum of the partial pressures p_{N_2} and p_{H_2} is practically identical with the total pressure, we may write as an approximation—

$$p^2_{NH_3} = K^2_p \times p_{N_2} \times p^3_{H_2} = K^2_p p^3_{H_2}(1 - p_{H_2}) = K^2_p(p^3_{H_2} - p^4_{H_2})$$

or—

$$p_{NH_3} = K_p\sqrt{p^3_{H_2} - p^4_{H_2}}$$

If, now, the partial pressure of the hydrogen has the value 0·99 atmosphere in one case and 0·01 atmosphere in another; that is, if in one case we use almost pure hydrogen and in another almost pure nitrogen, the ratio of the quantities of ammonia which we obtain in the two cases is—

$$\frac{p_{NH_3'}}{p_{NH_3''}} = \sqrt{\frac{0{\cdot}99^3 - 0{\cdot}99^4}{0{\cdot}01^3 - 0{\cdot}01^4}} = \backsim 100$$

The use of nearly pure nitrogen then depresses the ammonia content of the gaseous product till it is about 100 times smaller than when we started with almost pure hydrogen. It consequently escapes observation.

On the other hand, if we wish to choose conditions such that the yield of ammonia from the hydrogen will be a maximum, we make use of the easy transformation—

$$y = \frac{0{\cdot}67K_p\sqrt{p_{N_2}^{\frac{1}{2}} \times p_{H_2}^{\frac{1}{2}}}}{1 + 0{\cdot}67\sqrt{p_{H_2}^{\frac{1}{2}} \times p_{N_2}^{\frac{1}{2}}}}$$

where the partial pressures as usual refer to the escaping gas. We see that this expansion gives a maximum value for y when the product $p_{N_2}^{\frac{1}{2}} \times p_{H_2}^{\frac{1}{2}}$ has a maximum value. This happens when the escaping gas contains as much nitrogen as hydrogen.

Recurring for the last time to our equation for the formation energy of ammonia, we observe that the value of the last term in this equation, 21·98T, is surprisingly large. It indicates that our assumptions regarding the specific heats are incomplete, and we surmise that a more exact knowledge of the specific heat of ammonia would so alter the expression that the constant would come out smaller. The numbers relating to the location of the equilibrium would thereby be somewhat changed, but the general ideas regarding the formation of ammonia which we have derived from our equation would suffer no alteration.

The peculiar inertness characteristic of nitrogen, which appears so prominently in the preparation of ammonia from the elements, reminds us of the condition which we described in connection with the formation of nitric oxide. There we found that nitric oxide decomposed but slowly into the elements at an intense white heat. This peculiar inertness of nitrogen will always place a serious handicap on the cheap technical preparation of ammonia, particularly so long as the tremendous amount of combined organic nitrogen with which Nature has provided us lasts. Nature brings about the transformation of elementary nitrogen into ammonia and nitric acid in her slow way with the greatest facility and on a most colossal scale. While she has accumulated nitrates at but one place on the earth's surface in sufficient quantities to mine—and its exhaustion is imminent—she has provided for us a much richer and more varied supply of combined nitrogen easily convertible into ammonia.

SIXTH LECTURE

THE DETERMINATION OF THE SPECIFIC HEATS OF GASES

THERE are both direct and indirect methods for determining the specific heats of gases. Of the direct methods only those which give the specific heat at constant pressure yield results of any accuracy. Of the indirect methods there are two which are particularly important, and results obtained by them have found frequent mention in the preceding lectures.

One of these indirect methods consists in measuring the pressure produced by explosions. Knowing the heat evolved by the explosive reaction, we can tell from this pressure what the mean specific heat is between the room temperature and the temperature attained during the explosion at constant volume. The other indirect method depends on the determination of the ratio between the true specific heats at constant volume and at constant pressure.

First determination of the specific heat of a gas.

Crawford[1] first investigated the specific heat of gases. He sought to measure the difference between the heat given off from a metallic vessel when evacuated and when full of gas, when the vessel was heated to a definite temperature and then plunged into a water-calorimeter. But in this case the heat capacity of the gas is so small compared with that of the containing vessel, that it is impossible to obtain results of any value by this method, unless we have a calorimeter of extreme delicacy. It was only much later that Joly[2] and Bunsen[3] constructed such an instrument. Indeed, with Crawford's apparatus, it was impossible to obtain even approximately correct values. Thus he found the specific heat of air eight times greater than it really is.

[1] Gehler's "Physikal. Wörterbuch," 2nd edit., vol. 10, section 1.

[2] Joly, *Proc. Royal Soc.*, 41 (1886), 352.

[3] Bunsen, *Wied. Ann.*, 31 (1887), 1.

The use of greater quantities of gas would, of course, increase the accuracy of the determination, but at the same time it necessitates a change in the method of procedure. A closed vessel large enough to contain a sufficient quantity of gas would require a calorimeter of impossible dimensions. Consequently we abandon closed vessels, and resort to the device of passing a hot gas through a colder calorimeter, and measuring the heat it there gives off. In this way we are no longer determining the specific heat at constant volume, but at constant pressure. Lavoisier and Laplace,[1] who first employed this method, using an ice-calorimeter in conjunction with it, immediately got better values than those of Crawford. Yet even their results are quite divergent from the true ones.

Experiments of Lavoisier and Laplace.

It is easy to obtain large and easily measurable heat changes by passing large quantities of gas through a metal spiral enclosed in such a calorimeter, but, unfortunately, several sources of uncertainty are thereby introduced. The gas must, of course, be brought in a tube from the heating chamber to the calorimeter. It is not difficult to determine the temperature in the heating chambers, but it is much more difficult to ascertain the temperature change between the exit from the heating chamber and the entrance into the calorimeter. It does not suffice to place a thermometer where the gas enters the calorimeter coil, for if the temperature of the gas does not coincide with that of the surrounding metal cell at this point, then the reading of the thermometer will be determined as much by the radiation from the walls of the tube as by the temperature of the gas itself. Its readings are, therefore, incorrect. By bringing the heating chamber right up to the entrance of the calorimeter, we can, to be sure, cause the gas to enter the calorimeter at the temperature of this heating chamber. But in this case the heating chamber heats the calorimeter at the junction, and may easily affect the calorimetric results. Regnault was the first to overcome these difficulties.

Gay Lussac's experiments.

Before perfecting this difficult method experimenters naturally looked about for a simpler one. Gay Lussac devised a method by which he could measure specific heats, at

[1] Regarding these and other older experiments, see Regnault's historical treatment of the subject in the *Memoires de l'Institute de France*, 26 (1862), 1 to 40.

least relative to air, by a purely thermometric comparison, without using the heat changes of a calorimeter. We have already become acquainted in the transpiration experiment with the first method which he tried. A holder full of air and enclosing a thermometer was suddenly put in connection with a similar evacuated vessel containing a second thermometer. When the gas rushed over, the temperature in the first vessel sank by the same amount that it rose in the second. The same thing happened when other gases were used instead of air. The magnitude of the two opposed temperature changes was different with different gases, and varied with the density. Gay Lussac at first tried to draw conclusions regarding the specific heat of the gases from these measurements, but soon convinced himself that the variations in behaviour of the gases used depended on the rapidity with which their temperature changes were transmitted to the thermometer. This, however, is a function of the thermal conductivity of the gas. He, therefore, changed his procedure to one which agrees in principle with the ordinary method of mixing. According to this latter method, we find the specific heat of an insoluble solid by simply heating a known weight of it to t°, and letting it fall into water at t'°. A determination of the final temperature suffices for a calculation of the specific heat of the solid in terms of that of water. Gay Lussac caused a current of air of known temperature and velocity to mix with a current of the gas under investigation at another known temperature and of known velocity. Knowing the final temperature, it must also be possible in this case to calculate the specific heat of the gas referred to air. As Regnault asserts, the method is theoretically free from criticism, yet involved difficulties in actual execution which Gay Lussac could not overcome.

Apjohn and Suermann's experiments.

Apjohn[1] and Suermann[2] again undertook to measure the specific heats of gases in another way without using a calorimeter. Their results were equally unsatisfactory. Their idea was to abstract a known quantity of heat by some device from a gas, and to measure the consequent fall in temperature. The devise they hit upon was to allow water to evaporate into a current of the dry gas. Knowing the weight of water

[1] *Philos. Mag.*, 13 (1838), 261 and 339.

[2] *Pogg. Ann.*, 41 (1837), 474.

evaporated, it was easy to calculate the heat absorbed, and from the measured temperature fall, the specific heat of the gas.

Experiment of Delaroche and Bérard.

Delaroche and Bérard,[1] returning to the use of the calorimeter, obtained the first really successful results. They caused the gas under investigation to escape from a bladder into a tube surrounded by a vapour jacket. The tube was connected to a water-calorimeter containing a metal spiral, through which the gas passed and was cooled. The cooled gas was collected in a second bladder. Simply turning a couple of cocks caused the gas to escape from the second reservoir and to pass through the heating coil back into the first reservoir. This could be repeated as often as desired, and so a small quantity of gas would suffice to communicate a large and easily measurable quantity of heat to the calorimeter. On the one hand, the specific heat of a large number of gases was measured relative to air, and on the other, the absolute specific heat of air was measured, several procedures being employed to obviate the uncertainties caused by the radiations of heat from the calorimeter to the surroundings, and by the conduction of heat to the calorimeter from the heating chamber along the connecting tube. Their results were much nearer the truth than any previous ones.[2]

Experiments of Delarive and Marcet.

Delarive and Marcet[3] later undertook a reversal of Delaroche's and Berard's method, yet with but slight success. They let a current of gas blow into a calorimeter which was specially protected against any heat exchange with the surroundings. The temperature of the calorimeter was higher than that of the entering gas, and the fall in temperature caused by the passage of the gas theoretically permitted a calculation of the specific heat of the gas relative to the air.

Regnault's experiments.

Regnault was the first to get really good results. He elaborated the arrangement of Delaroche and Bérard with the greatest attention to detail. His experimental arrangements are shown in Fig. 5.

The gas was taken from a copper vessel of 35 litres capacity,

[1] "Annales de Chimie par Guyton de Morveau," etc., 85 (1813), 72.

[2] Experiments by Haycraft with an apparatus even more complete in principle yielded no serviceable results (Gilbert's *Ann.*, 76 (1824), 289).

[3] *Ann. Chim. Phys.*, 75 (1840), 113.

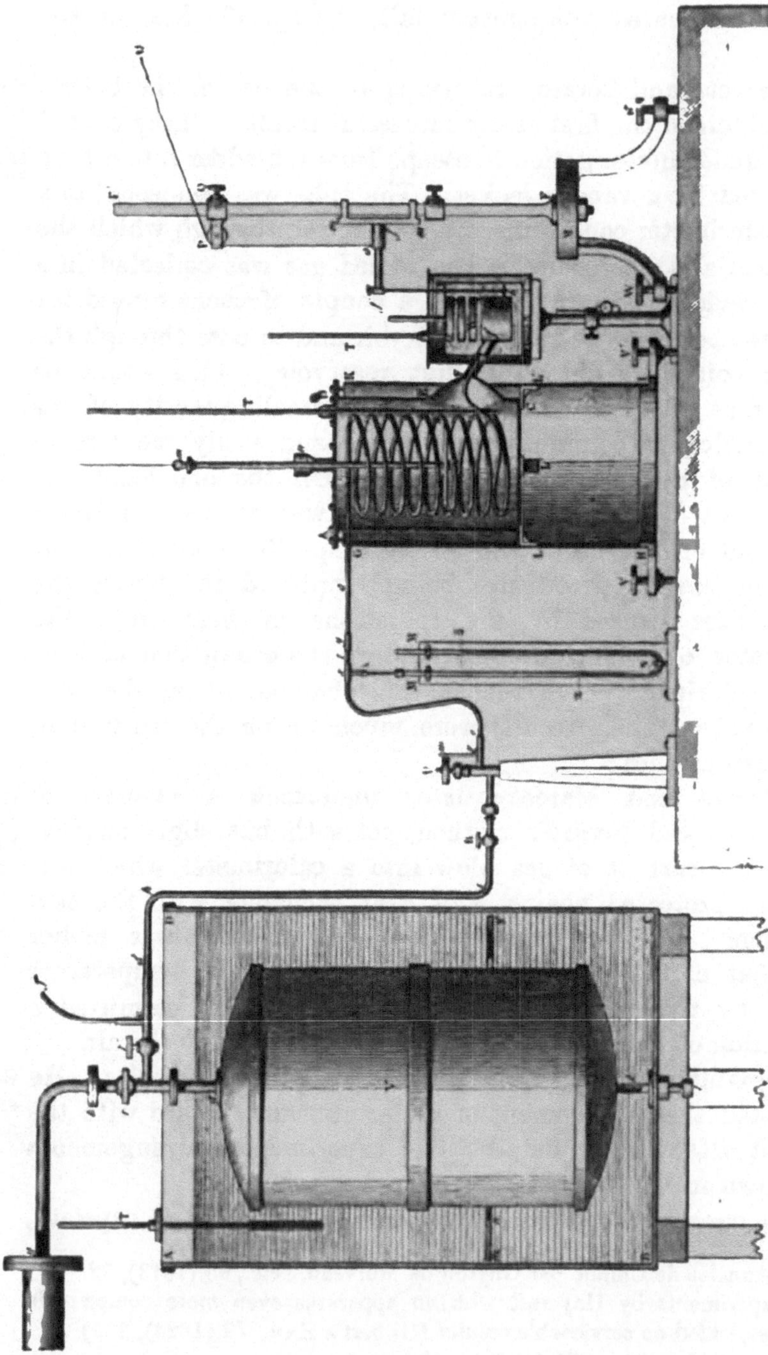

FIG. 5.

into which it had been previously forced through the tube *gba* by means of a force pump. The copper vessel was contained in a water-bath, ABCD, whose temperature was measured by the thermometer T, while the stirrer *mnq* kept the water in motion. The pressure in the copper vessel was determined by the attached manometer *ef*. The bottom tap *k*, which was closed hermetically below, allowed him to bring the copper vessel in communication with the atmosphere.

In an experiment, the gas shut up in V, where its temperature and pressure were known from *f* and T, escaped through *dhi*R and the fine reducing valve U into the circuit *oxlya*, leading to the heating coil. A capillary glass tube was cemented in at *t*. The screw of the reducing valve U was graduated, and the straight edge *vw* made it possible to return to any initial point. During the experiment the valve was regulated by hand, so that the manometer MM′ showed always the same difference of pressure $\alpha\beta$. The gas, therefore, entered the capillary *t* always at the same pressure, and consequently passed through it and the remaining circuit at the same speed, provided the atmospheric pressure did not change in the mean time.

The current of gas then passed through a metallic spiral, 8 mm. in bore and 10 m. long. This coil lay in a box, ABCD, which was heated by a burner from below. This box, in turn, rested in a larger lead box GLL″K, which protected it from the outer air, and increased the constancy of the temperature. The box ABCD was filled with oil, which was constantly stirred. The temperature was measured by the thermometer T.

Regnault's most unique device was his method of attaching his heating apparatus to the calorimeter. The bath ABCD was made to belly out at one place, and the heating coil extended 1 cm. out through and beyond this protrusion. The end of the coil is fitted by means of a cork into a thin-walled glass tube ending in the calorimeter vessel *f*. The calorimeter was a brass pot, in which were placed three brass boxes *v*. There was a metallic spiral in each box prolonging the path which the gas must follow through the box. The calorimeter rested upon cork wedges in a protecting mantle. The temperature of the water in the calorimeter was measured with a thermometer graduated to $\frac{1}{20}°$, and permitting $\frac{1}{200}°$ to be estimated. A stirrer moving up and down kept the water well mixed.

Regnault's accessory determination.

When the gas had passed through the last box it escaped freely into the atmosphere. The cross-section of the gas-pipe in the calorimeter was such that there was no appreciable change of pressure nor any expansion of the gas which might do work against the atmosphere, and thereby involve a loss of heat.[1]

In order to obtain serviceable values by this method of experimentation, it was necessary to carry out certain accessory measurements with great accuracy. In the first place, the quantity of gas which the copper reservoir gave off when its pressure fell from the initial value p_a to the final value p_e during an experiment, must be measured with the greatest care. Then, too, it was necessary to find the most suitable velocity with which the gas should be passed through the system. If the velocity was very small, the gas had time, after its exit from the heating coil and before its entrance into the calorimeter, to give off a perceptible quantity of heat. When its velocity was very great, the time spent in the calorimeter was not sufficient for the gas to cool to the temperature of the calorimeter. Finally, it was necessary to carefully ascertain before and after each experiment what heat changes were occasioned in the calorimeter by outside disturbances.

Regnault's method of experiment with corrosive gases and with vapours.

Regnault constructed coils and calorimeters of platinum for use with chlorine and other gases which acted on brass. In this case the gas was led out of the calorimeter into some absorption apparatus, and its mass determined from the gain in weight. For determining the specific heat of easily condensible vapours, Regnault adopted a procedure which we have already described in the Fourth Lecture. The results Regnault obtained are collected in the following table :—

[1] Leduc, *Comp. Rend.*, 106 (1896), 1860, calls attention to the fact that Regnault's statements in this regard are not necessarily conclusive, and by taking account of a slight deviation which Regnault called equal to zero, the specific heat of air comes out $\frac{1}{150}$ smaller than Regnault calculated.

REGNAULT'S SPECIFIC HEATS AT CONSTANT PRESSURE.[1]

Gas.	Formula.	Specific heat per gram.	Molecular weight.	Specific heat per mol.	Temperature interval ° C.[2] Z.T. = room temperature.	No. of atoms in molecule.
Air	—	[0·2375][3]	—	—	−30 to +200	2
Nitrogen	N_2	0·2438[4]	28·08	6·846	Z.T. to 200	
Oxygen	O_2	0·2175	32·00	6·960	Z.T. to 200	
Hydrogen	H_2	3·4090	2·016	6·873	Z.T. to 200	
Carbon monoxide ...	CO	0·2450	28·00	6·860	Z.T. to 190	
Nitric oxide	NO	0·2317	30·04	6·960	Z.T. to 170	
Chlorine	Cl_2	0·1210	70·9	8·579	Z.T. to 200	
Hydrochloric acid ...	HCl	0·1852	36·458	6·752	Z.T. to 200	
Bromine	Br_2	0·0555	159·92	8·876	83 to 228	
Carbon dioxide	CO_2	0·1843	44·00	8·109	−30 to +10	3
		0·2025		8·910	10 to 100	
		0·2169		9·544	100 to 200	
Nitrous oxide	N_2O	0·2262	44·08	9·970	Z.T. to 180	
Sulphur dioxide	SO_2	0·1544	64·06	9·891	Z.T. to 200	
Hydrogen sulphide ...	SH_2	0·2432	34·086	8·290	Z.T. to 210	
Water	H_2O	0·4805	18·016	8·657	125 to 215	
Carbon bisulphide ...	CS_2	0·157	76·12	11·95	70 to 194	
Ammonia	NH_3	0·5084	17·064	8·675	Z.T. to 218	4
Phosphorus trichloride	PCl_3	0·1347	137·35	18·50	111 to 246	
Arsenic trichloride ...	$AsCl_3$	0·1122	181·35	20·35	159 to 268	
Methane[5]	CH_4	0·5930	16·032	9·106	Z.T. to 208	5
Chloroform	$CHCl_3$	0·1567	119·36	18·70	117 to 228	
Silicon tetrachloride ...	$SiCl_4$	0·1322	141·80	18·75	90 to 234	
Stannic chloride ...	$SnCl_4$	0·0939	260·80	24·49	149 to 273	
Titanium tetrachloride	$TiCl_4$	0·129	189·90	24·50	162 to 272	
Methyl alcohol	CH_3OH	0·458	32·08	14·67	101 to 223	above 5
Ethylene[6]	C_2H_4	0·404	28·03	11·32	Z.T. to 200	
Ethyl chloride[7]	C_2H_5Cl	0·273	64·49	17·61	Z.T. to 203	
Ethyl bromide	C_2H_5Br	0·190	109·0	20·71	78 to 196	
Ethyl mercaptane[8] ...	C_2H_5SH	0·401	62·11	24·91	120 to 223	
Ethyl cyanide[9]	C_2H_5CN	0·426	55·08	23·46	114 to 221	
Ethylene choride ...	$C_2H_4Cl_2$	0·229	98·93	22·65	111 to 221	
Ethyl alcohol	C_2H_5OH	0·453	46·05	20·86	114 to 222	
Acetone	CH_3COCH_3	0·4125	58·05	23·95	129 to 233	
Benzene	C_6H_6	0·375	78·05	29·27	116 to 218	
Ethyl ether	$C_4H_{10}O$	0·480	74·08	35·56	65 to 230	
Ethyl acetate[10]	$C_4H_8O_2$	0·401	88·06	35·31	115 to 219	

[1] Regnault usually gave several more decimal figures, although the fourth, and with many substances the second and third, decimal figures were uncertain.

[2] In round numbers.

[3] Not per gram, but per litre at 0° and 76 cm. pressure.

[4] Not directly determined, but computed from values obtained for air and oxygen.

[5] Prepared from sodium acetate and lime. Analysis showed that it was not quite pure.

Wiedemann's experiments.

Eilhard Wiedemann[1] was able to considerably simplify Regnault's method by replacing his heating coil and calorimeter vessel by others of smaller dimensions and more manageable construction. This heating apparatus consisted of a copper box, M (Fig. 6), 20 cm. high, 18 cm. broad, and 21 cm. long, having the protrusion P devised by Regnault on its right side. Instead of the heating coil, it contained a copper cylinder, G, 4 cm. in diameter and 11 cm. long, packed full of fine copper turnings. The gas entered through the tube *mn*, and escaped through the tube *o*. As far as *s* the tube was of copper, but the small piece (17 mm. long) extending beyond the tip of the protruding part of the calorimeter was of German silver, which is a relatively poorer conductor of heat. As in Regnault's apparatus, the calorimeter was connected with the heating apparatus by means of a cork. The liquid in the heating bath was kept agitated by the stirrer V, and its temperature was read on the thermometer *t*. The temperature was constant during an experiment to within 1°. The calorimeter was a very small silver vessel 5½ cm. high and 4·2 cm. in diameter. Inside of it were arranged three small vertical silver tubes 41 mm. long and 9 mm. in diameter, filled with silver turnings. Passage through a single one of these tubes was sufficient to cool the hot gas down to the temperature of the calorimeter.

The advantage of Wiedemann's modification.

The use of metal turnings in the heating chamber and in the calorimeter was the most important improvement devised by Wiedemann. The extraordinarily increased rapidity of heat exchange between gas and wall thus obtained made it possible to construct the whole apparatus on a smaller and more compact scale than was possible for Regnault. A second fortunate alteration in the construction consisted of an arrangement by which the temperature of the gas could be measured

[1] *Pogg. Ann.*, 157 (1876), p. 1.

---

[6] Prepared from alcohol and sulphuric acid; washed with sulphuric acid and potassium hydroxide solution; doubtless impure.

[7] Prepared from concentrated hydrochloric acid and absolute alcohol, washed with water, liquefied, dried over calcium chloride, redistilled.

[8] From potassium sulphide and calcium ethylsulphate; washed with water and distilled from calcium chloride.

[9] From potassium cyanide and calcium ethylsulphate.

[10] From sodium acetate and calcium ethylsulphate.

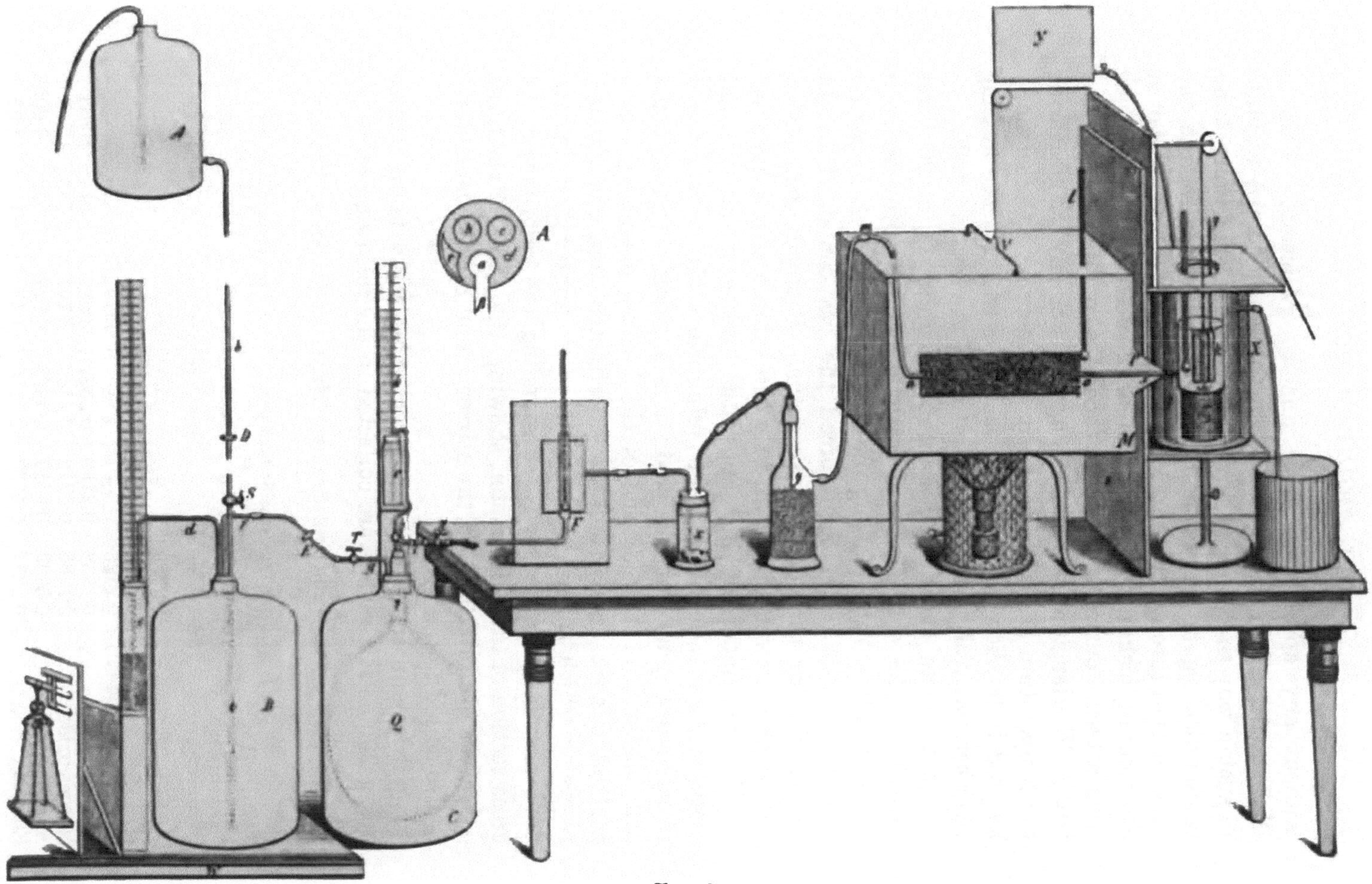

FIG. 6.

without any appreciable error due to radiation. This is illustrated by F in the figure. Two tubes are joined together in such a way that the gas first flows through the inner tube, communicating its heat to the thermometer inserted therein, and then escapes through the outer tube, which serves in this way as a protecting jacket to the inner tube. (To eliminate any effect from external currents of air, the tubes are enclosed in a wooden box.) Wiedemann used this arrangement chiefly for measuring the temperature of his gas as it left the reservoir. Occasionally he used it to show that the gas as it moved from the heating chamber was really at the temperature of the heating bath.

Wiedemann substituted a rubber bladder, Q, of 20 litres capacity, enclosed in the large bottle C for the copper reservoir of Regnault. Water flowing into the air-chamber B forced air over into C. This in turn forced the gas contained in the rubber bladder through the apparatus. B stood upon scales, and its gain in weight told directly the volume of gas expelled.

The remainder of his apparatus presented no unusual features. *x* is a bubble counter (sulphuric acid); y is a tower to catch any spray from the acid; *z* is a shield to protect the calorimeter from the radiation of the heating chamber; *k* is a perforated cork supporting the calorimeter; and *y* is a water-reservoir which furnishes water at room temperature for the double-walled brass jacket X surrounding the calorimeter. A cross-section of the calorimeter, showing the three silver tubes *a*, *b*, *c*, the gas inlet tube *β*, the stirrer *r*, and the thermometer *t*, is also added to the figure.

Wiedemann, like Regnault, restricted himself to temperatures below 200°. His results are collected in the following table:—

WIEDEMANN'S VALUES FOR THE SPECIFIC HEATS AT CONSTANT PRESSURE.

For	True specific heat per gram at			True specific heat per mol at		
	0°	100°	200°	0°	100°	200°
Air	0·2389	—	—	—	—	—
Hydrogen	3·41	—	—	6·874	—	—
Carbon monoxide ...	0·2426	—	—	6·793	—	—
Carbon dioxide ...	0·1952	0·2169	0·2387	8·589	9·544	10·50
Nitrous oxide ...	0·1983	0·2212	0·2442	8·741	9·750	10·76
Ammonia	0·5009	0·5317	0·5629	8·547	9·073	9·605
Ethylene	0·3364	0·4189	0·5015	9·43	11 74	14·06

Holborn and Austin[1] recently so altered the method that an initial temperature of 800° could be used, and so made it possible to measure the specific heat of gases over a much larger range. They heated their gases electrically in a nickel tube 8 mm. in diameter and filled with turnings. The gas was taken from commercial bombs, led through a long drying tube, then through the heating tube, and into a short connecting tube of porcelain. This last takes the place of the cork connection of Regnault. The gas then escapes into a silver calorimeter containing 1½ litres of water. The temperature of the gas was measured by a platinum-rhodium thermoelement (according to Le Chatelier) just before it entered the calorimeter. The wires of the thermo-element were 0·25 mm. thick, and isolated by thin quartz capillaries. The junction was protected by a special device against radiation from the sides of the tube, and against heat conduction. The gas finally entered a rubber bag contained in a bottle. As the bag expands water is expelled from the bottle, and the change in weight is noted. The mean specific heat between— **Experiments of Holborn and Austin.**

20° and 440° C.
20° „ 630° C.
20° „ 800° C.

was determined. The following values per gram for air, nitrogen, and a mixture of oxygen with 9 per cent. nitrogen were obtained. Regnault's values are appended in brackets. The values for pure oxygen are calculated from the observed values of the mixture of oxygen with 9 per cent. nitrogen and that of pure nitrogen.

Mean specific heat between	N_2	O_2 with 9 % N_2.	O_2 pure.	Air.	Air calculated from N_2 and O_2.
10° and 200°	(0·2438)	—	(0·2175)	(0·2375)	—
20° „ 440°	0·2419	0·2255	0·2240	0·2366	0·2377
20° „ 630°	0·2464	0·2314	0·2300	0·2429	0·2426
20° „ 800°	0·2497	—	—	0·2430	—

Holborn and Austin remark that their numbers for air and nitrogen agree satisfactorily with Regnault's, while their

[1] *Sitzungsber. der Kgl. preuss. Akad.*, 1905, p. 175.

number for oxygen differs considerably from that of Regnault. They conclude, from a control experiment which they made with air between 25° and 250°, that Regnault's value for the specific heat of oxygen must be raised to 0·2206 if the number 0·2375 for air is to be taken as correct. They estimate the absolute accuracy of their observation as about ±1 per cent. The observed change of the specific heat with the temperature is so slightly greater than ±1 per cent. that the change of the specific heat with the temperature cannot be determined from the data given for the simple gases.

Taking this result in connection with that of Stevens' mentioned before (p. 132), we may conclude that the specific heats of the permanent gases increase less rapidly with the temperature than corresponds to the formula—

$$4{\cdot}76 + 0{\cdot}00122t$$

which Mallard and Le Chatelier computed from their experiments with the crusher. On the other hand, Mallard and Le Chatelier's former assumption—

$$4{\cdot}8 + 0{\cdot}0006t$$

which Langen accepted, and which Schreber's re-calculation of Langen's results altered but slightly, is in agreement with the results of Holborn and Austin.[1]

Holborn and Austin's values for carbon dioxide.

As one would expect, Holborn and Austin found that the specific heat of carbon dioxide was markedly dependent on the temperature. Their values (with Regnault's appended in parentheses) are collected in the following table:—

Specific heat at constant pressure between	Specific heat per gram.	
	Observed.	Calculated.
20° and 200°	(0·2168)	0·2173
20° „ 440°	0·2306	0·2312
20° „ 630°	0·2423	0 2410
20° „ 800°	0 2486	0·2486

[1] According to Kalähne's criticism of Stevens' experiments (see later), the increase of the specific heat of the simple gases, instead of being greater, is really smaller.

The "calculated values" were obtained from the formula—

$$c_{(0,t)} = 0{\cdot}2028 + 0{\cdot}0000692t - 0{\cdot}0000000167t^2$$

which, in turn, was derived from the actual measurements. We can obtain the true specific heats referred to a gram of carbon dioxide and at constant pressure by multiplying by t and differentiating with respect to t. This gives—

$$c_{true} = 0{\cdot}2028 + 0{\cdot}0001384t - 0{\cdot}000000005t^2$$

Calling $t = -273°$, we get for the true specific heat at absolute zero—

$$c = 0{\cdot}1613$$

If, finally, we replace the gram by the gram molecule as a unit of mass, we get—

$$c_{(0,T)} = 7{\cdot}097 + 0{\cdot}003045\mathrm{T} - 0{\cdot}000000735\mathrm{T}^2$$

If we attempt to express the increase of the mean specific heat with the temperature between 200° and 800° by a linear expression, we find that the term representing the increment agrees very well with the number obtained by Langen from his explosion experiments.

Holborn and Austin give the following table, comparing true specific heats of carbon dioxide at constant pressure determined in various ways. In it are given the values derived from Regnault's, Wiedemann's, and their own experiments; also values calculated by help of the formula of Mallard and Le Chatelier for the specific heat at constant volume—

$$c_{v(0,t)} = 6{\cdot}3 + 0{\cdot}0006t - 0{\cdot}00000118t^2$$

and by help of Langen's formula—

$$c_{v(0,t)} = 6{\cdot}7 + 0{\cdot}0026t$$

° C.	Régnault.	Wiedemann.	Mallard and Le Chatelier.	Langen.	Holborn and Austin.
0	0·1870	0·1952	0·1880	0·1980	0·2028
100	0·2145	0·2169	0·2140	0·2100	0·2161
200	0·2396	0·2387	0·2390	0·2220	0·2285
400	—	—	0·2840	0·2450	0·2502
600	—	—	0·3230	0·2690	0·2678
800	—		0·3550	0·2920	0·2815

If we estimate the accuracy of the numbers given in this table for 600° and 800°, on the basis of the limit of error given by Holborn and Austin, we find that it is too small to justify any further discussion of the deviations between the numbers given in the table.

Holborn and Henning's values for water-vapour.

Recently the specific heat of water-vapour at the constant pressure of one atmosphere has been studied by Holborn and Henning[1] up to 820° C. in a similar manner. The calorimeter through which the steam, superheated to temperatures between 270° and 820°, passed was maintained in the neighbourhood of 115° C. in order that the steam should escape without condensing. Thus, contrary to Regnault's procedure, the specific heat of the water-vapour was found directly. Only the specific heat referred to air was determined. The relation of the mean specific heat of one gram of water-vapour at the constant pressure of one atmosphere to the specific heat of one gram of air at the same pressure, was found as follows :—

Between	110° and 270° C.	... ...	1·940
„	110° „ 440° C.	... ...	1·958
„	110° „ 620° C.	... ...	1·946
„	110° „ 820° C.	... ...	1·998

If the mean specific heat of air (per gram at constant pressure between 0° C and t° C) is assumed to be—

$$(a) \qquad c_t = c_0(1 + 0{\cdot}00004t)$$

or

$$(b) \qquad c_t = c_0(1 + 0{\cdot}00008_5t)$$

then we get for one gram of water-vapour from the above determinations—

$$(a) \qquad c_t = 0{\cdot}446(1 + 0{\cdot}00009_6t)$$

$$(b) \qquad c_t = 0{\cdot}441(1 + 0{\cdot}00014_5t)$$

These equations referred to the mol as unit have already been given on p. 126.

Influence of pressure on the specific heat at constant pressure.

We cannot dwell upon the investigations of the effect of pressure on specific heat at constant pressure. According to the results of Lussana[2] and of Amagat,[3] it is certain that the

[1] Holborn and Henning, *Ann. der Physik.* (IV.), 18 (1905), 739.

[2] Lussana, *Fortschritte der Physik*, i. J. 1896, p. 345, and i. J. 1897, p. 331.

[3] Amagat, *Compt. Rend.*, 122 (1896), 66 and 121; comp. also Witkowsky, *Fortschritte der Physik*, i. J. 1896, p. 343.

effect of pressure becomes less the nearer the gas approaches an ideal gas in its behaviour. We therefore find that pressure has a very slight influence even at ordinary temperatures on the so-called permanent gases, and in the easily condensible gases the effect of pressure vanishes as we approach the high temperatures at which gas reactions mostly take place.

Joly[1] has undertaken the direct determination of the specific heat of gases at lower temperatures at constant volume. **Joly's experiments**

The principle of his method is as follows. A metal vessel is hung by a fine metal filament from the pan of a delicate balance, and its weight determined. The vessel hangs freely in a cavity through which a rapid current of steam from some sort of boiler may be forced at any desired moment. As soon as the weight of the vessel has been ascertained the steam is admitted. When the steam strikes the cold metallic vessel, it condenses upon it and heats it with its heat of condensation to the steam temperature of 100°. Further condensation does not take place, because the vessel, being freely suspended in a space filled with steam whose walls are kept constantly at 100,° can lose no heat by radiation. The vessel becomes heavier because of the water condensed upon its surface. This increase of weight rapidly approaches a permanent maximum, and may be readily determined. It permits us to find directly how much heat is necessary to raise the vessel from its initial temperature to 100°, since the latent heat of the vaporization of water is known. If we first use the vessel exhausted, and then filled with gas, the difference between the two increments of weight gives directly the quantity of heat necessary to heat the enclosed gas to 100°. Hence we get the specific heat of the gas at constant volume between this initial temperature and 100°.[2]

This method may very well be used to test how much the specific heat of a gas at constant volume depends on the density of the gas. For this purpose it is only necessary to use a gas first at a small pressure and then at a greater one. Still, it is impossible to determine the values at high temperatures, which are of special importance to us, with a steam calorimeter.

[1] Joly, *Phil. Trans.*, 182 (1892), 73.

[2] Air, carbon dioxide, and hydrogen were investigated. For a criticism of these results, see Wüllner, "Experimentalphysik," vol. ii. (Leipzig, 1896) p. 538.

Apparently no one has ever tried using a higher boiling substance than water.

Indirect methods: explosion experiments.

Let us now proceed to a discussion of the indirect methods of measuring the specific heats of gases. The explosion method is of especial interest to us. We have already discussed in the previous lectures the results obtained by this method, and have in part explained how the measurements were made.

Bunsen's experiments.

Historically, it is to be observed that Bunsen [1] first measured the pressure developed in the explosion of gases. He used explosion mixtures of oxygen and hydrogen, and of oxygen, hydrogen, and carbon monoxide. They were contained in glass tubes 8·15 cm. high and 1·7 cm. in diameter, sealed at one end, and closed at the other by a plate resting freely upon the aperture. The plate was loaded according to a method long used for the technical testing of explosives,[2] and the maximum load was determined which the pressure of the explosion was just able to lift. Knowing this load and the size of the opening of the tube, the pressure of the explosion could be calculated. If a part of the gas had already cooled off somewhat before the explosion had traversed the rest of the gas, the pressure would be found too low. In order to obviate this danger, Bunsen ignited his gas with a chain of sparks along the whole length of the tube.

Perfecting Bunsen's method.

Bunsen's determinations of the explosion pressure may be improved by the use of some kind of a self-registering manometer attached to the explosion bomb. The so-called "indicator" is an instrument of this kind, and is often used nowadays in the study of gas and other explosion engines.

Indicators.

The indicator consists of a cylinder connecting with the explosion chamber. A piston slides with as little friction as possible inside the cylinder, carrying with it a pencil-point. This pencil-point registers the motion of the piston upon a strip of paper which glides rapidly past it. The piston is pushed along the cylinder by the explosion pressure against a standardized spiral spring. We get in this way a record on the paper of the static pressure. There is a difficulty involved in the application of this method to those explosions which are used for measuring specific heats. The explosion must take place very rapidly, in

[1] *Pogg. Ann.*, 131 (1867), p. 161.

[2] Guttmann, *Explosivstoffe*. Braunschweig, 1895.

order that the heat loss, before the explosion is completed, may be as small as possible. But when the pressure changes very rapidly, the pencil-point, because of the momentum of the moving parts, describes oscillatory motions, and writes a wavy instead of a smooth curve for the static pressure. It consequently requires some graphic treatment in order to get the true value from this curve.

A second method of measuring explosion pressures is based on the effect of impacts. The piston is allowed to be thrown a short distance by the action of the explosion, while a pencil-point attached to it writes upon a moving paper. In this way we get a *trajectory*, and from it one can determine the distance traversed by the piston in successive fractions of a second. The differences between these distances measure the acceleration which the piston experienced, and consequently the force which acted upon it. The maximum acceleration tells us the maximum force which the explosion has exerted. The quotient of this force, divided by the area of the piston head upon which it acted, represents the maximum force of the explosion.

Measurement of impacts.

Mallard and Le Chatelier used the first of these methods; Vieille and Berthelot and Vieille the second.

Experiments of Berthelot and Vieille, and of Mallard and Le Chatelier.

Berthelot and Vieille[1] carried out their experiments in metal bombs of various sizes (0·3, 1·5, 4 litres). The bombs were provided with an accessory tube in which moved a well-fitting piston. The pencil-point wrote upon blackened paper on a revolving drum. An electrically operated tuning-fork traced its vibrations on the same paper, so that a comparison of the lines afforded a knowledge of the distance travelled by the piston, and the time consumed in the movement. While Bunsen set off his gas by a chain of sparks extending the whole length of his explosion tube, Berthelot and Vieille exploded their gases at but one end. Mallard and Le Chatelier[2] used a cylindrical bomb of 3·86 litres capacity, and ignited the gas-mixture by a spark at its centre. This was decidedly better than Berthelot and Vieille's method of igniting at one end. They obtained, by means of their pressure indicator, a curve for

[1] *Comp. Rend.*, 95 (1882), 1280; 96 (1883), 116, 1218, 1358. *Ann. chim. phys.*, 4 (1885), 13.

[2] *Ann. des Mines*, 4 (1884), 379.

the static pressure from the beginning of the explosion till the bomb had cooled off. They could determine the rate of cooling of the gas-mixture from this curve, and could therefore apply a correction of the observed maximum pressure to account for the slight amount of heat which the gas lost during the process of combustion, and which consequently contributed nothing to the rise of pressure. In Vieille's method it was only possible to determine this correction indirectly and incompletely.

Clerk's experiments.

Clerk[1] later performed several experiments, analogous in principle to the experiments of Mallard and Le Chatelier, with a cylindrical bomb, a spark igniter at the bottom, and a steam-engine indicator to record the pressure.

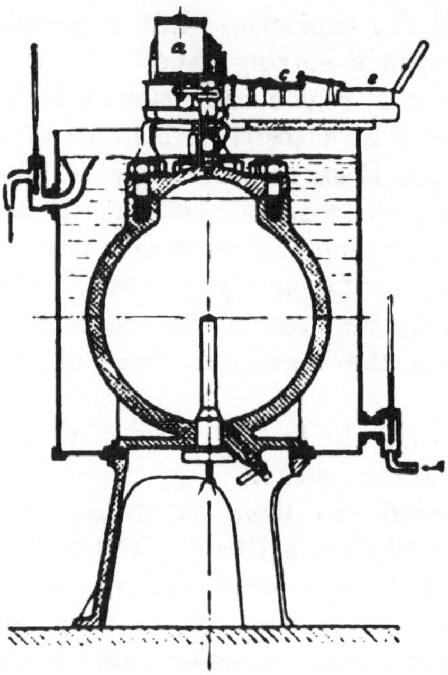

FIG. 7.

Langen's experiments.

Langen[2] employed the most highly perfected apparatus of this type. He used a spherical explosion bomb of steel with a capacity of 34 litres, thus reducing the loss of heat during combustion to a minimum. The gas was electrically ignited at the centre of the bomb. The loss of heat during combustion, and consequently the correction to be applied to the maximum pressure, was smaller in this large apparatus than before, and at the same time the gas cooled off more slowly, and consequently the correction could be more easily determined.

His bomb is illustrated in Fig. 7. It is enclosed in a large vessel through which water is kept flowing. The bomb may be evacuated through a cock attached to its lower part. The cover of the bomb has three openings—one for leading in the gas to be studied; a second for attaching an open, two-armed

[1] Dugald Clerk, "The Gas and Oil Engine." London, 1897.

[2] "Mitteilungen über Forschungsarbeiten aus dem Gebiete des Ingenieurwesens" (Berlin, 1903), Heft 8.

mercury manometer; the third for attaching the indicator *a*, which was practically identical with the common and widely known instrument of commerce. Its pencil recorded the static pressure on a drum, which revolved at a regular but adjustable velocity. The electrical contacts *c* and *e* were so arranged that by pressing a button the recording apparatus and the apparatus for measuring the velocity of revolution of the drum were simultaneously started, and directly afterwards the spark igniter was made to act. The purpose of the mercury manometer was chiefly to measure the initial pressure before explosion, and the final pressure after the exploded gas had completely cooled off.

It cannot be denied that measurement of pressure with this apparatus presents no serious difficulties. The correction for the maximum pressure applied by Langen is, perhaps, susceptible of some improvement, because, while the heat conduction by the walls was corrected, the heat radiation was not.[1]

This correction, however, amounted to only a few per cent. in the experiments of Mallard and Le Chatelier, and would be even less in these experiments; consequently any inaccuracy in it could not greatly affect the results. If the reaction proceeds without perceptible dissociation taking place, then the calculations of the specific heat of gases from these experiments by means of the gas law ($pv = RT$) is certainly the simplest way to find out anything about the specific heats of gases at high temperatures.[2]

The crusher manometer.

The other method, which we have considered in this same connection, makes use of the crusher manometer. This method has the advantage that we can employ higher temperatures without fearing that dissociation will ensue.

The following figure (Fig. 8) will serve to illustrate this.[3] A charge of high explosive is freely suspended in the cavity

[1] Compare Nernst, *Physikal. Zeitschr.* (1904), 777.

[2] Langen's experiments have been repeated lately by Hatisser with a very much smaller apparatus ("Mitteilung über Forschungsarbeiten auf dem Gebiet des Ingenieurweeen," Heft 25 (Berlin, 1905). The results seem to be affected seriously by imperfectness of the apparatus (see Haber, *Zeit. für Gasbeleuchtung und Wasserversorgung* (1905), p. 1046).

[3] Nach Heise, "Sprengstoffe und Zündung der Sprengschüsse." Berlin, 1904.

A, where it may be electrically exploded by means of the spark wires b. The pressure of the explosion is exerted against the steel cylinder a, which in turn presses upon the copper cylinder r (8 mm. in diameter and 13 mm. high). The screw d fits tightly into the lower head-piece B′, and serves as a backing for the copper cylinder.

The great pressures developed by the explosion deform the copper cylinder. By comparing this deformation with that produced by known hydraulic pressure, we estimate the magnitude of this explosion pressure. It is self-evident that the chemical composition of the copper cylinders, as well as their previous mechanical treatment, must be and have been exactly identical, else the comparison would mean nothing. With this proviso, it appears that, within certain limits of pressure, the deformation ε (in millimetres) may be represented in terms of deforming pressure p (in kilograms) by the expression—

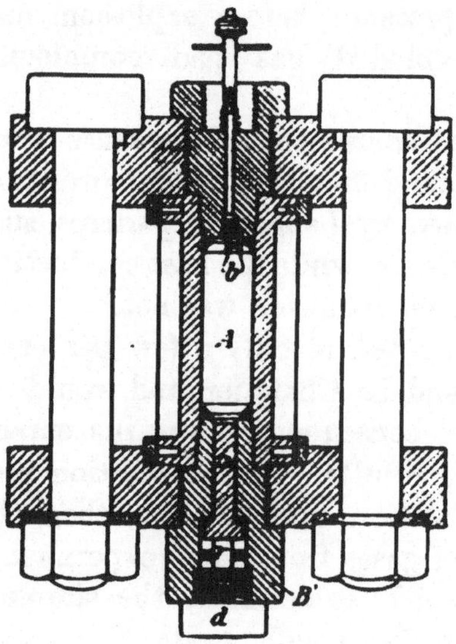

FIG. 8.

$$p = a + b\varepsilon$$

where a and b are constants which may be determined from the hydraulic measurements.

Theory of the crusher manometer.

But, as Sarrau and Vieille[1] have shown both theoretically and experimentally, it is by no means permissible to put the maximum pressure of the explosion P directly equal to the static pressure p, calculated from the observed deformation on the basis of the previously determined constants a and b. This procedure would be correct only where the explosion takes place with comparative slowness, and when the moment of inertia of the piston is comparatively small. If the explosion is very rapid, and the momentum of the piston great,

[1] *Comp. Rend.*, 95 (1882), 26, 133, 181; also 102 (1886). 1054, and 104 (1887). 1759.

the maximum pressure P of the explosion approximates to the limiting value.[1]

$$P = a + b\frac{\varepsilon}{2}$$

Example illustrating the theory of the crusher manometer.

It therefore becomes necessary, if we would calculate the maximum pressure of the explosion from the deformation, to know how great an acceleration, if any, the pressure piston experiences. Sarrau and Vieille give several very instructive examples of experiments in which they were able to determine these accelerations by attaching a pencil to the piston and causing it to record the piston's movements upon a rotating strip of paper. They then measured with a microscope the distances travelled by the piston along the paper from moment to moment. A few of their numerical results are given in tabular form below—

A					B				
t	*r*	K	*p*	P	*t*	*r*	K	*p*	P
1	0·29	+20·6	541	561	1	0·96	+1607	480	2087
2	0·92	+9·3	696	705	2	2·37	+1013	1099	2112
3	1·69	−1·8	1033	Not perceptibly different from *p*.	3	3·27	+384	1775	2159
4	2·43	−4·6	1445		4	3·61	−454	2337	1903
9	4·31	−2·4	2750		5	3·20	−1083	2844	1761
12	4·66	−1·2	2936		6	2·25	−1537	3219	1682
16	4·89	−0·6	3136		7	0·89	−1537	3219	1682
19	5·04	−0·6	3232						

[1] If u represents the path of the piston, $\frac{du}{dt}$ its velocity, and m its mass, then—

$$P - p = m\frac{d^2u}{dt^2}$$

and—

$$(P - p)du = m \times \frac{d^2u}{dt^2} \times du$$

But the sum of all the accelerating forces from $u = 0$ to $u = \varepsilon$ is zero, so that

$$\int_0^\varepsilon m\frac{d^2u}{dt^2} \times du = 0$$

and hence, since $p = a + b\varepsilon$ —

$$\int_0^\varepsilon (P - p)du = \int_0^\varepsilon (P - a - bu)du = 0$$

Under A are placed numbers referring to a particular grade of powder called "C," used in the French army. The ratio of the weight of the powder to the volume of the bomb was 0·650. (This number is called the "density of charge." It represents kilograms per litre.) The times, t, are multiples of the arbitrary unit of time chosen 0·000317 sec. The velocity of the piston is given in metres per second, and the accelerating action K on the piston in kilograms. Further, under p are given the pressures calculated from the formula $a + b\epsilon$. The sum of p and K[1] represents the total pressure P of the explosion. We see that in this case K has a value so small as to be negligible, and hence the explosion pressure is correctly represented by the formula $a + b\epsilon$. We have here, then, the simple case where the explosion develops its maximum pressure so slowly that the piston experiences almost no acceleration.

Similar measurements made with gun-cotton are given under B. The density of the charge was 0·2. The unit interval of time is here 0·0003242 sec.[2] We see that the maximum pressure is almost instantly attained here, and the acceleration of the piston plays an extraordinary important *rôle*. The explosion does not push the piston, it *hurls* it against the copper cylinder, and the kinetic energy of this motion brings about a deformation of the cylinder greater than an hydraulic press exerting the same pressure as the explosion could ever produce.

In order, therefore, to make use of any data obtained with

Assuming with Sarrau and Vieille that P is constant, integration gives—

$$P = a + \frac{b}{2}\epsilon$$

This limiting formula therefore applies only when the explosion develops its maximum pressure P in an immeasurably short period of time, and when this pressure remains constant during the entire though very brief course of the piston's movement.

[1] According to the fundamental rules of mechanics—

$$K = m\frac{d^2u}{dt^2}$$

where m represents the mass, u the path, and t the time. $\frac{d^2u}{dt^2}$ was determined graphically from the diagrams of the piston movements, as above explained; m, that is the mass of the piston, was 60 grams.

[2] The piston weighed 3·601 Kg., but the same results were obtained with a piston weighing but 60 grams.

such a crusher manometer in the deduction of specific heats, it is essential that we know the nature of the explosive we employ. Apparently no calculations of this sort, other than those of Mallard and Le Chatelier just mentioned, have ever been undertaken, although the apparatus finds everyday use in the manufacture of high explosives, and the possibility of obtaining data regarding the specific heats of gases at high temperatures with this apparatus is apparent. It should further be noted that surprising chemical facts have come to light in these experiments under high pressure.[1]

The ratio of specific heats.

We will now discuss how the ratio κ, between the true specific heats at constant pressure C_p and at constant volume C_v may be determined. Knowing this ratio, we can obtain the true specific heat at constant volume per mol, because—

$$\kappa = \frac{c_p}{c_v} = \frac{c_v + R}{c_v}$$

and hence—

$$c_v = \frac{R}{\kappa - 1}$$

The true specific heat per mol at constant pressure would be—

$$c_p = \frac{\kappa \times R}{\kappa - 1}$$

The determination of the value κ for gases depends upon certain phenomena accompanying an adiabatic expansion. Thus, if a gas has a pressure p at a volume v, then after an adiabatic change the pressure p_1 and the volume v_1 are connected by the relation—

$$pv^{\kappa} = p_1 v_1^{\kappa}$$

If, instead of the volume v, we substitute the concentration $\frac{1}{v}$, or as it is usually called in this connection, the density (d), we get—

$$\frac{p_1}{p} = \left(\frac{d_1}{d}\right)^{\kappa}$$

Procedure of Clément and Desormes.

A method originating with Clément and Desormes permits

[1] Sarrau and Vieille, *Comp. Rend.*, 105 (1887), 1223, find that the reaction—

$$2CO + 2H_2 = CH_4 + CO_2$$

is greatly favoured by high temperatures and pressures.

us to measure the pressures and volumes before and after an adiabatic expansion. They filled a large bottle or flask with the desired gas at the same temperature as the surrounding air, but at a slightly higher pressure P_a. Or sometimes they connected this first flask with a second one containing the gas under investigation at a somewhat lower pressure. Connection with the atmosphere or the second flask was established by opening a stop-cock. The pressure inside and outside, or in the two flasks, was then rapidly equalized. The part of the gas escaping into the air or into the second vessel, did work either against the atmosphere or against the gas in the second vessel. After the pressure was fully equalized the cock was closed, and the gas in the first vessel allowed to regain the temperature of the outside air. The pressure P_e existing in this vessel is then determined. Since the initial and final temperatures coincide, the densities of the gas before and after the expansion are to one another as the pressures, that is—

$$d_1 : d = P_a : P_e$$

But the density corresponding to the final pressure P_e was already established at the end of the process of expansion when the cock was closed. For no gas escaped after that, nor did the volume change, and consequently the mass per unit of volume or the density did not change. The pressure before the expansion was P_a. We will call B the pressure at the closing of the stop-cock. κ can then be found from the equation—

$$\frac{P_a}{B} = \left(\frac{P_a}{P_e}\right)^\kappa$$

or—

$$\kappa = \frac{\log P_a - \log B}{\log P_a - \log P_e}$$

The difficulty of the method of Clément and Desormes.

Strictly speaking, the method assumes that the cock is opened for only an infinitely short interval of time, and yet that the pressures inside and outside are, nevertheless, equalized in that time. But in practice we cannot be sure that the equalization has really taken place when the cock is open for but a very brief interval, because oscillation takes place during the first instant the cock is open. Yet if the cock is left open for a perceptible time, heat is given off during this interval from the walls of the vessel, so that the rise of pressure after closing the cock comes out too small.

Cazin[1] has carried out an investigation to determine how long one must wait in order that these oscillations of pressure shall have just ceased. He closed the cock the very moment this condition was reached. In this way he got the following values:— **Cazin's experiments.**

Gas	Air	O_2	N_2	H_2	CO	NH_3	CO_2	N_2O	SO_2	C_2H_4	Ether
κ	1·41	1·41	1·41	1·41	1·41	1·328	1·291	1·285	1·262	1·257	1·079

All the values refer to ordinary temperatures. Some of the gases did not possess the desired purity. Ethylene is, for this reason, especially uncertain.

It is clear that heat communicated from the walls is more dangerous the smaller the vessel one uses, for just so much more unfavourable does the ratio of area to volume become. Cazin was mindful of this, and used vessels of 30 and 60 litres capacity.

We must conclude, however, from a critical discussion of Caxin's results by Röntgen, that he left the cock open too long, and consequently got too small final values. Röntgen[2] materially improved the method, using a vessel of 70 litres capacity, and a very delicate manometer. He determined both the initial pressure P_a, the pressure B after a brief opening of the cock, and the final pressure P_e. He applied a special correction for the heat communicated to the gas by the walls of the flask. He obviated the effect of oscillations of pressure by the arrangement of his manometer. In this way he got the very accurate values for— **Röntgen's experiments.**

$$\text{Air} = 1\cdot4053$$
$$\text{Carbon dioxide} = 1\cdot3052$$

at ordinary temperature. The high thermal conductivity of hydrogen vitiated the results obtained for this gas.

Maneuvrier,[3] partly in collaboration with Fournier, attempted a somewhat different procedure. He compressed the gas contained in a spherical vessel of ca. 50 litres capacity **Maneuvrier's experiments.**

[1] Cazin, *Ann. Chim. Phys.*, iii. 66 (1862), 243; see also iv. 20 (1870), 243.

[2] *Pogg. Ann.*, 148 (1873), 580.

[3] *Ann. Chim. Phys.* (7) 6 (1895), p. 321; also p. 377, where a table is given showing all previous values of κ.

by the rapid impact of a piston. The volume change may be easily derived from the area of the piston-head and the distance it travels. It is easy to calculate what the change of pressure would be in an isothermal compression of the same magnitude. The actual pressure change is measured by means of a hydrostatic manometer, which is put into connection with the compressed gas for a very brief period just after the compression has taken place. The opposing pressure on the manometer may be varied, and so, by repeating the experiment often enough, that pressure may be found which just balances the pressure developed by the gas, and hence allows no movement of the manometer. By carrying out the experiment with different volumes, and consequently different changes of pressure, a curve may be obtained from which it is possible to compute the ratio between the adiabatic and isothermal change of pressure for the limiting case of a vanishingly small change of volume. This ratio represents, according to the theory, the value κ—

$$\frac{d_{p\,(\text{adiab.})}}{d_{p\,(\text{isoth.})}} = \kappa$$

In this way Maneuvrier and Fournier[1] found—

Gas	Air	CO_2	H_2	C_2H_2
κ	1·395	1·298	1·380	1·26

Lummer and Pringsheim's experiments. We owe to Lummer and Pringsheim still another method of using an adiabatic process to determine the value of κ.[2] It depends on the evaluation of T_1, T_2, p_1, and p_2 in the adiabatic equation—

$$\frac{T_1}{T_2} = \left(\frac{p_1}{p_2}\right)^{\frac{\kappa-1}{\kappa}}$$

which merely represents a transformation of the expression previously given. Lummer and Pringsheim filled a vessel of 90 litres capacity with a gas at known temperature and a known pressure, which exceeded that of the atmosphere. They let the excess pressure blow off at the right moment through a wide orifice. While the older methods were very sensitive toward a prolonged opening of the cock, their method has the

[1] *Comp. Rend.*, 123 (1896), 228, and 124 (1897), 184.

[2] Lummer and Pringsheim, *Wied. Ann.*, 64 (1898), p. 582.

great advantage that the cock can be left open as long as desired. A complete equalization with the atmospheric pressure is thus made certain, and the final pressure is the atmospheric pressure. All now depends on determining quickly enough the cooling in the body of the gas incident to its expansion, and to find its value free from any disturbing effect due to the transfer of heat from the walls of the vessel to its interior. Lummer and Pringsheim solved this problem by arranging a platinum wire of the very greatest tenuity in the centre of the gas, and using its change of resistance as an indication of the temperature change. Makower[1] modified this arrangement slightly, and measured the ratio of specific heats of steam a few degrees above 100°, using a vessel containing but 9·3 litres, having first ascertained that even with this size vessel the effect of the heat transfer from the walls was slight. His value is appended to those of Lummer and Pringsheim in the following table:—

	κ	Temperature.
Air	1·4026	15° (about).
Oxygen[2]	1·3977	
Carbon dioxide	1·2995	
Hydrogen	1·4084	
Water-vapour	1·305 ± 0·002	105° (about).

The values of Lummer and Pringsheim are recognized as being especially trustworthy.

Experiments of Müller. All the methods so far described have employed but a single adiabatic change of the gas. In order to measure accurately the changes thus induced before the transfer of heat shall have altered the adiabatic condition, it is necessary to use large vessels and large quantities of gas. To obviate this difficulty, Afsmann[3] proposed studying a rapid periodic adiabatic change. Müller,[4] at O. E. Meyer's suggestion, has carried out a great number of determinations according to Afsmann's principle. He used a U-tube of about 4 cm. bore which contained some mercury, and at the beginning was open at both ends. By tilting the tube

[1] *Phil. Mag.* (6) 5, (1903), p. 233.

[2] Commercial oxygen from a bomb. It probably contained 9 per cent. nitrogen.

[3] *Pogg. Ann.*, 85 (1852), 1.

[4] *Wied. Ann.*, 18 (1883), 95.

momentarily in the plane of the U, oscillations were set up in the mercury, whose period Müller measured to within a thousandth of a second by means of an electrical recording device. In the cases he studied the period was usually about ½ second. The mercury came to rest after some 50 such oscillations. If now a glass globe holding somewhat over a litre of the gas be placed upon each arm of the U-tube, we find, on repeating the experiment, that the period of the oscillation becomes shorter. The cause of this is the alternate compression and expansion of the gas in each globe. Müller's apparatus showed this difference to be about 0·1 second. When exactly measured to the thousandth of a second, this difference permits us to compute the ratio of the specific heats, provided the volume changes are adiabatic. It is doubtful whether this is really the case. Yet the results seem serviceable, even though they are certainly not so accurate as those of Röntgen and of Lummer and Pringsheim.

All his experiments were carried out between about 0° and 50°. Carbon dioxide was investigated from 9·6° to 33·4°, and ammonia from 10·7° to 30·1°, and yet no change due to the temperature was discovered, although a very accurate method should have disclosed one. Such a change was observed with several vapours, and the result obtained with them are placed at the end of the following table; yet here the value of κ is also dependent on the pressure.

Gas.	Temperature interval, °C.	κ	Gas	Temperature interval, °C.	κ
Air	10 to 22	1·405	CH_3Cl	16 to 17	1·199
O_2	16 to 21	1·402	$CHCl_3$	23 to 39	1·110
CO_2	9·6 to 33·4	1·265	CH_2Cl_2	24 to 42	1·12
HCl	18 to 41	1·398	C_2H_4	15 to 30	1·243
HBr	10 to 38	1·365	CH_3COH	23	1·146 (?)
SO_2	16 to 34	1·256	CH_3CCl_3	44	1·04
H_2S	10 to 40	1·276	C_2H_5Cl	21 to 30	1·126
CS_2	21 to 40	1·189	$C_2H_4Cl_2$	42	1·086
NH_3	17 to 30	1·262	CH_3OCH_3	6 to 30	1·113
CH_4	19 to 30	1·316	$C_2H_5OC_2H_5$	22 to 45	1·029

Velocity of sound in gases.

Now, the passage of sound through gases generates very rapid adiabatic changes. Besides, observations of the velocity of sound may be made at both low and high temperatures. This method is, therefore, more widely applicable, and also more convenient in execution, than the other methods for determining κ so far mentioned.

According to the theory, the velocity of propagation of a longitudinal wave depends upon the elasticity e and the density d of the propagating medium, as stated by the equation—

$$\gamma^2 = \frac{e}{d}$$

Newton put the elasticity equal to the pressure p. La Place showed that in an adiabatic compression and expansion the elasticity is represented by the product $p\kappa$.

If we replace the density d at the temperature and pressure of the experiment by the density at 0° (273° abs.), and of 760 mm. of mercury pressure (p_0), we have—

$$d = d_0\frac{273}{T} \times \frac{p}{p_0}$$

and hence—

$$\gamma^2 = \kappa\frac{T}{d_0} \times \frac{p_0}{273}$$

Here d_0 depends only upon the chemical nature of the gas under consideration. $\frac{p_0}{273}$ is a constant for all gases, and only the values κ and T are determined by the conditions of the experiment. Conversely, if we know κ we can measure temperature by determining the velocity of sound. An apparatus for measuring the velocity of sound therefore constitutes an acoustic thermometer. Conversely, knowing T, we can determine κ from a measurement of the velocity of sound. The formula applies primarily to the propagation of sound in an unlimited medium. In tubes the friction of the gas against the walls and the exchange of heat between the compressed or expanded gases and the walls create disturbances, the theory of which has been worked out by Helmholtz and Kirchhoff. Observations have shown that, in agreement with the theory, these disturbances become smaller the wider the tubes and the higher the pitch of the sound used. This is especially true when the gases are not too dilute. Tubes of 3 cm. bore show no noticeable deviation from conditions in a free gas space.[1]

Dulong[2] was first able to derive good values of κ from the phenomena of acoustics. He made use of the old idea of blowing a whistle in the various gases and determining the pitch. The number of vibrations in the note emitted is proportional to the velocity of the sound. Knowing the density of the gas under standard conditions and the temperature, the value of κ relative to air can be calculated. Dulong obtained the absolute value of κ for air by direct measurements of the velocity of sound in the open. Masson[3] made a large number of measurements by the same method. We have placed the results of both these men in a table of Wüllner's given somewhat farther on.

Kundt's method. The use of these whistling notes is always affected with a slight uncertainty due to the mouthpiece. It has therefore been generally abandoned in favour of the method of Kundt,[4] which permits us to directly measure a wave-length. Kundt generates stationary waves in a tube containing a little fine dust. The dust in this "wave tube" collects at the nodes, each of which is a half wave-length from its neighbour. If λ is the half wave-length, and x the number of half-vibrations which the note makes in a second, then—

$$\gamma = \lambda x$$

and hence—

$$\lambda^2 = \kappa \frac{T}{d_0} \times \frac{p_0}{273} \times \frac{1}{x^2}$$

Kundt always compares the wave-length with the same tuning-fork in the gas and in air—that is, always using the same frequency, x. It then follows, if the index L refers to air (Luft), that—

$$\frac{\lambda^2}{\lambda^2_{(L)}} = \frac{\kappa}{\kappa_{(L)}} \times \frac{T}{T_{(L)}} \times \frac{d_{0(L)}}{d_0}$$

The use of high notes not only eliminates the disturbing effect of the walls, but gives a larger number of nodes in a moderate distance whose mean distance apart may be very accurately determined.

[1] For the literature and new experiments, see Sturm (*Drudes Ann.*, 14. (1904), 823). Compare also Vieille's paper on the velocity of sound ("Rapports presentées au congrès international de Physique" (Paris, 1900), vol. i. p. 228)

[2] Dulong, *Pogg. Ann.*, 16 (1829), 438, *Ann. Chim. Phys.*, 41, 113.

[3] Masson, *Ann. Chim. Phys.*, 53 (1853), 277.

[4] *Pogg. Ann.*, 135, 337 and 527 (1868).

Wullner [1] has executed a series of measurements at 0° and 100°, using this acoustic method. He tabulated his results with those of Dulong and Masson, and we reproduce the table here:— Wüllner's experiments.

Gas.	Dulong.	Masson.	Wüllner 0°	Wüllner 100°
Air	1·405	1·405	1·40526	1·40513
Oxygen	1·402	1·405	—	—
Nitrogen	—	1·405	—	—
Hydrogen	1·394	1·405	—	—
Carbon monoxide ...	1·410	1·413	1·4032	1·3970
Carbon dioxide	1 326	1·277	1·3113	1·2843
Nitrous oxide	1·331	1·270	1·3106	1·2745
Ammonia	—	1·304	1·3172	1·2791
Ethylene	1·228	1 260	1·2455	1·1889
Nitric oxide	—	1·394	—	—
Marsh gas	—	1·319	—	—
Sulphur dioxide	—	1·248	—	
Hydrochloric acid ...	—	1·392	—	
Hydrogen sulphide ...	—	1 258	—	—

The fourth decimal figure in these results is quite uncertain; the fifth is wholly so.

Capstick [2] has also made many determinations by this method. Capstick's experiments.

Gas.	Formula.	κ	Gas.	Formula.	κ
Methane	CH_4	1·313	Carbon tetrachloride	CCl_4	1·130
Methyl chloride ...	CH_3Cl	1·279	Ethylene chloride...	$C_2H_4Cl_2$	1·137
Methyl bromide ...	CH_3Br	1·274	Ethylene	C_2H_4	1·134
Methyl iodide ...	CH_3I	1·286	Vinyl bromide ...	C_2H_3Br	1·264
Ethane	C_2H_6	1·182	Allyl chloride ...	C_3H_5Cl	1·198
Ethyl chloride ...	C_2H_5Cl	1·187	Allyl bromide ...	C_3H_5Br	1·137
Ethyl bromide ...	C_2H_5Br	1·188	Ethyl formine ...	$HCOOCH_3$	1·145
Propane	C_3H_8	1·130	Methyl acetate ...	CH_3COOCH_3	1·124
n. propyl chloride ...	C_3H_7Cl	1·126	Hydrogen sulphide	SH_2	1·137
Iso propyl chloride	C_3H_7Cl	1·127	Carbon dioxide ...	CO_2	1·340
Iso propyl bromide	C_3H_7Br	1·131	Carbon disulphide	CS_2	1·308
Methylene chloride	CH_2Cl_2	1·219	Silicium tetrachloride	$SiCl_4$	1·129
Chloroform	$CHCl_3$	1·154			

[1] *Wied. Ann.*, 4 (1878), 321; corrected results in Wüllner's "Handb. der Physik," 5th edit., vol. ii. p. 553.

[2] *London Phil. Trans.*, 185 (1894), 34; and *Trans. Royal Soc., London*, 57 (1895), 323.

Dust figures at high temperatures.

This "dust figure" method may be used even at temperatures of several hundreds of degrees. Kundt and Warburg[1] used it near 300° in a celebrated research where it was wished to determine the value of κ for mercury vapour. The arrangement of their apparatus may be easily seen from Fig. 9. Strecker[2] later used the same method to determine the ratio of the specific heats of iodine, bromine, chlorine, and a series of compounds containing other elements. We see a glass tube in the figure about a metre long, closed at both ends and clamped at *s*. This is the "sounding tube." The point *s* is about 30 cm. from its left-hand end. Its right-hand end is firmly

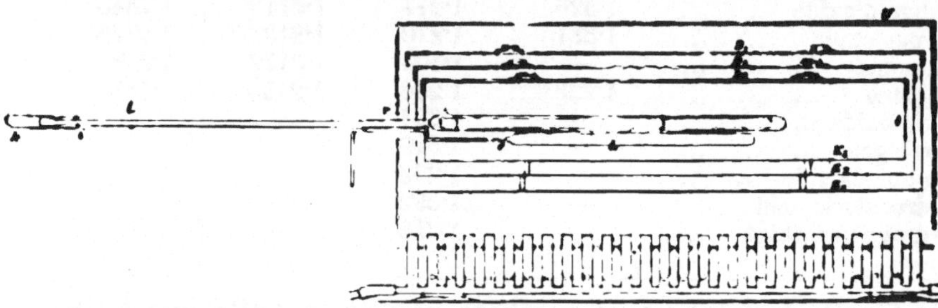

FIG. 9.

attached to the glass tube, about 3 cm. wide, in which the gas to be studied, as well as a little finely powdered silica, is contained. Inside this so-called "wave tube" there is another tube, open at both ends, in which the wave figures are formed. An inner tube of this kind is necessary, for the vibrations of the glass of the outer tube interferes with their formation there. The left-hand end of the sounding tube projects at *i* into another glass tube, *k*, closed at the opposite end, and containing air and lycopodium powder. This tube serves as a comparison tube. The heating apparatus in which the experimental tube is contained is really nothing but a large air-bath heated by a row of bunsen burners. It has triple walls and floor, K_1, K_2, and K_3, with corresponding covers, D, D_2, and D_3. Thermometers are introduced through the end walls at *r* and *t*. Alongside the wave tube is an air-thermometer, L, with a thin stem *e*. When the sounding tube is rubbed to the right of *s* with a

[1] *Pogg. Ann.*, 157 (1875), 353.
[2] *Wied. Ann.*, 13 (1881), 20; 17 (1882), 85.

damp cloth, so that a longitudinal vibration and a high note are produced, the air in the comparison tube and the gas in the wave tube are set vibrating. The length of these stationary waves is shown by the lycopodium and silica powders. Kundt and Warburg,[1] as well as Strecker, take the value of κ for air which Röntgen obtained in his above-mentioned experiments. The accurate measurement of the wave-lengths in the experimental tube is rendered somewhat difficult by the fact that the glass tube contracts slightly on cooling, and the distance between nodes appears somewhat shorter than it should.

Strecker's experiments.

Strecker found the following values :—[2]

Cl_2	Br_2	I_2	ClI	BrI	HCl	HBr	HI
1·323	1·293	1·294	1·317	1·33	1·394	1·431	1·397

The acoustic method of measuring κ.

At high temperatures we can no longer use glass, and quartz is not applicable because of the large dimensions necessary. This exhausts our supply of available transparent substances, and, since we must see the node deposits in order to measure them, prevents us using the method under these conditions. Nevertheless, the acoustic determinations of wave-lengths at least in air have been twice carried out at temperatures up to nearly 1000° by another method. The method in each case was based on a principle originated by Quincke. When a tube closed at one end is filled with stationary sound waves, there is a wave-crest at the closed end. At a quarter of a wave-length from the end comes the first node, and at a distance of a half wave-length comes another crest, and so on. By using a "hearing" tube, which may be shoved along the tube containing the gas, it is possible to recognize and locate the change from crest to node. Quincke made use of this phenomenon in his acoustic thermometer.

This acoustic thermometer of Quincke consisted of two straight tubes, one a wider interference tube and the other a narrower hearing tube. The interference tube was from

[1] *Wied. Ann.*, 63 (1897), 66.

[2] Strecker's value for chlorine is almost exactly identical with the value which Martini (*Phil. Mag.* (5), 39 (1895), 143) and Beiblätter (*Weid. Ann.* (1881), 565) found at the same time and by another method at 0°. This value was 1·327. Martini measured the lengths of tubes at which the tubes when filled with the gases were in reasonance. These lengths are directly proportional to the velocity of sound in the gases. The experiments were extended to CO_2 and N_2O. (Beibl. *Wied. Ann.*, *l.c.*)

40–150 cm. long, 1–5 cm. wide, and closed at one end. Into this the glass, metal, or clay hearing tube was inserted. It was from 1–2 m. long, 4–6 mm. clear bore, and its walls were from 0·75–1·5 mm. thick. The hearing tube was open at both ends. One end was connected with the ear by a rubber tube 1·2 m. long, 5 mm. in inside diameter, and with walls 2 mm. thick The open end of a sounding box attached to a tuning fork was placed before the mouth of the interference tube, and the hearing tube was shoved in as far as possible. If the hearing tube was now drawn slowly out, maxima and mimima were observed distant a quarter of a wave-length from each other. The distance between any two of the minima could be read off by means of a millimetre scale placed alongside the hearing tube. In this way the ear was made to take the place of the dust figures of Kundt and Warburg.

Stevens' experiments. The method gives good results at ordinary temperatures. The observations are more difficult to carry out at high temperatures. Stevens found the following values for air in this way:—

At	0°	100°	950° C.
κ ...	1·4006	1·3993	$1{\cdot}34 \pm 0{\cdot}01$

The ratio of the numbers for 0° and 100° is more certain than the numbers themselves. They are given to four places, in order to show the decrease of 0·1 per cent. in the value of κ between 0° and 100°. According to Kalähne,[1] Stevens' measurement of temperature is in error at 950° by some 27°, and κ is consequently 0·03 smaller than it otherwise would be.

It is a serious disadvantage, in using the acoustic thermometer, that we are obliged to use tuning forks which will keep sounding for a long while. Only forks of rather low pitch fulfil this requirement. These slowly vibrating forks give very long waves at high temperatures, and it is consequently necessary to use very long interference tubes, and to keep them very exactly at the same temperature; which means, of course, increased difficulty of experimentation. Thus, according to the different tones used, Kundt and Warburg found a half wave-length of 2 cm. in mercury vapour, Wüllner a half wave-length of 6·5 cm. in air and Stevens one of 60 cm. in air at 950°.

Kalähne's experiments. Later Kalähne[2] employed a "resonance" method originated

[1] Drude's *Ann.*, 11 (1903), p. 231. [2] *Ibid.*

by Quincke and developed by Seebeck[1] and Low[2]. It is based on the following principle. A straight tube open at both ends is provided with a movable piston. The piston head is nearly as large as the cross-section of the tube, and therefore shuts off an air-column of variable length extending from it to the mouth of the tube. A telephone is placed directly in front of the tube's mouth, and its membrane is set vibrating by an intermittent electric current at the rate of perhaps a thousand vibrations to the second. These vibrations enter the tube and set up stationary waves in it. At certain definite positions of the piston, always a half wave-length distant from one another, pronounced maxima of resonance are obtained. These are best observed by using a rubber tube opening near the mouth of the tube and leading to the ear. It is advisable to introduce a Quincke interference tube into this rubber tube, thereby dampening the fundamental tone and the undertones, and leaving only the high overtones, with which the maxima may be determined with especial precision.

This arrangement too, according to Kalähne's experience, becomes more difficult to use the higher the temperature. Kalähne found no perceptible change of κ with the temperature up to 450°. He found, with Stevens, that at higher temperatures κ becomes smaller, though he did not find so great a difference as did Stevens. Thus, according to him κ decreases to but 1·39 at 900°, and not to 1·34; that is, it only decreases by about 0·6–0·7 per cent. of its value at ordinary temperatures.

Low's experiments.

To illustrate the applications which this method has found, we will add the results which Webster Low obtained for a number of gases, using it in a somewhat modified form and at ordinary temperatures.

Gas.	κ
Air	1·3968
CO_2	1·2914
Ether vapour	1·0244
(Hydrogen)	1·3604

The value for hydrogen is inexact.

[1] Seebeck, *Pogg. Ann.*, 139 (1870), 104.
[2] Low, *Wied. Ann.*, 52 (1894), 641.

Experiments of Thiesen and Steinwehr.

The last-mentioned acoustic forms of the method are not only difficult of execution, but limited in application, for they involve the use of open tubes, which at high temperatures are difficult to keep filled with gases other than air. In this respect the method which Thiesen and Steinwehr have just announced [1] seems to be a distinct improvement. In this method the number of impulses per second required to set up sympathetic vibration in a closed tube filled with the gas under investigation at a given temperature is determined.

The foregoing survey does not include all the researches of an experimental nature which have been made in this field. It does, however, afford a complete summary of the methods now at our disposal for determining specific heats.

It is evident that everything in this field pertaining to high temperatures is still in a most unsettled condition. The twenty years of quiescence which followed the explosion experiments of Mallard and Le Chatelier and of Berthelot and Vieille has been just brought to an end by the researches of Langen, Stevens and Kalähne, Holborn and Austin, and Theisen and Steinwehr.

At the end of the Fourth Lecture we discussed the possibility of deducing the specific heats at high temperatures from equilibrium measurements. Regnault's method as amplified by Holborn and Austin, the measurement of explosion pressures, and this chemical method offer the best prospects for obtaining accurate results.

[1] *Z. f. Instrumentenkunde*, 23 (1903), 114; and 24 (1904), 133.

SEVENTH LECTURE

THE DETERMINATION OF GASEOUS EQUILIBRIA. THEORETICAL AND TECHNICAL OBSERVATIONS REGARDING RELATED SUBJECTS

THIS last lecture will be devoted to a consideration of matters important in equilibria determinations, especially at high temperatures. In connection with this we shall find opportunity for treating many theoretical and technical questions which we have not heretofore considered.

Physical methods of determining gaseous equilibria.

We may investigate gaseous equilibria in several ways.[1] If a change in the number of reacting molecules takes place during the reaction which leads to the equilibrium, as, for instance, in the reactions—

$$2NO_2 \rightleftarrows N_2O_4$$
$$2SO_2 + O_2 \rightleftarrows 2SO_3$$

we may determine the composition by measuring the changing density of the mixture. We became acquainted in the Fifth Lecture with several examples illustrating this method. Again, if the colour changes, as it does in the dissociation of nitrogen tetroxide, we may also investigate the equilibrium colorimetrically, as Salet[2] has done. The Natanson brothers[3] have realized a third possibility by investigating the velocity of sound in nitrogen dioxide and tetroxide, using Kundt's

[1] The physical investigation of equilibria by means of density determination is treated very fully by Windisch in "Bestimmung des Molekulargewichts" (Berlin, 1892). An extended bibliography is given in the same place.

[2] *Compt. Rend.*, 67 (1868), 488.

[3] *Wied. Ann.*, 24 (1885), 454.

dust-figure method. In this way they studied the effect of changing pressure on the condition of equilibrium.

Chemical determination of the equilibrium.

Physical methods of this sort do not disturb the equilibrium. On the other hand, chemical-analytical methods cannot be applied without doing this. For suppose we abstract one of the gaseous constituents by means of absorption; the other substances present will immediately strive to made good the loss. If this re-formation proceeds with any appreciable rapidity, our analytical methods yield worthless results, for what we get is not the amount of the desired constituent actually present in the equilibrum mixture, but rather that amount increased by what is re-formed during the process of absorption. The use of an analytical method therefore assumes that the velocity with which readjustment of the equilibrium takes place has been greatly reduced.

There are many cases where a gaseous equilibrium is very slowly attained. For instance, Bodenstein was obliged to allow a mixture of iodine and hydrogen to react for five and a half *months* at 283° C. in order to fully attain an equilibrium which had been very nearly attained by heating to a temperature 150° higher for but a short while. It is clear that in such a case we need not fear any perceptible change of composition if the gas mixture is rapidly cooled to ordinary temperature; and when its mixture has once been brought to the ordinary room temperature, the "paralysis" of the reaction is complete, and we may determine the single constituents of the mixture by our ordinary analytical methods. But this paralysis of the reaction by cooling cannot always be carried out. In the first place, there are equilibria, such as that of nitrogen tetroxide, where the adjustment of the equilibrium takes place rapidly so long as the substances concerned remain gaseous. In the second place, the readjustment is a very rapid one in all cases if the temperature is very elevated, and we are unable to cool the mixture rapidly enough to prevent a certain amount of chemical change during the refrigeration.

We will consider this matter of velocity somewhat more closely, taking the hydriodic acid gas equilibrium as an example. According to the fundamental principles of mass action, the velocity of reaction is determined by the difference between the quantities formed and decomposed in unit time. We may measure the formation of hydriodic acid from iodine and hydrogen according to the formula—

$$I_2 + H_2 \rightarrow 2IH$$

by determining the decrease in the concentration of the iodine, C_{I_2}, in unit time. For this we get—

$$-\frac{dC_{I_2}}{dt} = k'C_{I_2} . C_{H_2}$$

where C_{H_2} signifies the concentration of the hydrogen and k' what the change in the concentration of the iodine in the unit of time would be if we lead in hydrogen in such quantity that the product $C_{I_2} \times C_{_2H}$ remained equal to unity. k' is called the velocity constant of the formation of hydriodic acid. The opposed reaction—

$$2IH \rightarrow I_2 + H_2$$

can similarly be represented by—

$$+\frac{dC_{I_2}}{dt} = k''C_{IH}^2$$

where k'' is the velocity constant for the decomposition of hydriodic acid. The two reactions run contrary to one another and bring about the change—

$$\frac{dC_{I_2}}{dt} = k'C_{I_2} . C_{H_2} - k''C^2_{IH}$$

When equilibrium is reached, the change—

$$\frac{dC_{I_2}}{dt} = 0$$

and—

$$k'C_{I_2} . C_{H_2} = k''C^2_{IH}$$

or—

$$\frac{k'}{k''} = \frac{C^2_{IH}}{C_{I_2} . C_{H_2}}$$

We have previously called the expression—

$$\frac{C_{IH}}{C^{\frac{1}{2}}_{I_2} . C^{\frac{1}{2}}_{H_2}} = K$$

the equilibrium constant of the reaction. We see from the preceding formulæ that—

$$K = \left(\frac{k'}{k''}\right)^{\frac{1}{2}}$$

The velocity of the formation and the decomposition of hydriodic acid.

Bodenstein (*loc. cit.*) has measured these velocity constants. He found the following values for the velocity constant of the formation of hydriodic acid, k', and for the velocity constant of its decomposition, k''.

Degrees C.	k'	k''
508	3·58	$1·059 \times 10^{-1}$
443	$3·75 \times 10^{-1}$	$0·670 \times 10^{-2}$
427	$1·72 \times 10^{-1}$	$0·31 \times 10^{-2}$
410	$6·59 \times 10^{-2}$	$0·137 \times 10^{-2}$
393	$3·79 \times 10^{-2}$	$0·588 \times 10^{-3}$
374	$1·40 \times 10^{-2}$	$0·230 \times 10^{-3}$
356	$6·76 \times 10^{-3}$	$0·809 \times 10^{-4}$
326	$1·46 \times 10^{-3}$	$0·326 \times 10^{-5}$
302	$3·53 \times 10^{-4}$	$0·942 \times 10^{-6}$
283	$1·19 \times 10^{-4}$	

The velocity constants, and consequently the amounts changed under comparable conditions increase ten thousand times for a temperature rise of 200°. This is in agreement with the general rule, according to which a temperature rise of 10° doubles the velocity of a reaction. But our cooling agents are subject to certain natural limitations. If, then, we push the temperature above a certain point, differing in different cases, we shall be unable to prevent a certain amount of recombination taking place during the refrigeration. The mixture of gases which we should analyze would not correspond to the equilibrium at the highest temperature, but to one at some lower temperature where the equilibrium had, as it were, been overtaken by the refrigeration.

Graphic representation of the effect of velocity.

We may represent these deductions graphically by picturing a case (see Fig. 10) where the equilibrium constant decreases with falling temperature. We will represent the change of the equilibrium constant by the arbitrary heavy line. Between a and b our determinations will give the true value of the constant K. If we raise the temperature above the point corresponding to b, we find values for K represented by the dotted line. They are too great, for our rate of cooling is too slow to *freeze* our gas, as it were, in an unchanged condition. If we lower the temperature to a point below *a*, we shall no longer attain an equilibrium, because now the rate of adjustment is too slow. We should then get any value in the dotted area depending on the composition of our initial mixture.

Views differ concerning the conditions in this dotted area. It is assumed that the longer we wait, the more nearly do the

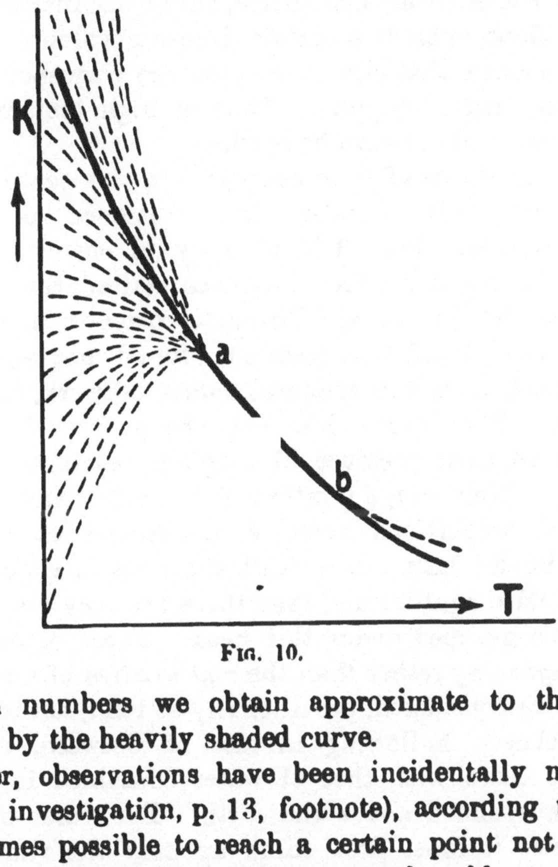

FIG. 10.

equilibrium numbers we obtain approximate to the values represented by the heavily shaded curve.

However, observations have been incidentally made (see Jouniaux's investigation, p. 13, footnote), according to which it is sometimes possible to reach a certain point not far from equilibrium quite quickly, and yet not be able to get any farther, even after prolonged waiting. This is often explained, especially in the writing of French investigators, on the basis of a theory propounded by Duhem. According to this theory, a "false equilibrium" has been attained in these cases, which is not to be considered simply as a *phase* in a slow reaction, but as a *true halting-place* of the reaction. The real point at

False equilibria and passive resistance.

issue is readily seen if we compare our case with the motion of heavy masses on an inclined plane of continually decreasing pitch. The usual view would consider the reaction course in the dotted area to be comparable to the flowing of a viscous liquid down this inclined plane. The rate of flow decreases the more the pitch of the plane decreases, but flow would not cease till the liquid had reached the lowest point. Duhem, on the other hand, would compare the reaction with the sliding down of solid bodies; which, of course, only continues so long as the angle of slope exceeds a certain limiting value. In both cases we can assume that rise of temperature corresponds to an increase in the angle of pitch. Then at high temperature a real equilibrium could always be reached.

While the existence of false equilibria in this sense may be questionable, we may consider their existence as beyond question in another. For while it is by no means generally admitted that a reaction which has once started can of itself come to a real stop before equilibrium is reached, there is no doubt, on the other hand, that reactions often do not *begin* when we would expect, from the chemical forces at work, that they really should. Phosphorus does not change at all in pure oxygen at atmospheric pressure, although it reacts easily with dilute oxygen. Similarly, a mixture of hydrogen and chlorine in equivalent proportions remains unchanged at ordinary temperatures in the dark. If we call the state in which these bodies exist a false equilibrium, then there are very many cases which could be grouped under this head. These observations concern the *beginning* rather than the *continuation* of a reaction, and may well be considered, provisionally at least, as belonging to a special class. Following an idea of Ostwald, we may compare these cases with that of water contained in a deep vessel freely suspended above the earth. The water does not run out, because it must first climb up over the rim of the vessel. This rim represents a "passive resistance." In the same way, we may suspect that there is a "passive resistance" in the chemical illustration cited, which van't Hoff suggested is due to the change in the orientation of the molecules which must precede the reaction.[1]

[1] "Vorlesungen über theoretische Chemie," 2nd edit., 1901, vol. i. pp. 208, 209, and 223.

The location of the points *a* and *b* in Fig. 10 evidently depends not only on the reaction studied, but also to a great extent on the method of experiment which we choose. Deville's method (p. 158) of allowing the gases from his flame to strike into a tube kept cool by water, pushed *b* toward a higher temperature, because it permitted a more sudden cooling. Bodenstein's observations with glass bulbs filled with hydriodic acid, extending over months at a time, pushed *a* to a very low temperature, because by so prolonging the duration of an experiment very small reaction velocities sufficed to attain equilibrium. On the other hand, heating a gas in closed vessels is not suitable for following gaseous equilibria to very high temperatures, for it is difficult to cool such vessels rapidly enough. Bodenstein, therefore, did not go above 508° in studying the dissociation of hydriodic acid. On the other hand, by heating a current of gas it is not possible to bring *a* to such a low temperature as we can by heating the gas in a closed vessel.

Return to the graphic representation of the reaction velocities.

These statements apply to a reaction which is unaffected by catalytic influences. If we have a solid substance at our disposal which greatly accelerates the gaseous reaction, not only are we able to very greatly lower the limiting temperature *a*, but at the same time we secure an important advantage as regards our method of cooling. For if we keep the gas in that region of temperature where the reaction velocity without the catalyst is very small, but with it is very great, then the equilibrium remains fixed *at a constant temperature* the moment it loses contact with the catalyst. The behaviour of mixtures of sulphur dioxide and oxygen, and of nitrogen and hydrogen, furnish us with examples of this. Sulphur dioxide and oxygen easily form sulphur trioxide at 450° in the presence of finely divided platinum, until equilibrium is reached. Yet away from the platinum this mixture of sulphur dioxide, trioxide and oxygen changes very slowly at this temperature. Ammonia is similarly formed from the elements at 1000° in contact with finely divided iron, until equilibrium is reached. Yet this gaseous mixture alters very slowly at the same temperature if the catalyst is absent.

Catalytic influences.

More accurately stated, *every* solid substance exerts some accelerating action on gaseous reactions. Reactions always go, relatively, most slowly in a free gas space. Yet there is the greatest difference in the catalytic activity of various solids.

Nature of catalytic action. The nature of catalytic action is not in all cases clear. Yet certainly a great many, and perhaps all of them, may be ascribed to intermediate reactions. For while Ostwald,[1] several years ago, very rightly emphasized that this explanation was merely *possible* and not *necessary*, it is still true that special investigations of the better-known cases have, in the mean time, always shown this explanation to be the correct or probable one.[2]

But it follows from this that the catalytic action of solids on gases is directly comparable with the reaction of solids with gases. Many investigators now consider that in such heterogeneous reactions there is an equilibrium established on the boundary of the two phases (solid to gas, solid to liquid, or liquid to gas), and that the velocity of the reaction only depends on the rapidity with which the gaseous (or liquid) mixture at the boundary can replenish itself by diffusion and convection. In support of this view, Nernst[3] has pointed out that were there no equilibrium at the boundary of the phase in a heterogeneous reaction, we should have finite differences of force between systems infinitely close together. Yet it is clear that finite forces acting at infinitely small distances must produce infinitely great velocities tending toward equalization. In answer to this, I have pointed out[4] that the same consideration would naturally be applied to the phenomena of solution and vaporation, and, indeed, that it has already been long applied in this way. It is, as a matter of fact, theoretically impossible, on the ground just mentioned, for any constituent of a gas mixture to be completely insoluble in any solid substance with which it is in contact. If we bring a mixture, say of oxygen, sulphur dioxide, and sulphur trioxide in contact with platinum, then Nernst's conception would require that we should consider these three substances

[1] *Physikal. Zeitschr.*, 3 (1902), 318. (Reprint of his lecture at the Naturforscherversammlung in 1901.)

[2] For examples, see Bredig and Haber, *Z. f. angew. Chem.*, 16 (1903), 558.

[3] *Zeitschr. f. physik. Chemie*, 47 (1904), 52.

[4] *Zeitschr. f. Elektrochem.*, 10 (1904), 156.

dissolved, to a certain extent at least, in the platinum. The same thing applies to any other solid, as glass or porcelain, and to any other gases which we may choose to consider in place of sulphur dioxide, trioxide, and oxygen. Now, we may certainly assume that equilibrium exists between a dissolved gas and the same free gas outside. But if both the gas outside in the gaseous phase and the gas inside dissolved in the solid phase be slowly reacting, then a chemical equilibrium will not exist either on the gaseous or the solid side of the boundary layer. Each constituent, however, of the system on one side of the boundary will be in equilibrium with the same constituent of the system on the other side.

Experimental knowledge of velocities in hetrogeneous systems.

The theoretical difficulty of conceiving that finite forces acting at infinitely small distances equalize themselves at the boundary surface, disappears as soon as we assume a partition equilibrium to exist between the substances on either side of this boundary surface. The idea of a slow reaction is by no means incompatible with this assumption. Now, to be sure, we have no experimental evidence whatever of such traces of solution. But this makes no difference with the theoretical discussion; for the sole value of this discussion is simply to demonstrate the impossibility of reaching any *à priori* conclusions regarding the velocity with which equilibrium is attained in the boundary zone of heterogeneous systems.

Since theoretical reasons for assuming an equilibrium at the boundary of a phase are lacking, let us see if experiment furnishes us with any information on the subject. Now, Brunner[1] has shown in several cases that the rate of reaction in the boundary layer solid to liquid actually does depend only on the rate of diffusion. Nernst has been able to show the quantitative relation of diffusion velocity and amount transformed by a very happy treatment of the matter, and Brunner (*loc. cit.*) has experimentally confirmed his conclusions. Yet when we examine more carefully the cases which Brunner studied, we see that so far as chemical reasons and not diffusion velocities are concerned, we could never expect that the reaction would take place with measurable slowness. For, in every case, they merely involved the addition of a *charge* to substances going over into the ionic condition, and the loss of a *charge* by ions

[1] *Zeitschr. f. physik. Chemie*, 47 (1904), 56.

leaving the ionic condition. We universally find that the simple loss or gain of a charge takes place "instantaneously," so far as the measuring instruments now at our disposal are concerned. In more complicated cases, where a "chemical change," in the strict sense, is involved in the reaction at the boundary, I have shown, in collaboration with R. Russ,[1] that no such "instantaneous" adjustment of equilibrium takes place, and that the reaction on the solid-liquid boundary depends on the chemical forces at work.

Nernst's theory of reaction velocity at surfaces of heterogeneity, which Brunner has confirmed, therefore applies to the limiting case, where all the specifically chemical influences which would make the reaction a slow one are absent. It teaches us, then, how the rate of diffusion is determining the velocity when other influences are excluded. This is the case whenever the chemical velocity is very great as compared with the velocity of diffusion. Nernst's theory by no means shows that the chemical velocity *must* be great as compared with the diffusion velocity, but, indeed, when rightly considered, leaves the point wholly undecided. It would certainly be a great simplification if we could *à priori* neglect the chemical influences affecting the velocities of reactions in boundary layers in comparison with the velocity of diffusion. Herr v. Jüptner,[2] for instance, has made use of this assumption in certain numerical computations he has published regarding water-gas. He imagines that on the boundary between carbon and hydrogen at incandescence the two equilibria

$$C + H_2O \rightleftarrows CO + H_2$$
$$C + 2H_2O \rightleftarrows CO_2 + 2H_2$$

exist simultaneously. We shall see, however, at the close of this Lecture, that this assumption, which requires that the gases shall be in equilibrium, both among themselves and with the carbon, does not hold.

Application to the combustion of coal.

The question as to what takes place on the surface of a glowing coal when exposed to air is related to this case. Even at a white heat coal burns completely to carbon dioxide, when

[1] *Zeitschr. f. physik. Chemie*, 47 (1904), 257.

[2] "Beiträge zur Theorie des Generator- und des Wassergases." Stuttgart, 1904, in Ahrens Sammlung chemischer und chemisch-technischer Vorträge.

the air is forced over it at high speed. Yet at equilibrium carbon monoxide ought to be the chief constituent at all temperatures above 800°. Above 1000° carbon monoxide ought to be practically the *only* oxide present! The important effect which the velocity of the air supply has on the products of combustion forms the basis of the Dellwick-Fleischer[1] process for the manufacture of water-gas. We may evidently make two assumptions regarding the phenomenon. We may imagine that the gas mixture formed on the surface of the coal contains originally carbon monoxide in an amount corresponding to equilibrium, but that when the gas mixture has once got out of contact with coal it reacts with the fresh supply of atmospheric oxygen to form the slightly dissociated carbon dioxide, according to the equation

$$O_2 + 2CO = 2CO_2$$

yet this would be the first case on record where oxygen reacts by a primary splitting up into the two atoms constituting its molecule. According to all our experience with autoxidation, the first product formed when a substance burns contains the whole O_2 molecule.[2]

In the case at hand the primary reaction would then be—

$$C + O_2 \rightleftarrows CO_2$$

and we should consider the adjustment of the equilibrium to be brought about by the secondary reaction—

$$CO_2 + C \rightleftarrows 2CO$$

This adjustment would or would not take place, depending on whether or not the carbon dioxide had time to react. As yet we know nothing about the reaction velocities which determine this. Still, there is nothing known which conflicts with the view that in a rapid current of air the primarily formed carbon dioxide does not have time to enter into a secondary reaction

[1] See in this connection *Journal f. Gasbeleuchtung und Wasserversorgung*, 41 (1898), 557 (Bunte); 528 (Leybold). Also 42 (1899), 593 (Lunge); 43 (1900), 355, 373, 575, 672, 694, 709, 957 (Strache and Jahodal); 44 (1901), 393 (Bueb); 47 (1904), 268 (Placidi and Kettner), and 1079 (Keppeler), etc.

[2] See *Zeitschr. f. Elektrochemie*, 7 (1901), 441, 446. In particular, see also *Engler u. Weissberg Autoxydation Braunschweig*, 1904.

with the coal and produce the carbon monoxide which, were there time enough for the establishment of equilibrium, would become the chief constituent of the gas mixture.

Reverting to our discussion of the catalysis of gas reactions by solids, we now see that, if the activity of catalysts depend solely on their relative rates of diffusion, the extent of their boundary surfaces ought to be the sole determinating factor. Observation fails to corroborate this. The catalytic effect of platinum on mixtures of sulphur dioxide and oxygen surpasses that of finely divided porcelain to a degree which is quite inexplicable on any possible assumption regarding the relative extent of the surfaces exposed. Such cases are numerous. It may be remarked that a rise of temperature does not favour diffusion and catalysis to the same degree. Thus, with the same catalyst chemical influences may be the determinating factors at low temperatures and diffusion at high ones. (See Appendix, No. I.)

The temperature coefficient of heterogeneous reactions.

It should be remarked that the temperature coefficient of diffusion has been compared with the temperature coefficients of heterogeneous chemical reactions, in order to decide whether the velocity of diffusion differs from the velocity of the chemical reaction. Rising temperature accelerates diffusion but slightly, while it greatly accelerates chemical reactions.

In the course of a discussion regarding this point between Bodenstein and Stock, the question came up as to what, in this sense, a slight or great acceleration was. Bodenstein considered that the accelerating effect of temperature was great when at high temperatures a rise of 10° increased the velocity constant by 20 per cent. Stock disagreed with him.[1] Bodenstein was certainly right, for as far as our information goes, the coefficient of gaseous diffusion increases no more than proportionally to the *square* of the absolute temperature,[2] that is, for a 10° rise of temperature it would increase by 6 to 7 per cent. at ordinary temperature, and by 2½ per cent. at 500°. The increase of the velocity constant, both of gas reactions and of gas diffusions,

[1] *Zeitschr. f. physik. Chemie*, 50, 112 (1904). Compare Berl, *Zeitschr. f. Anorg. Chemie*, 44 (1905), 267, whose results, in my opinion, favour the reaction velocity rather than the diffusion velocity explanation.

[2] O. E. Meyer, "Kinetische Theorie der Gase," 2nd edit. (Breslau, 1899), § 101.

becomes smaller as the temperature rises, but both preserve approximately the ratio of 10 to 1.

Returning to our graphical representation of the matter in Fig. 10, we see there is still one other way imaginable of determining an equilibrium at low temperatures, to the left of *a*, without actually reaching it; that is, as Nernst has pointed out, the *reaction velocities* may be studied.[1] (See Appendix, No. II.)

Nernst's indirect determination of equilibria. Nernst made observations on the formation of nitric oxide from air at 1538°. He used a platinum tube of 13·5 cm. length, of 0·85 cm. bore, and with walls 0·17 mm. thick, with which to heat his current of gas. This tube was kept hot by means of an alternating current of low voltage. A thermo-element was inserted in the platinum tube, and a magnesium tube, which served to lead the gas out of the zone where the platinum was brightly glowing into a cold tube. When simply air was forced through, Nernst got the following results:—

Quantity of air.	Time (minutes).	c.c. NO.	c.c. NO per minute.
1·3	31·5	c. 0·2	c. 0·006
1·0	70·0	0·4	0·006
1·6	145·0	0·96	0·0066

Rate of formation of nitric oxide. So we see that in the least rapid stream there was 0·96 c.c. NO in 1·6 litre, or 0·6 c.c. per litre. When the velocity was greater the yield of nitric oxide per litre of air was smaller, and yet the rate of formation was practically the same in every case.

Rate of decomposition of nitric oxide. Nernst similarly conducted a mixture containing 3 per cent. NO and 97 per cent. air through the same tube at the same temperature. When t represents the number of minutes required for a litre of this mixture to pass through, he found the following quantity, x, of nitric oxide in the escaping gas (c.c. per litre):

t	x
0	30·0
44	19·9
198	8·2

[1] "Göttinger Nachrichten" (1904), p. 269.

that is, in the slowest current 8·2 c.c. NO remained undecomposed. Evidently equilibrium was neither reached during the formation nor the decomposition. It must have lain between the limits of 0·6 c.c., and 8·2 c.c. per litre. The exact place where it did lie may be found by a further consideration of the numbers given.

What we said above relative to hydriodic acid applies also to the reaction—

$$2NO \rightleftarrows N_2 + O_2$$

namely—

$$-\frac{dC_{NO}}{dt} = k'C_{NO}^2 - k''C_{N_2} \cdot C_{O_2}$$

All concentrations may be expressed in cubic centimetres per litre of gas, for it does not matter what unit we adopt. Suppose that in one case we allow a few cubic centimetres of NO to form in a litre of pure air. In another case suppose we allow a little NO, in a mixture containing 3 per cent. NO and 97 per cent. air, to decompose. Evidently the relative amounts of oxygen and nitrogen change but very slightly in each case. The product $C_{N_2}C_{O_2}$ then remains nearly constant. Consequently, we are not surprised that the rate of formation of NO, represented by the product $k''C_{N_2}C_{O_2}$, was found to approximate closely to 0·0066 c.c. per minute, whatever the velocity of the current of gas happened to be. If we further substitute x for the concentration of the nitric oxide (measured in c.c. per litre) we obtain at the temperature used in this experiment—

$$-\frac{dx}{dt} = k'x^2 - 0{\cdot}0066 \quad . \quad . \quad . \quad . \quad . \quad (a)$$

At equilibrium x will have the particular value x_0, and $\frac{dx}{dt}$ will vanish ; hence—

$$k'x_0^2 = 0{\cdot}0066 \quad . \quad . \quad . \quad . \quad . \quad . \quad (b)$$

Now we are able to calculate the constant k' from the observations of the rate of decomposition by integrating the kinetic equation (a) just given. Nernst found in this way that—

$$2x_0k' = 0\ 0036$$

combining this with *b*, we get—

$$x_0 = 3{\cdot}7$$

That is, when equilibrium is reached at 1538°, there would be 3·7 c.c. NO per litre of air. (See Appendix, No. III.)

Let us now return to a discussion of those cases where it is possible to reach an equilibrium by some known experimental arrangement.

Proof that equilibrium has really been established.

It is evident that we must have some method of telling when, or better whether, the equilibrium has been reached. By far the best method for doing this is to approach the equilibrium from both sides. If we obtain the same composition in each case, we may be sure that equilibrium has really been attained. To apply the method, we start in one case with the substances on the left-hand side of the reaction equation, and in the other with the substances on the right-hand side. We ought to get the same value in each case for the equilibrium constant calculated from the composition of the escaping gases. We must not, however, expect that equilibrium is necessarily reached with equal *rapidity* from both sides. The equilibrium constant represents simply the *ratio* of the velocity constant of direct and counter reactions. If it does not approximate to unit value, then the velocities of the direct and counter reactions will be quite different.

Equilibrium may be reached from both sides in one and the same experiment.

It is possible, under certain circumstances, to attain equilibrium from each side simultaneously and in the same experiment. A method used by van Oordt and myself[1] in studying the ammonia equilibrium may serve as an illustration. The arrangement of the apparatus is shown in Fig. 11. We see at the extreme right a flask containing the compound of ammonia and ammonium nitrate easily decomposed by heat. The flask rests upon a ring bent in one end of a straight copper wire, whose other end is kept hot by an adjoining burner. A constant supply of heat is thus maintained, and in consequence a regular stream of ammonia is developed. This stream of ammonia is dried by passing over lime. It then passes through a bubble counter, A, containing a movable drop of mercury, and into the porcelain tube labelled *eins*, which lies in the electrically heated furnace. There are several wads of

[1] *Zeitschr. f. anorg. Chemie*, 44 (1905), p. 341.

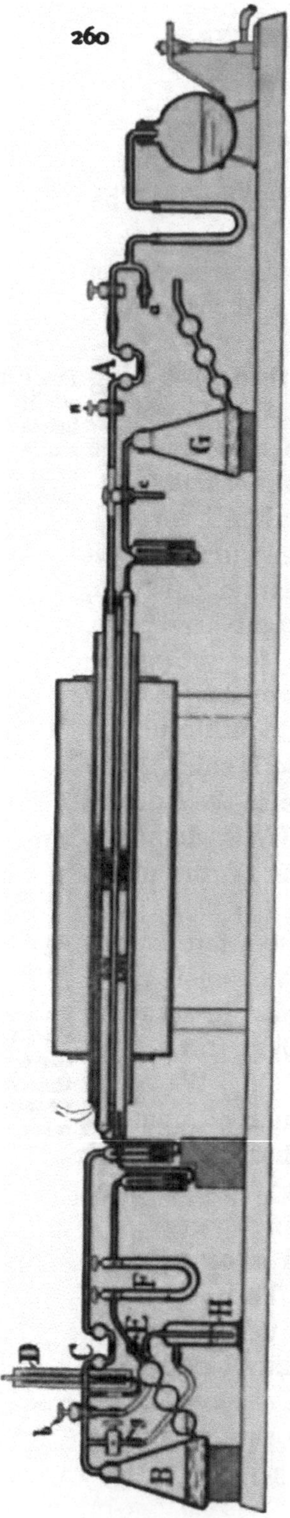

Fig. 11.

asbestos in the middle of this tube. These had been digested in hydrochloric acid till they no longer gave a test for iron, then ignited in a stream of hydrogen, given another thorough treatment with hydrochloric acid, and then finally, after washing with water and drying, impregnated with some ferrous oxalate and placed in the tube. At the beginning of the experiment the ferrous oxalate was first reduced to metallic iron by heating it in a current of hydrogen or of ammonia. The gas passed the iron asbestos and entered the Vollhard absorption flask B, containing sulphuric acid, and seen at the extreme left of the figure, without ever coming into contact with rubber or cork. A glass sleeve made tight with red lead served to attach this flask to the outlet of the porcelain tube. Near this junction was a movable mercury seal such as is used in ozone experiments. Below it and the absorption flask came a bubble-counter C, a mercury manometer D, and two stopcocks. The purpose of these stopcocks as well as the other details of manipulation need not be given here. On leaving the Vollhard flask the gases passed through a three-way cock E, a tube F filled with quicklime and provided with a mercury seal, and entered the second porcelain tube, which was identical in every way with the first. From here they again escaped. without ever coming in contact with rubber, into a second Vollhard absorption flask G, to which a gasometer or an experimental gas-meter was attached, in order to measure the quantity of gas passed through the apparatus.

In cases where an equilibrium cannot be reached from both sides, there are several other ways of deciding whether or not the final condition really represents an equilibrium. In the first place, the duration of the reaction may be increased in order to see whether any further change is thereby brought about. Again, one can investigate whether the equilibrium constant obtained is quite independent of the initial composition of the reacting mixture. If these conditions are varied over wide limits and yet no change is produced in the equilibrium constant, it is reasonable to consider that equilibrium has really been attained. Again, if we know the heat of the reaction, we may then compute the change of the equilibrium constant for a certain moderate change of temperature, using the formula of van't Hoff. If we actually obtain the previously calculated value at the higher temperature, it is strong evidence that the equilibrium constant was correctly determined. Besides, **Indirect evidence that an equilibrium is attained.**

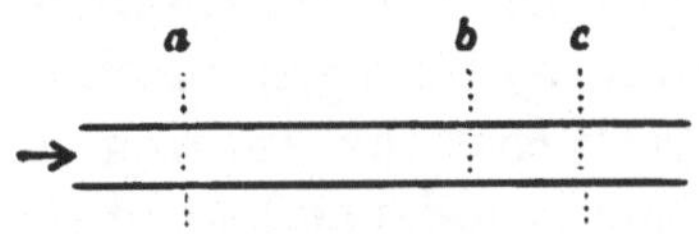

FIG. 12.

if we know the specific heats of the substances involved, then a single correct equilibrium determination permits us to find the free energy for all temperatures and all compositions. We may therefore, in such cases, use observations at widely different temperatures to check our results.

It is often more difficult to further test whether or not the equilibrium has been displaced *during the cooling*. If our oven has a tubular shape, we may, with Nernst (*loc. cit.*), use the above diagram in our discussion (Fig. 12). The gases flow through the tube in the direction of the arrow. We assume that the temperature *t* prevails throughout the length *ab*, and that the gases reach equilibrium in this region. To promote this as much as possible, the cross-section of the tube should be large, thus diminishing the velocity of the gas. Wherever possible we should put catalytic agents in this part of the tube. We will further assume that at *c* there prevails such a temperature as would wholly "paralyze" the reaction. Our **Tests whether the equilibrium has changed during the period of cooling.**

object then would be to so regulate the flow of the gas and so adjust the length *bc*, that the gas would pass from *b* to *c* without in the least changing its composition and yet be cooled from $t \rightarrow t_0$. Our first thought is to decrease the cross-section of the tube at this point to capillary dimensions, and so hasten the passage of the gas through the dangerous region. Using such an arrangement, one finds in general that the composition of the gases escaping at *c* at first depends on the rate of flow; but when this is increased beyond a certain point, the composition remains constant. The natural and agreeable conclusion from this is, that at low velocities the composition changes in passing from *b* to *c*, but that when the velocity is sufficiently increased this displacement no longer occurs, and the escaping gases correspond to the equilibrium condition, provided the equilibrium condition has been reached between *a* and *b*. Yet this conclusion may well be a wrong one, because we know that at high velocities this last provision is not fulfilled. Most of the uncertainty may be removed by trying to reach the equilibrium from the other side. In each case the limiting velocity is found above which the composition is independent of the velocity, and the equilibrium constants which are then determined ought to be equal. But another source of uncertainty still remains. As Nernst emphasizes, an increase in the velocity does not necessarily mean an increase in the rate of cooling. The more rapid current carries more heat along with it, and so prolongs the stretch in which the temperature is falling from $t \rightarrow t_0$. Such an experiment would only yield conclusive results when the *apparatus is changed*, as, for instance, by the addition of a cooling-jacket, so that the length of the tube in which the temperature is falling remains constant in spite of the increasing velocity.

This uncertainty is eliminated if we are able to use such a low temperature in the region *ab* that we may assume that the reaction practically ceases when the gas escapes from it and loses contact with the catalyst. But with gases at high temperatures we must always take the above fact into consideration, for all substances at sufficient high temperature act catalytically on gas reactions.

We may be most certain of our results when, as Bodenstein did in the case of hydriodic acid, we measure not only the equilibrium constant, but also the velocity constants of the direct and counter reactions. For if in the equilibrium determination the point we find lies out beyond *b* (Fig. 10) on the dotted line, it will be disclosed very clearly by an abnormal variation of the velocity constant.

Nernst's theory of the displacement of the equilibrium along a temperature gradient.

We will finally mention an instructive idea of Nernst's,[1] which will show the significance of the rate of cooling in a somewhat different light. We will first assume, for the sake of illustration, that a mixture of nitric oxide, nitrogen and oxygen in equilibrium at 4200° is escaping from a flame. The flame represents a plane surface, and the gas mixture is moving in a direction normal to this surface, and so suffers a very rapid cooling along a very short distance. Let us assume, that the cooling amounts to 1000° per millimetre, and that the equilibrium corresponding to each temperature in every thin lamella along the path is instantaneously adjusted. Then the composition of the mixture will be decidedly different, even at a distance of a single millimetre from the flame. The partial pressure of the NO will have sunk to about half its original value, while the partial pressure of the oxygen and nitrogen will have increased very slightly to counterbalance this. Diffusion will therefore tend to force NO out of the flame into the cooler region, while oxygen and nitrogen, on the other hand, will tend to diffuse back into the flame. The equilibrium in the various layers will, nevertheless, remain undisturbed, so long as the chemical reaction takes place rapidly enough to replace or remove these diffused gases, which would otherwise disturb the equilibrium. But as soon as the reaction becomes so slow[2]

[1] "Boltzmann-Festschrift" (Leipzig, 1903), p. 905.

Mathematical relation between the diffusion constant and the velocity constant of reaction, according to Nernst.

[2] The mathematical theory given by Nernst (*loc. cit.*) for the conditions in each separate layer may be stated as follows: Consider a layer of unit volume (1 litre), in which the mass of the nitric oxide (measured in mols, *i.e.* c_{NO}) experiences a certain increase in the time dt, which may be represented by—

$$\frac{dc_{NO}}{dt} \times dt$$

Since unit mass (in mols) exerts the pressure RT in unit volume (1 litre) (R = 0·0821 litre atmospheres), then the increment of the pressure in the layer is—

$$RT \times \frac{dc_{NO}}{dt} \times dt$$

that it can no longer do this, diffusion forces more and more nitric oxide out of the flame, and more and more oxygen and

Similarly the mass increases in a layer of thickness dx and area q by the amount—

$$q \times dx \frac{dc_{NO}}{dt} \times dt$$

and the partial pressure of the nitric oxide by—

$$RTqdx \times \frac{dc_{NO}}{dt} \times dt$$

Since this increase is due to diffusion, those principles apply which were first enunciated by Fourier (Mach, "Prinzipien der Wärmelehre" (Leipzig, 1900), p. 83 ff.) for the conduction of heat, and later used by Fick (*Pogg. Ann.*, 94 (1855), 59) in the study of diffusion. According to these principles the increase of the partial pressure p_{NO} is—

$$Dqdx \times \frac{d^2p_{NO}}{dx^2} \times dt$$

when D is the diffusion constant. We may best get a conception of the significance of this diffusion constant if we imagine two immense closed spaces at the same temperature throughout. There is nitric oxide in one of the spaces, and a mixture of nitrogen and oxygen in the other, both at atmospheric pressure. In the first there is no oxygen or nitrogen, in the second no nitric oxide. We join the two spaces by a tube 1 cm. long, and of 1 sq. cm. cross-section. The amount of nitric oxide diffusing through in one second represents the diffusion constant.

If, then, the chemical reaction destroys exactly as much nitric oxide as diffusion brings to that particular layer, then the condition is a stationary one. Now, in accordance with well-known principles, the rate of decomposition of nitric oxide at any given temperature is represented by—

$$-\frac{dc_{NO}}{dt} = k'c^2_{NO} - k''c_{N_2} \times c_{O_2}$$

The decrease of its partial pressure—

$$-RTqdx \frac{dc_{NO}}{dt} dt$$

may then, in so far as it is due to decomposition of NO into the elements, be represented by—

$$RTqdx \times dt\,(k'c^2_{NO} - k''c_{N_2} \times c_{O_2})$$

This expression must be equal in the stationary condition to the increase of the partial pressure due to the diffusion

$$Dqdx \frac{d^2p_{NO}}{dx^2} d$$

It therefore follows that—

$$D\frac{d^2p_{NO}}{dx^2} = RT\,(k'c^2{}_{O} - k''c_{N_2} \times c_{O_2})$$

It is easier to evaluate the right-hand members of the equation than the

nitrogen into it. Consequently, the composition of the gas mixture in the cool zone approximates nearer and nearer to that in the hottest zone the less the velocity of the reaction becomes and the shorter the length through which diffusion can act. The diffusion velocity of gases increases rapidly with the temperature. It should, according to the kinetic theory of gases, increase with the 1·5 power of the absolute temperature. Actual experiment at relatively low temperatures indicates that it actually increases with the 1·7 to 2·0 power of the absolute temperature. This explains why, in cooling a gas from a very high temperature, the effect of diffusion outweighs that of reaction velocity, and makes the composition of the mixture in the cooling zone approximate to that of the equilibrium mixture in the flame itself.

The formation of nitric oxide in the explosion of oxy-hydrogen mixtures containing nitrogen.

We shall now examine a little more closely, in the light of these observations, the equilibrium attained in the formation of nitric oxide from its elements, as studied by Nernst. This equilibrium is of especial interest, because the temperature at which oxygen and nitrogen react is very high. Nernst made use of some explosion experiments of Bunsen[1] to determine the position of equilibrium at this high temperature.

Bunsen exploded equivalent mixtures of oxygen and hydrogen to which air had been added, and found the values given in the following tables :—

	Volume of oxy-hydrogen mixture per 100 vols. air.	Residual air (per cent.).	Per cent. NO.	Temperature, degrees C.
1	64·31	99·90	0·07	2200
2	78·76	99·43	0·38	2500
3	97·84	96·92	2·05	2700
4	226·01	88·56	7·63	3200

left, because the diffusion constant depends, in the first place, on the temperature of the gas mixture, and, in the second, on the composition of the gas mixture. Similarly the increase in rate of the partial pressure change—

$$\frac{d^2 p_{NO}}{dx^2}$$

depends on these same variables, though in a different way.

[1] "Gasometrische Methoden" (2nd edit.), p. 73. Braunschweig, 1877.

From this Nernst calculated the percentage of nitric oxide produced in the explosion. The fact that the nitric oxide formed reacted with the oxygen to form nitrogen peroxide was taken into consideration in the calculation. When this has been absorbed we get the "residual air" given by Bunsen in the table. The formation of nitrogen peroxide uses up half a volume O_2 for every volume NO, so that two-thirds of the deficit between the 100 c.c. of air taken and the "residual air" equals the volume of the entire nitric oxide produced. The temperatures appended to the table represent the maximum temperatures attained in the explosion. It was, of course, impossible to measure them directly. They may be calculated, however, from the specific heats of air, nitric oxide, and the oxy-hydrogen mixture, and from the heat of formation of water-vapour and of nitric oxide. Nernst used Langen's values for these quantities. Since the specific heats of hydrogen, oxygen, nitrogen, and nitric oxide are nearly identical, and have in addition the same temperature coefficients characteristic of all diatomic, permanent gases, this calculation is not so uncertain. It naturally makes some difference in this case what the heat loss due to radiation was, and what allowance was made for it, but Nernst[1] says nothing about it.

Conclusions from the explosion experiments.

Without knowing more about the equilibrium we could draw no conclusions from the numbers given in the table. But Nernst's observations in his hot tube at lower temperatures show clearly that the value 0·07 for 2200° is smaller than corresponds to the equilibrium. (See the record of later experiments in Appendix, No. V.) Even at 2500° the equilibrium is certainly not reached. It therefore appears that the exceedingly short interval of time in which the explosion of the oxy-hydrogen mixture heats the air to something like 2200° does not suffice for the establishment of equilibrium. We could determine the length of this interval of time if we knew the velocity of propagation of the explosion, and the rate at which the gas cooled. But neither of these values can be obtained directly from Bunsen's experiments. The rate of

[1] In regard to the radiation from flames, see R. v. Helmholtz, "Über die Licht- und Wärmestrahlung verbrennender Gase" (prize essay presented to the Vereins zur Beförderung des Gewerbefleisses in Preussen, Berlin, 1890). See also, Nernst, *Physikalische Zeitschr.*, 5 (1904), p. 777.

cooling depended on the width of the eudiometer tube. It was much greater in narrow than in wide vessels. The velocity of propagation in the eudiometer which he used was not constant, but increased as the flame advanced. We may perhaps assume that in Bunsen's experiment the maximum temperature was maintained for about the one-thousandth part of a second. In any case, we may conclude that, using the explosion method, the point *a* (Fig. 10) in this case must lie at least as high as 2500°. On the other hand, experiments three and four lie in the region *ab*, where equilibrium is attained; for the value of equilibrium constant determined by Nernst at lower temperatures extrapolated to their temperatures shows a satisfactory agreement with them (p. 106).

The technical preparation of nitric oxide.

We may here appropriately discuss the preparation of nitric oxide from the air on a technical scale. (See Appendix, Nos. IV., V., and VI.) This process, which we may call the "burning of air," is of the very greatest technical importance, because from nitric oxide, air, and water we can make nitric acid, and from this the nitrates which are of such fundamental importance in agriculture and the manufacture of explosives. It is estimated that the saltpetre beds of Chili will be exhausted by 1940.[1] No other natural deposits of importance are known. Attempts to fix the atmospheric nitrogen by heating calcium carbide and to then obtain a fertilizer which could take the place of saltpetre have been successful so far as the process itself is concerned, but is doubtful how far this fertilizer can be substituted for saltpetre, and its transformation, first into ammonia salt and further into nitrates, does not stand perhaps on so favourable an economic basis as does the burning of the air. It is possible to get nitric acid by the oxidation of ammonia with oxygen, and ammonia may be obtained in great quantities from the distillation (Mond) of coal. But it seems that this change is always connected with an appreciable loss of ammonia transformed into nitrogen, so that here too the economic basis is less favourable than in the burning of air, where all that is needed to produce nitric acid is air, water, and power.

The possible yield of

Let us now examine from this theoretical standpoint the actual attempts which have been made to effect this combustion

[1] Sigfrid Edström, *Transactions of the American Electrochemical Society*, vol. vi. p. 16 (1904).

nitric acid per unit of electrical energy.

of the air. The yield of nitric oxide per unit of applied electrical energy is, of course, the crux of the whole matter. Our first problem will then be to find how much energy should be theoretically necessary to effect the combustion. Here we may base our calculations on nitric acid, for this can be made by a series of spontaneous reactions directly from nitric oxide, water, and the oxygen in the air, without the expenditure of any more energy. These reactions may be collected in the following empirical equation :—

$$2NO + 1\tfrac{1}{2}O_2 + aq. = 2HNO_3 \text{ aq.}$$

The case of the hot arc.

We will consider the case of a quietly burning, high-tension electric arc whose temperature may be placed at something like 4200°. From our previous equation we see that at this temperature the equilibrium constant must equal—

$$\frac{p_{NO}}{p_{O_2}^{\frac{1}{2}} \times p_{N_2}^{\frac{1}{2}}} = 0.29$$

and if air were the starting material the composition would be—

NO	O_2	N_2
10 per cent.	16 per cent.	74 per cent.

Ten mols of NO will yield 630 grms. HNO_3 with the help of air and water. To get this amount of nitric acid we must, therefore, heat 100 mols of permanent gases to 4200°, and besides, furnish the 10 × 21,600 cal. of heat necessary to form the 10 mols NO. Of course, we cannot state accurately the heat needed to raise the 100 mols of permanent gas from room temperature to 4200°, yet the formula—

$$6.8 + 0.0006t$$

for the mean specific heat per mol will not be very much in error. It does not, of course, make much difference whether we take 0° or 20° as our starting-point. We then find for w, the heat needed to heat the gas—

$$w = 100(6.8 + 0.0006 \times 4200)4200 = 3,914,400 \text{ cal.}$$

To this we must add 216,000 cal. for the heat of formation, so that altogether we should have to expend 4,130,400 cal. In electrical units this amounts to 17,286 kilowatt seconds, or

4·71 kilowatt hours for 630 grms. HNO_3. The yield then, per kilowatt hour, would be 134 grms. HNO_3. In practice it is usual to take the kilowatt year as a unit, it equalling $365 \times 24 = 8760$ kilowatt hours. It follows that the possible yield from one kilowatt year would be about 1174 kgs. of nitric acid.

Let us, in the second place, assume that it would be possible to have the arc burn at a temperature 1000° lower. The nitric oxide concentration at equilibrium would then be but half as great as before. The energy needed to heat 100 mols of the gas would be— **The case of the cold arc.**

$$w = 100(6{\cdot}8 + 0{\cdot}0006 \times 3200)3200 = 2{,}790{,}400 \text{ cal.}$$

The formation of 5 mols NO would require 108,000 cal. The total consumption of energy would then be 2,898,400 cal. In electrical units this would be 3·37 kilowatt hours. But our yield of nitric acid would only be 315 grms. The theoretical yield in this case has therefore sunk to 93·5 grms. per kilowatt hour, or 819 kgs. nitric acid per kilowatt year. Lowering the temperature thus diminishes the possible yield. It also involves increased difficulty in the subsequent treatment of the gas, for while it is relatively easy to transform concentrated nitric oxides quite completely into nitric acid by the use of air and water, it is difficult to effect the same conversion with dilute vapours. **Advantages of a high temperature.**

On the other hand, the use of low temperatures offers two advantages. The explosion experiments of Bunsen show that no matter how short the period of reaction is made, there would still be time enough for the adjustment of the equilibrium. It would be practically impossible to force air through an arc at a rate which would exceed that of the explosion of an oxy-hydrogen mixture. But experiment has repeatedly shown that the equilibrium attained at 3200° is permanently paralyzed the instant it enters the cooling zone, while at 4200° the readjustment of the gaseous system is so rapid a one that a part, at least, of the nitric oxide formed certainly reverts to nitrogen and oxygen, unless some special device be used to increase the rate of cooling. One advantage of a low temperature, then, is that it better avoids the reconversion of the nitric oxide. **Advantages of a low temperature.**

Another advantage lies in the fact that the loss of heat energy by radiation is less at the lower temperature. The radiation of heat by gases increases with the temperature much more rapidly than the temperature itself. We can get some idea of the matter by recalling that the energy radiated by a solid body increases with the fourth to fifth power of the absolute temperature. The intensity of the visible radiation from a steadily burning arc is so great that in the fraction of a second it completely blinds the eye. These rays are chiefly absorbed by the walls of the vessel, and hence their energy is lost so far as the formation of nitric acid is concerned. We must evidently take both these points into consideration in designing our apparatus.

There is also a third factor to be considered. Any portions of gas which enter the arc a second time waste just so much electrical energy. For since the equilibrium was reached during the first passage of the gas, a second passage can do no more than reach a former condition, which in the mean time may have been altered by loss of heat or reversion into the elements.

Older attempts to prepare nitric oxide in the arc.

Cavendish was the first to observe the combination of oxygen and nitrogen in electric sparks. Indeed, he was able to prepare some milligrams of saltpetre per hour with the modest appliances of his time. His method was simply to pass sparks through air confined over a caustic alkali solution. In 1892, Sir William Crooks showed that a high-tension arc was especially well suited to the preparation of nitric acid from the air. In 1897, Lord Rayleigh determined the yield of nitric acid obtainable from a 2000-volt arc burning in a closed vessel of 50 liters capacity, into which a jet of caustic alkali, and a gas-supply containing eleven parts oxygen and nine parts air were led. These are the stoichiometric proportions of oxygen and nitrogen required for the formation of nitric dioxide. A conversion of 21 litres of gas per hour was obtained. This corresponds to a yield of 50 grms. nitric acid per kilowatt hour.

Experiments of McDougall and Howles.

After the problem had been shown to be experimentally accessible in this way, McDougall and Howles[1] undertook a more thorough investigation. They burnt the air in a high-tension arc inside of vessels of various shapes. They found that a great deal depended on forcing the burnt gas as rapidly as possible out of the arc and preventing it from getting back

[1] *Memoirs and Proceedings of the Manchester Literary and Philosophical Society* (IV.), 44 (1900), No. 13.

again. The form of vessel they finally chose is illustrated in Figs. 13 and 14. The air enters the vessel at D at the rate of 20 liters per hour. A and A′ are the electrodes, and E a glass-covered observation-hole in the earthenware vessel. McDougall and Howles obtained 34 grms. of nitric acid per kilowatt-hour under the most favourable conditions, using air. Using Lord Rayleigh's mixture of oxygen and air they obtained 67 grms. McDougall and Howles instinctively ascribed their double yield from a mixture containing 67 per cent. O_2 and 33 per cent. N_2, 20·9 per cent. O_2 and 79·1 N_2 (air) instead of to mass action. Since the rate at which the nitric oxide is formed is dependent on the product—

$$C_{N_2} \times C_{O_2}$$

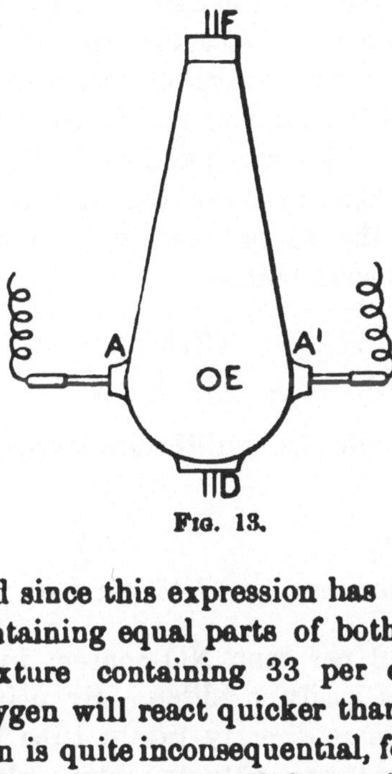

FIG. 13.

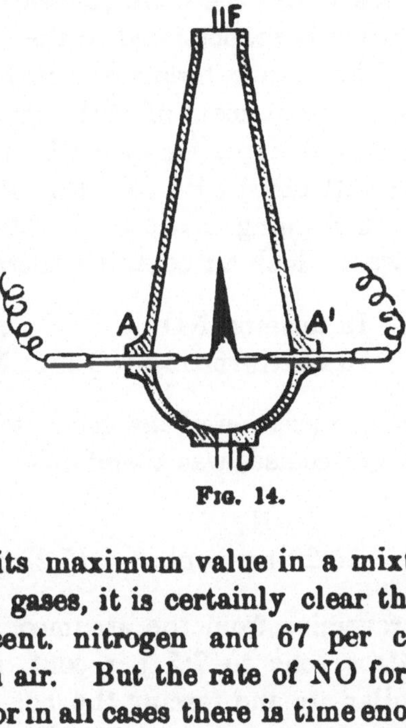

FIG. 14.

and since this expression has its maximum value in a mixture containing equal parts of both gases, it is certainly clear that a mixture containing 33 per cent. nitrogen and 67 per cent. oxygen will react quicker than air. But the rate of NO formation is quite inconsequential, for in all cases there is time enough for the attainment of the equilibrium at the temperature of the arc. McDougall and Howles have therefore overestimated the importance of mass action here, just as others have done in the preparation of sulphuric acid (p. 199).

One might at first lay stress upon the view that the substitution of a mixture of one-third nitrogen and two-thirds oxygen for air would increase the NO concentration at equilibrium, The supposed mass action of the oxygen.

since the temperature and consequently the equilibrium constant is not changed. The arc which McDougall and Howles used was certainly very hot. We shall not be very greatly in error if we place its temperature at 4200°. With air this would correspond to a NO content of 10 per cent.; with the mixture containing more oxygen it would correspond to 12·5 per cent. This evidently does not explain a doubling of the yield. The only conceivable explanation from this point of view is that when oxygen is used the equilibrium is "frozen," or fixed, much more quickly in the cooling zone than when air alone is used. But there is absolutely no reason for believing this would be the case. Indeed, with the same rate of cooling, the decomposition of the nitric oxide would be decidedly more rapid in the mixture where it was the more concentrated, for the rate of its decomposition is proportional to the *square* of its concentration.

The nitric oxide content of the gas mixture in McDougall and Howles' experiments.

The matter becomes a little more comprehensible when we note the statement of McDougall and Howles that in their most successful experiments with air (34 grms. yield of HNO_3 per kilowatt hour) 51·5 per cent. of the oxygen of the air was used up in forming nitric acid. Of the 21 per cent. by volume of oxygen which air contains, there were used—

In forming NO	3·716 per cent.
In controverting NO into N_2O_5 . .	5·574 „

The composition of the gas at which the equilibrium stopped as the gas cooled, was therefore—

N_2	O_2	NO
75·28 per cent.	17·28 per cent.	7·44 per cent.

A reversion from the assumed 10 per cent. NO content in the hottest zone to 7·5 per cent. is easily credible. By using a similar arc and forcing the hot gases directly into a tube kept cool with running water, Brode[1] was able to obtain almost the same amount. Muthmann and Hofer[2] with similar arcs found a 6·7 per cent. content of NO without using any such cooling device. To explain the increased yield of NO in McDougall and Howles' experiments on the basis of mass action

[1] "Habilitationsschrift" at Karlsruhe (1905).

[2] *Berl. Ber.*, 36 (1903), 438.

would necessitate assuming that 15 per cent. NO was not only formed but also retained in the gas, neither of which assumptions is at all plausible. The improved yield must rather have been due to some unintentional change in the arc. The great influence of the particular conditions under which the arc burns is clearly shown by the experiments of McDougall and Howles on this point.

Influence of the shape of the arc as found by McDougall and Howles.

Indeed, working under the same conditions they obtained in the same time almost identical amounts of nitric acid, using in one case 174 watts per second and in another 302 watts (current strength 0·18 to 0·34 amp.). 130 watts was therefore wasted in the second case simply in radiation. If we use the fixed arc of McDougall and Howles, we are unable to produce long arcs without a large consumption of electricity. We get short and thick arcs, in which the gas lingers too long, and needlessly radiates its heat away. We could accomplish much more if some arrangement were used in which the arc is kindled and expands through a great volume of air, and when it goes out is quickly kindled again. In this way we should obtain long, thin, thread-like arcs which would heat every part of the air through which they pass for an excessively short period of time. The amount of useless radiation and consequent loss of energy would be reduced. At the same time the rate of cooling in this way would be very rapid. Bradley and Lovejoy[1] on the one hand, and Birkeland and Eyde[2] on the other, have followed out these considerations. (See Appendices, Nos. IV., V., and VI.)

Technical apparatus of Bradley and Lovejoy.

Bradley and Lovejoy used mechanical, Birkeland and Eyde magnetic, means to realize the above conditions Bradley and Lovejoy constructed an iron cylinder 1·54 metres high and 1·23 metres in diameter. A steel tube was placed in the longitudinal axis of the cylinder and was rotated at the rate of 500 turns per minute by an electric motor. This tube was connected electrically with the positive pole: 23 brass collars were fastened one above the other along it. Six metal posts were attached to the outer circumference of each of these collars at an angular distance of 60° from one another, and all pointing in the same horizontal plane. Each post was tipped by a platinum needle

[1] *Siehe Zeitschr. f. Elektrochemie*, 9 (1903), 382.

[2] For data see *Zeitschr. f. angew. Chemie* 18 (1905), 217.

0·1 mm. in diameter. Corresponding to these crowns of thorns, 6 rows of 23 opposite electrodes were led in through the walls of the iron cylinder. They were insulated by short porcelain tubes, and each was connected to the negative pole of the dynamo. Each of the 23 collars, or crowns, was turned 2·5° around its neighbour immediately below. Turning the steel tube 2·5° about its axis, therefore, brought each needle exactly above the place where the corresponding needle of the next lower collar had been before. The opposite electrodes were small platinum-iridium hooks, and whenever a point got near a hook the spark struck across and the arc was lighted. The rapid turning of the axis steadily lengthened the arc till it finally broke at a length of some 15 cms. Meanwhile, the arc had been kindled on the points of other collars, and these arcs in turn were stretched and broken. In this way some 6900 arcs were formed every second. There were between 250 and 300 arcs burning at the same time. A direct current of 10,000 watts fed these arcs. It was supplied at a pressure of 10,000 volts. The heating current was then 1 ampère and the current per arc about 3×10^{-3} ampère. These tiny current strengths, and the great average length of the arcs, naturally resulted in making the arcs extremely thin. The air blown into the vessel could, therefore, only remain for the briefest instant in the arc long enough to allow for the heating of the gases and the adjustment of the equilibrium, but affording much less opportunity for recombination or loss by radiation than did the apparatus of McDougall and Howles. Although the apparatus makes it easy for the gases to enter a second arc and so be heated again, still the yields with this apparatus are much better than in the arc of McDougall and Howles. 88·3 grms. of nitric acid per kilowatt hour, or 770 kgs. per kilowatt year, was obtained.

Technical apparatus of Birkeland and Eyde.

Birkeland and Eyde have devised an apparatus which is distinctly superior to that of Bradley and Lovejoy. By its means they have been able to obtain a still better yield of nitric acid per kilowatt. The principle of the apparatus is illustrated in Fig. 15. An alternating-current arc burns under high tension[1] between copper poles. A strong magnetic field,

[1] Birkeland and Eyde's method is in practical operation at the works of the Actieselskabet det Norske Kvaelstofcompagni, in Arundel, Norway. Bradley and Lovejoy's method does not appear to be in operation.

set up by a direct current, constantly "blows" the arc away from its starting-point, so that it is forced outward into semicircular form and finally broken off. One arc follows the other. Those belonging to one phase of the alternating current are blown upward, those belonging to the opposite phase are blown downward. The condition of affairs indicated by dotted lines in the figure is thus brought about. The whole plate-shaped arc is contained in a flat chamber through which the air is passed. The yield is said to amount to 900 kilograms per kilowatt year, and sometimes to as much as 950 kilograms.

The form of apparatus devised by Birkeland and Eyde, like that of Bradley and Lovejoy, furnishes an excess of air to the gases after they have passed through the arc. Therefore the

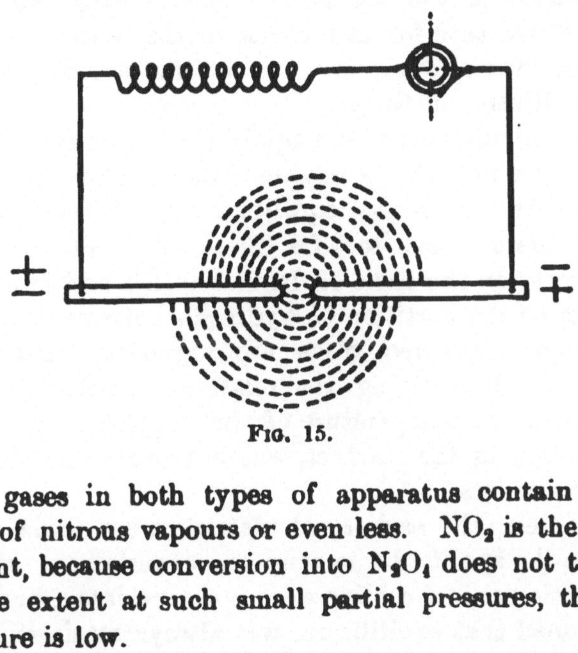

FIG. 15.

escaping gases in both types of apparatus contain only 2·3 per cent. of nitrous vapours or even less. NO_2 is the principal constituent, because conversion into N_2O_4 does not take place to a large extent at such small partial pressures, though the temperature is low.

Comparing the actual yields with the theoretical requirements, we see that it will be difficult to make much further progress in this direction. The 900 kgs. per kilowatt year already amounts to four-fifths of the theoretically possible yield when we instantaneously cool gases heated to 4200°. If we heat the gas beforehand in some cheaper way it will improve matters, provided the rapidity of cooling is not lessened by so doing. Howles and McDougall got poorer yields when they used heated

The technical results considered from a theoretical standpoint

air. Besides, the improvement of the yield per unit of energy is no longer the point of greatest importance : 900 kgs. of nitric acid corresponds to 200 kgs. of nitrogen in the combined condition. Combined nitrogen has a market value of about 25 cents per kg. A kilowatt hour per year, or a kilowatt year costs in favourable localities about $10.00. This means that the yield of nitric acid pays for the power five times over.[1]

Under these circumstances we see that the technical solution of the problem depends on how simply and completely we can convert the oxides of nitrogen into nitric acid. This part of the procedure is materially assisted by keeping the percentage content of oxides of nitrogen as high as possible in the gases from the arc. It follows from this that a further perfection of the electrical part of the process should keep this in view, and hence strive both for an increase in the temperature of the arc, and an increased rapidity of cooling, avoiding also any subsequent dilution of the gas.

A further discussion of explosions.

Nernst's calculations of the equilibria in Bunsen's explosion experiments are not wholly unique. Calculations by Hoitsema dealing with the water-gas equilibrium have been mentioned before, and there exists, too, Le Chatelier's[2] treatment of the Deacon process on the basis of Hautefeuille's and Margottet's[3] experiments on the partition of hydrogen between chlorine and oxygen when the three gases are exploded together. The idea itself, of determining the relative concentration in the equilibrium at the temperature of the explosion by measuring the distribution in the product, was enunciated by Horstmann at a very early date.

Nevertheless, the earlier calculations were made either in cases where the specific heats were too uncertain, or where the rates of reaction and of cooling were not considered; but, instead, it was assumed that equilibrium was always attained, and that it was not displaced during the cooling.

The measurement of extreme temperatures.

If we would gain an insight into equilibrium conditions at

[1] It seems, however, that the result of 900 kgs. per kilowatt year is not the regular one. The guarantee given by Birkeland and Eyde does not exceed, up to date, a yield of 550 kgs. per kilowatt year working on a commercial scale.

[2] *Compt. Rend.*, 109, 664.

[3] *Ibid.*, 109, 641

high temperatures, it is evidently of prime importance to be able to measure these high temperatures with certainty. We saw, to be sure, in the Fifth Lecture, that theoretically it is possible to deduce both the temperature and the location of the equilibrium simultaneously from measurements of the explosion pressures. But we also learned there what uncertainties beset this method of Le Chatelier and Mallard. It therefore becomes of especial interest to know how temperatures above the range of our ordinary instruments can be optically measured.[1]

Older attempts to measure temperature optically.

As early as the sixties Ed. Becquerel[2] used the radiation from incandescent solids as an index of their temperature. Le Chatelier first made the method practically available. From this it was but a short step to measure[3] photometrically the intensity of the light radiated from a body at relatively low and independently known temperatures, and having then found an empirical formula expressing the total brightness per unit area as a function of the temperature, to extrapolate from it into regions of high and unknown temperatures. But these optical methods have only found general acceptance within the last few years, when the study of radiation phenomena has given us a clearer conception of the fundamental laws of radiation.

Bolometric investigation of the laws of radiation.

The laws of radiation have been investigated bolometrically; that is, an apparatus has been employed which absorbs incident rays of all lengths which contribute towards raising the temperature. The rise in temperature produced can be followed very accurately by measuring the change in resistance of the substance absorbing the radiation, provided this substance is a metallic conductor. A compact system of fine platinized platinum strips is most suitable for this purpose. The bolometer is an invaluable appliance for measuring radiation, because it is sensitive both for the longest wave-lengths of heat and the shortest wave-lengths of light. Its use, however, requires such care and perseverance as to almost entirely preclude its use as an accessory method for measuring temperature. The result of bolometric investigations have not, therefore, sufficed to establish a practical bolometric system of temperature measurement.

1 W. C. Heraus has recently constructed a thermoelement which can be used up to 2000°. The appliance, however, has not yet been fully perfected.

2 Ed. Becquerel, *Ann. Chim. Phys.*, 86 (1863), 49.

3 Le Chatelier, *Compt. Rend.*, 114 (1892), 214.

Yet the real *significance* of the optical method depends very largely on the results of bolometric measurements.

Basis of the theory of radiation. Kirchoff's first Proposition.

All the measurements of radiation are based on two principles enunciated by Kirchhoff.[1] The shortest statement of the first of these is that every body sends out the same radiation which it absorbs at that temperature. A body which is not in the least transparent, and which does not reflect at all, but absorbs all incident radiation at every temperature, is called *absolutely black* by Kirchhoff. Its radiation, for the sake of brevity, is called "black" or "dark" radiation. It is, at the same temperature, greater than that of every other body which is to a certain extent reflecting or absorbing, both in sum total and in the value of every single wave-length. Generally, we call the intensity of its radiation unity, on which basis the radiation of any other body is expressed by a fraction representing that part of the "black" radiation to which it is equivalent.[2] In this case the radiative power E of any substance is equal to its absorptive power A. On the other hand, if we designate the emissive power of the absolute black body as S, then—

$$\frac{E}{A} = \frac{S}{1} = S$$

Temperature-radiation, and luminescence.

This statement holds with one important reservation. The radiation must be a pure *temperature radiation*. That is, it must be caused solely by heat, and not by chemical or electrical effects. Radiation due to such chemical or electrical effects is called "luminescence." A luminescent body can emit far more radiation than the absolute black body at the same temperature. A gas in a Geissler tube glows because of the luminescence produced by the passage of electrical discharges through it. Phosphorus shines because of the luminescence produced by the process of oxidation. At higher temperatures we have the

[1] For the older literature, see Wüllner, "Lehrbuch der Experimentalphysik," 5th ed., vol. iv., Strahlung, 3rd chapter (Leipzig, 1899). Lummer, in his "Ziele der Leuchttechnik" (Munich, 1903), gives a review of the newer literature, especially of his own work. For the theoretical side of the question, see Drude's excellent "Lehrbuch der Optik" (Leipzig, 1900).

[2] The radiation of the absolute black body varies as the square of the index of refraction of the surrounding medium. The refractive indices of gases, however, are so nearly equal to that of air that this variability can be completely neglected.

phenomenon of luminescence in the green inner cone of a Bunsen flame burning with an abundant air-supply, and more generally in every explosion of gases. (See Appendix, No. VII.) Every gas flame from which solid particles (of carbon) or the vapours of salt are absent shines chiefly, if not exclusively, because of luminescence. On the other hand, solid bodies exhibit this phenomena at high temperatures with comparative rarity.

Solid bodies. Bodies which are not absolutely black either transmit radiation or reflect it. If the coefficient of reflection[1] and of transparency[2] for a certain temperature is given, then the coefficient of absorption[3] and at the same time the coefficient of emission are known (taking the coefficient of emission of the absolute black body as unity). We see immediately that the transparency of many solid substances is zero. No heat or light rays can pass through a piece of carbon, of platinum, or any similar material of any appreciable thickness at any temperature. The difference between their radiation and that of the absolute black body is based, then, solely on their different powers of reflection.

Flames. Flames in which solid particles, or the vapours of a salt are glowing, leaving luminescence out of account, present in general quite a different case. An ordinary flame of illuminating gas in which solid particles of carbon are glowing, is still highly transparent to radiation, as evidenced by our ability to see through it. The more glowing particles it contains, the more opaque it becomes. It must be considered quite opaque when a mirror or a second similar flame placed behind it no longer increases its luminosity. On the other hand, the reflecting power of such flames is almost always very small. We can convince ourselves of this by directing a beam of sunlight or light from an electric arc transversely through the flame, and again observing the previously measured radiation. If this has been perceptibly increased, it is evident that the flame has reflected a part of the incident radiation. If it does not increase, then the flame can have no perceptible reflecting power.

[1] That is, the fraction of the radiation which is reflected.

[2] That is, the fraction of the radiation which is allowed to pass through.

[3] That is, the fraction of the radiation which is absorbed.

The sum of the coefficient of reflection R, the coefficient of absorption A, and the coefficient of transparency D are always 1. That is—

$$A + R + D = 1$$

If the body is opaque, then D = 0, and—

$$A = I - R$$

Grey and coloured radiation. If the coefficient of reflection of the opaque body is not equal to zero (as in the case of the absolutely black body), but is the same for all wave-lengths and temperatures, then its coefficient[1] of absorption, and consequently its coefficient of emission, bears a fixed ratio to that of the absolutely black body. Such a body is called, from a bolometric standpoint, grey. If the coefficient of reflection of a body, either for different wave-lengths at the same temperatures or in its sum total, changes with changing temperatures, then the body differs more or less in its radiation from the absolutely black body, depending on the temperature. It is then a selective radiator, and is frequently called "coloured" in a bolometric sense. The designations "grey" and "coloured" must not be thought to refer to the appearance of these bodies to the eye. A body can have all the colours imaginable at ordinary temperatures, and yet approximate extraordinarily close to the behaviour of the absolute black body at higher temperatures. Conversely, it can be highly coloured in a bolometric sense, and yet at ordinary temperatures appear wholly white to the eye. The reason for this lies partly in the fact that the colours which a body shows at ordinary temperatures have nothing directly to do with its colour when incandescent. But another circumstance is of at least equal importance.

The small range of optically active wave-lengths. Our sense of colour depends wholly upon the phenomena of absorption within the narrow range of wave-lengths between 0·4 and 0·8μ ($\mu = 0{\cdot}001$ mm.) visible to our eyes, while the spectrum of radiation measurable by the bolometer extends

[1] Here we tacitly assume that the incident radiation always strikes the body in question at the same angle, most simply at a right angle. The coefficient of absorption, and therefore also the coefficient of radiation, is independent of the angle of incidence only in the case of the absolutely black body.

from the very shortest waves up to 19μ—often, indeed, further. And it is precisely the non-visible radiation which is of paramount importance from a bolometric point of view, because at all temperatures attainable in practice, its energy of radiation is tremendously greater than that of the visible radiation.

Kirchhoff's second principle.

The second principle enunciated by Kirchhoff upon which the theory of radiation is based concerns the actual realization of an absolutely black body. Kirchhoff pointed out that black radiation must exist in every completely enclosed cavity whose walls were opaque and at the same temperature. Every bundle of rays in such a cavity would be identical both in quality and in intensity with the radiation from an absolutely black body at the same temperature.[1]

The realisation of the absolutely black body.

Starting from this principle, it was but a short step to actual realization of the absolutely black body. Yet this step was not made till some forty years afterwards by Wien and Lummer.[2]

If, indeed, every bundle of rays in a closed isothermal cavity surrounded by opaque walls exactly corresponds to radiation from an absolutely black body, then the rays which issue from a small opening out of such a cavity would not differ by a measurable amount from black radiation. If we would be very cautious, we may line the inside of the cavity with a layer of non-reflecting material. Lummer and Pringsheim[3] chose a mixture of chromium, nickel, and cobalt oxide for this purpose. Their later experiments showed that such niceties were not necessary. "Black radiation" is emitted from every cavity where the opening is not altogether too large.

Experiments of Lummer and Pringsheim.

Lummer and Pringsheim used a porcelain tube closed at one end as a cavity. It was open in front and encased in sheet platinum, which was kept heated to incandescence by an electric current. The temperature inside this tube was measured by means of a thermoelement (platinum—platinum-rhodium, after Le Chatelier). By this means the relation between the black radiation and the temperature could be followed up to

[1] For the proof of this see Pringsheim, *Verhandl. d. Deutschen physik. Gesellschaft*, 3 (1901), 83.

[2] Wien and Lummer, *Wied. Ann.*, 56 (1895), 451.

[3] *Verhandl. d. Deutschen physik. Gesellschaft*, Bd. 1 (1899), 23 and 215.

1500°. The observations confirmed in a most striking fashion a law which Stefan had originally obtained empirically, but which Boltzmann had later deduced[1] from theoretical considerations with the necessary limitation to radiation from an absolutely black body.

This law says that the sum total of the radiation emitted by an absolutely black body at the absolute temperature T is proportional to the fourth power of the absolute temperature.[2]

If we call E the intensity of radiation, then $E d\lambda$ represents the radiant energy for an infinitely narrow strip of the spectrum. (The letter λ here represents the wave-length.) The

[1] *Wied. Ann. der Physik*, 22 (1884), 31 and 291.

[2] The law can be viewed as a simple consequence of the relation (p. 22) which stated that—

$$U = A - T\frac{dA}{dT}$$

Radiation, being a progressive wave-motion, exerts a pressure upon the structure it strikes (like every progressive wave on a water surface). Now, the pressure which gas molecules exert on the plain walls of a containing vessel because of their impacts from all directions can be mathematically replaced, as Joule first showed (O. E. Meyer, *Kin. Theorie der Gase* (Breslau, 1899), § 10), by just one-third the number of exactly perpendicular impacts. In the same way we may imagine the pressure exerted by the diffuse radiation against an absolutely black surface as having been caused by a precisely normal radiation of one-third the intensity. But it is the peculiarity of the absolutely black body that it absorbs *all* the rays which strike it. It therefore follows that one-third of all the radiation which the walls absorb appears as pressure, and would be capable of doing work were the absolutely black body connected to some suitable machine. Equilibrium, with a maximum production of work, would therefore only be obtained on the surface of an absolutely black body at a constant temperature and with radiation E incident from all directions, when one-third of the radiant energy E absorbed is given off as mechanical work. Then the quantity of the total energy U given off is equal to − E, and the work done A is equal to $\frac{1}{3}$E, and it follows that—

$$-E = \tfrac{1}{3}E - \frac{T dE}{3 dT}$$

or

$$4E = T\frac{dE}{dT}$$

Integration gives us the Stefan-Boltzmann law.

total radiation of the absolutely black body is therefore, according to Stefan-Boltzmann—

$$\int_{\lambda=0}^{\lambda=\infty} E d\lambda = s \,.\, T^4$$

where s is a constant to be experimentally determined. Measurements by Lummer and Pringsheim at known temperatures up to 1500° agreed excellently with this formula. Knowing s from these measurements, it was then possible to reverse the procedure and to calculate temperatures up to 2000° from observations of the radiation of an absolutely black body by means of the bolometer. The absolutely black body consisted of a carbon tube with thickened ends, to which the current leads were attached.

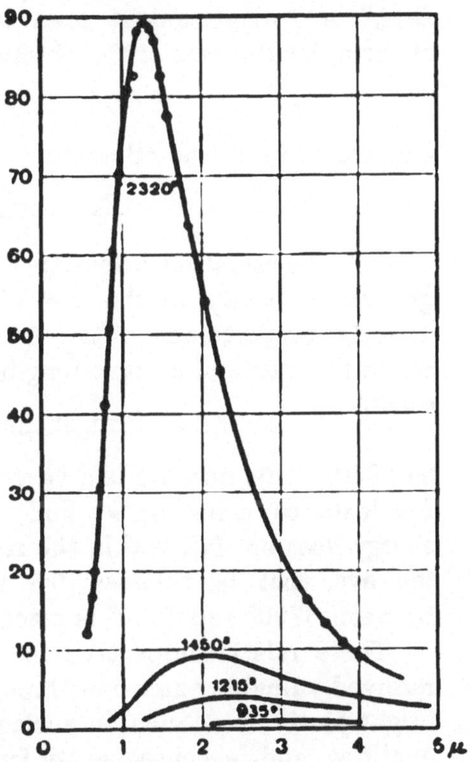

FIG. 16.

The relation between radiant energy, wave-length, and temperature, at various temperatures of the absolutely black body, is represented in Fig. 16. The wave-lengths in μ's (thousandths of a millimetre) are plotted horizontally, and intensities of radiation, measured in arbitrary units, are plotted vertically. It is remarkable how small the visible fraction (between 0·4 and 0·8 μ) of total radiation is, even at the highest temperature (absolute) given in the diagram.

The single observations seen upon the highest of the radiation curves in Fig. 16 were obtained by measuring the radiation with the bolometer after the radiation had been resolved into a spectrum. The resolution is best brought about by using a prism of sylvin, which is transparent for very long waves (19μ) (fluor-spar is only transparent up to 12μ).

Maximum radiation and the displacement law of Wien.

The measurement of the radiant energy of the single wave-lengths possesses significance independently of its importance in determining the curve for total brightness. Indeed, it is possible in this way to find for every temperature a wave-length at which the radiant energy has a maximum. This wave-length is connected with the temperature by two relationships discovered by Wien, usually spoken of as the "Verschiebungs Satz," or "displacement laws of Wien." According to one of them, for the case of the absolutely black body—

$$\lambda_m . T = \text{const.}'$$

and according to the other—

$$E_m = \text{const.}'' T^5$$

That is, the radiation intensity E_m for that wave-length λ_m of greatest intensity, in the case of an absolutely black body, is directly proportional to the fifth power of the temperature, while this particular wave-length is also given by the quotient $\frac{\text{const.}'}{T}$. Lummer and Pringsheim found the value of const.' to be 2940. This number is a very valuable one for many reasons. For instance, using it, we find that the maximum of radiant energy does not fall within the region of wave-length visible to the eye, that is, between 0·8 and $0·4\mu$, until a temperature between 3700° and 7400° is reached.

Bright platinum as a minimum radiator.

These relationships, true for absolutely black bodies, have received valuable extension through a study of the behaviour of bright platinum. Such a surface has very strong reflecting qualities, and is consequently far removed, optically, from the black body. Without this investigation the laws of black radiation would be merely limiting laws, from which an ordinary solid substance freely glowing might diverge very widely. The investigation of bright platinum, on the other hand, afforded a knowledge of the behaviour of an opaque solid substance which, because of its high reflecting power, possessed the character of a minimum radiator. It is here assumed that luminescence does not set in, and that the reflecting power of carbon, ferric oxide, etc., does not surpass that of bright platinum at high temperatures, which is certainly reasonable. The interesting fact is, that the radiation of bright platinum still approximates very closely to that of the absolutely black body. The form of their radiant energy curves is quite the same. Only the total energy of radiation does not vary as the fourth, but rather as the fifth power of the absolute temperature. The displacement of the maximum of radiation satisfies the formula—

$$\lambda_m T = 2630$$

which differs from the formula applying to the absolute black body only in having a somewhat smaller constant. The fact that so slight a change is produced by a substitution of bright platinum for the absolutely black body is of great importance in determining temperatures from radiation phenomena. It alone makes it possible for us to apply without great error methods which in principle are only permissible with absolutely black bodies, to solid, opaque substances of perceptible reflecting power, such as the carbon filaments of an incandescent lamp, or any other glowing conductor.[1]

Optical measurement of temperature on basis of the laws of radiation.

Limiting our discussion for the moment to black radiation, we see that the laws of radiation furnish a basis from which we may measure the temperature of the black body optically, when we know either the change of intensity of radiation of a single wave-length in the visible region, or the change of the total visible radiation with the temperature. The photometer could then be conveniently used in place of the bolometer as a measuring instrument, and the light radiated by one square millimetre of the black body's surface (the so-called "surface brilliancy,") would either tell us directly, or after preliminary resolution into a spectrum and selection of one particular wave-length, the temperature of the black body.

The Wien-Planck formula.

Wien has developed an expression for the intensity of any given wave-length, which Planck[2] has improved by the results of experimentation. This formula is—

$$E = \frac{a\lambda^{-5}}{e^{\frac{b}{\lambda T}} - 1}$$

[1] Consult in this regard the discussion at the Naturforscherversammlung at Hamburg. *Physik. Zeitschr.*, 3 (1901–1902), 97.

[2] *Verhand. der Deutschen physik. Gesellschaft*, 2 (1900), 202.

where a and b are two empirical constants. It can readily be transformed into—

$$E = a\lambda^{-5} . e^{-\frac{b}{\lambda T}}\left(\frac{1}{1 - e^{-\frac{b}{\lambda T}}}\right)$$

The fraction enclosed in brackets would always be very nearly equal to unity for short wave-lengths and any temperature practically attainable. Actual experiment shows that this is true, and it can therefore be omitted. Taking logarithms, we get—

$$ln E = ln a - 5 ln \lambda - \frac{b}{\lambda T}$$

If we measure the intensity of radiation from the same body at two different temperatures, T' and T'', and for the same wave-length, we should find that—

$$ln \frac{E'}{E''} = \frac{b}{\lambda}\left(\frac{1}{T''} - \frac{1}{T'}\right)$$

The constant b can be calculated from the investigations of Lummer and Pringsheim on the radiation of the black body, using the Wien-Planck formula. It lies in the neighbourhood of 14,600.

Wanner's spectrophotometer. A convenient and valuable optical thermometer devised by Wanner,[1] depends in its action upon the above principles. With this instrument we simultaneously observe the surface brilliancy of a black body and that of a ground-glass plate illuminated by an incandescent light. Each half of the field of view corresponds to one of these two sources of radiation. Within the instrument is a direct-vision spectroscope which resolves the incident light into a spectrum. Diaphragms of proper dimensions then cut out all the light except that of wave-lengths near to $0{\cdot}6563\mu$. A pair of Nicol prisms are inserted in the path of the rays coming from the ground glass. By turning one of these (the analyzer) about its axis, the brilliancy of the corresponding half of the field of view may be varied at will. By reading off the angle through which we turn the Nicol, we can obtain a quantitative measure of this change in brilliancy. If we adjust the

[1] *Physik. Zeitschr.*, 3 (1901), 112.

brilliancy of the incandescent field so as to equal that of an absolutely black body, in one case at the temperature T′, and in another at the temperature T″, then the ratio of the two angles as read will represent that of the intensities of radiation of the absolutely black body at these two temperatures. (This, of course, applies only to the wave-lengths $0 \cdot 6563\mu$ and the same radiating surface.) If we know the temperature of the black body in one case, then, by the help of the above formula, we can calculate the temperature in the other case. In order to be independent of any decrease in the brilliancy of the incandescent light after long use, such a temperature T′ of the black body is chosen that its intensity is exactly equal to that of a "Hefner" lamp for the wave-length $\lambda = 0 \cdot 6563$. The von Hefner amyl acetate lamp is in general technical use as a unit of light intensity. It is easily reproduced, and corresponds to a definite temperature of the absolutely black body. What this particular temperature may be is determined once for all by a simple blank measurement with each instrument.[1]

This instrument permits us to measure temperatures from 900° up to 2000°. We cannot go higher, because the incandescent filament then disintegrates, and the glass bulb becomes darkened. But there is no reason why some other source of light should not be used in its place.

Application of the Wanner optical pyrometer

The interior of technical blast furnaces and glowing tubes sends out nearly pure black radiation, so that the Wanner pyrometer may be unhesitatingly used in both these cases. But the temperature even of solid, coherent surfaces glowing in the open may be measured by its means without perceptible error, provided they are intermediate in their behaviour between bright platinum and the absolutely black body. Lummer and Pringsheim have compared bright platinum and the absolutely black body at 1100° abs. and 1800° abs., using another pyrometer based on the same principles, and found a deviation of but 40° at the lower temperature, and 100° at the higher. With poorer reflectors than platinum, with carbon filaments, for instance, the error would be much smaller. A special advantage of the instrument consists in the fact that the comparison of the two halves of the field of view need not be a very accurate one.

[1] See for the use of this instrument, Nernst and v. Wartenberg, *Verhandl. der physik. Gesellschaft*, 8 (1906), 48.

The intensity of radiation increases so extremely fast that an error of 10 per cent. in the adjustment at the temperature of molten platinum causes only an error of 10° in the temperature read.

Measurement of total brightness.

Finally, instead of measuring the intensity of any particular wave-length, we may measure the intensity of the total visible radiation, using an ordinary photometer. The expression given by Wanner applies to every particular wave-length in the visible spectrum. The simplest extension of this expression, and, so far as I know, one which has never before been pointed out, would be to consider the total visible spectrum which is included between 0·4 and 0·8μ as simply an ill-defined band of a mean wave-length, *e.g.* 0·6μ; and, compared with the whole spectrum of wave-lengths which affect the bolometer, that is all it really is. We should then obtain for the photometrically measurable surface brilliancy H′ and H″ of the black body at the temperature T′ and T″, without resolution into a spectrum, the relation—

$$ln\frac{H'}{H''} = \frac{b}{0{\cdot}6}\left(\frac{1}{T''} - \frac{1}{T'}\right) = \text{const.}\left(\frac{1}{T''} - \frac{1}{T'}\right)$$

If, with Wanner, we call b equal to 14,500, which is somewhat uncertain, $\frac{b}{0{\cdot}6}$ equals 24,170. If we denote by $T_{H=1}$ that temperature at which the surface brightness of the black body equals that of one light-unit (one Hefner candlepower), then this expression becomes—

$$ln H_1 = \text{const.}\left(\frac{1}{T_{H=1}} - \frac{1}{T'}\right)$$

or, if we divide the constant by $T_{H=1}$ and call the quotient a,

$$ln H_1 = a\left(1 - \frac{T_{H=1}}{T'}\right)$$

Total brightness and temperature.

Rasch[1] obtained this expression in another and certainly less simple manner, and tested it by experimental data which Lummer and Pringsheim on the one hand, and Nernst[2] on the other, had obtained from photometric measurements of total brightness. Rasch puts a equal to 12·943 and $T_{H=1}$ equal to

[1] Drude's *Ann.*, 14 (1904).
[2] *Physik. Zeitschr.*, 4 (1903), 733.

2068·4° abs., that is, 1795° C. On this basis we obtain the surface brilliancies labelled "calculated" in the following table:—

	Degrees C.	Surface brilliancy in H.C. per sq. mm.	
		Observed.	Calculated.
1	1175	0·0042	0·0039
2	1325	0·0220	0·0222
3	1435	0·0636	0·0653
4	1690	0·50	0·500
5	1690	(0·47)	0·500
6	1780	0·91	0·909

Nernst apparently got number 5 by extrapolation. The temperature of 1780° represents, according to Holborn and Wien, the melting-point of platinum. If we calculate from this and a the value of the radiation constant b, using the relation—

$$a \,.\, \lambda \,.\, T_{(H=1)} = b$$

and putting $\lambda = 0{\cdot}6$, we get—

$$b = 16{,}062$$

while we should have expected (p. 83)—

$$b = 14{,}500$$

or, according to Lummer and Pringsheim, 14,600. The difference is not large, and may be ascribed to physiological causes. The eye is not equally sensitive to light of various colours, and though we consider the boundaries of the visible spectrum to be 0·4 and 0·8μ, we may better put the mean visual wavelength, not equal to 0·6μ, but somewhat shorter, perhaps 0·55μ.

Experiments of Lummer and Kurlbaum.

We may further test this relationship by means of Lummer and Kurlbaum's[1] data regarding the total radiation of bright platinum. They have expressed their results in the following equation:—

$$\frac{H'}{H''} = \left(\frac{T'}{T''}\right)^x.$$

[1] *Verhandl. d. Deutschen physik. Gesellschaft*, 2 (1900), 91.

where x varies with the temperature. Taking logarithms, then—

$$ln\frac{H'}{H''} = xln\frac{T'}{T''}$$

We note that T′ and T″ must be very nearly equal, if we would have x replaceable by a fixed value in this expression, and if we represent the difference between T′ and T″ by ΔT, it follows that—

$$ln\frac{T'}{T''} = ln\frac{T'' + \Delta T}{T''} = ln\left(1 + \frac{\Delta T}{T''}\right)$$

But, according to the rules governing calculation with small quantities—

$$ln\left(1 + \frac{\Delta T}{T''}\right) = \frac{\Delta T}{T''} = \frac{T' - T''}{T''}$$

Thus the Lummer-Kurlbaum formula becomes—

$$ln\frac{H''}{H'} = x\frac{T' - T''}{T''}$$

This expression is identical with the above when we put $x = \frac{\text{constant}}{T'}$, for we then obtain—

$$ln\frac{H''}{H'} = \text{const.}\frac{T' - T''}{T'' \times T'} = \text{const.}\left(\frac{1}{T''} - \frac{1}{T'}\right)$$

Lummer and Kurlbaum have prepared a table giving the values of x for various temperatures of bright platinum, and I reproduce it below. The products xT are added. They should represent the constant in the expression obtained above, and ought not to vary if this expression has been correctly deduced.

Absol. temp. T .	900	1000	1100	1200	1400	1600	1900
Exponent x . .	30	25	21	19	18	15	14
Product xT .	27,000	25,000	23,100	22,800	25,200	24,000	26,600

We see that the products xT are, indeed, nearly constant. The different values vary around 24,920, as Rasch has pointed out. On the basis of the mean wave-length $0{\cdot}55\mu$, we then obtain 13,706 as the value of b. The ratio of this value of the constant b for bright platinum to the value 14,500 for the black body, corresponds quite closely to the difference between these two extreme cases as found by Lummer and Pringsheim, namely—

$$\lambda_m T = 2940 \text{ (black body)}$$
$$\lambda_m T = 2630 \text{ (bright platinum)}$$

Example showing the application of the optical method of temperature measurement.

The calculation of absolute temperature of solid incandescent bodies by means of the formula—

$$ln\text{H}' = 12{\cdot}943\left(1 - \frac{2068{\cdot}4}{\text{T}}\right)$$

therefore, appears justifiable up to very high temperatures, particularly if the radiation issues from a tube or cavity. Nernst found the total radiation from 1 sq. mm. surface in a cavity at the temperature of melting iridium to equal 12·1 ± 0·6 Hefner candle-power, assuming that a in the Lummer-Kurlbaum formula remained constant and equal to 14 at a bright, white heat, or in any case equal to 13 above the temperature of melting platinum. From this he calculated the melting-point of iridium to be 2203° and 2338° respectively. On the other hand, Rasch, using the formula first explained, calculated from Nernst's observations that this temperature was 2285°, or, taking a slightly greater value for a (13·02), 2287°. Heræus' value of 2400° for the melting-point of iridium is certainly too high.

Application of this method to the measurement of chemical equilibria.

If, then, it is possible for us to measure the temperature of solid bodies, and particularly the temperatures of the interiors of tubes, with considerable accuracy, we naturally ask what advantage we may derive from this for the study of equilibria at elevated temperatures. We are greatly hampered here by the fact that our supply of substances impermeable to gases above 1500° is very scanty. Carbon vessels cannot be made gas-tight, and metallic vessels, among which those of iridium are the most important, can only be used with certain limitations, for at an intense white heat this substance preserves neither the refractory chemical properties nor the impermeability to gases which characterize it so markedly at ordinary temperatures.

Flames, on the other hand, are available without restriction, and for this reason measurements of their temperatures by optical means is of especial interest to us.

Measurements on luminous flames. It is true that our experimental data in this field are still somewhat incomplete. Kurlbaum[1] first described a method for determining the temperature of carbon flames (candle, illuminating gas, and acetylene flames of the ordinary sort), and this method is very simple and direct. He used the spectro-photometer previously constructed by Holborn and himself, the principle of which simply was to simultaneously observe through a red glass an incandescent lamp filament and the image of a radiating surface thrown into the plane of the incandescent filament by a lens. If the red radiation of the incandescent lamp and of the surface were equally strong, then the filament appeared neither bright nor dark against the shining surface. Kurlbaum then substituted an absolutely black body for the radiating surface, and compared the brilliancy of the incandescent filament with it at various temperatures. When the filament was no longer visible, he inserted a candle-flame between the opening of the absolutely black body and the instrument. If the radiation of the flame was less intense (in the region of red rays) than that of the absolutely black body, the candle-flame cast a shadow, the field of view was darkened, and the carbon filament appeared. But if the radiation from the candle-flame was more intense, then the field of view became brighter and the filament appeared dark against it. The temperature of the absolutely black body at which the insertion of the candle-flame produced no change then represented the temperature of the flame, assuming that the flame possessed no reflecting power, and that its coefficient of absorption equalled that of the black body, that is, equalled one.

The conditions under which this photometric method of temperature measurement in any region of wave-lengths is justifiable may be easily defined with more precision if we recall the two formulæ (see pp. 278 and 280)—

$$A + R + D = 1 \quad . \quad . \quad . \quad . \quad (a)$$

$$\text{and } E : A = S_2 : 1 \text{ or } E = AS_2 \quad . \quad . \quad . \quad (b)$$

Here E represents the radiation coefficient of the flame in the

[1] *Physik. Zeitschr.*, 3 (1901–1902), 187, 332.

region of wave-lengths under consideration; A the absorption coefficient of the flame for the same wave-lengths; S_2 the radiation coefficient, and 1 the absorption coefficient of the absolutely black body at the same temperature and also for the same wave-lengths. If, then, we assume that at any one temperature of the absolutely black body insertion of the flame into the path of the rays produces no change in the field of view, then—

$$E + DS_1 = S_1 \quad . \quad . \quad . \quad . \quad . \quad . \quad (c)$$

Here E has the same significance as before. D is the coefficient of transparency of the flame for the wave-lengths used in the red, and S_1 the radiation coefficient of the absolutely black body for the same wave-length and at the temperature at which insertion of the candle-flame produces no change in the field. This formula is derived from the simple consideration that when the candle is inserted, the radiation coefficient E of the flame plus the transmitted radiation DS_1 are active, while when the candle is removed only the unimpeded radiation S_1 of the absolutely black body produces any effect. From this equation and (*b*) it follows that—

$$AS_2 + DS_1 = S_1$$

and with the help of (*a*) that—

$$S_2 = S_1 \frac{1 - D}{A} = S_1 \frac{A + R}{A} = S_1\left(1 + \frac{R}{A}\right)$$

When, therefore, we assume that the radiation S_1 of the absolutely black body at the actual temperature prevailing when we adjusted to precise equality, is equal to S_2, the radiation which the black body *would* send out at the temperature of the flame, we commit a mistake the magnitude of which depends solely on the ratio of the reflecting to the absorbing power of the flame—

$$\frac{R}{A}$$

and which may therefore be practically zero if this ratio is very small. Indeed, the method would still give correct results even if the absorptive power were considerably less than that of the absolutely black body, provided only that its reflecting power were also sufficiently small.

Uncertainty of temperature measurements of luminous flames.

Kurlbaum's observations carried out as above described, gave 1431° for the temperature of the candle-flame. Lummer and Pringsheim[1] have criticized this result on the ground that the ratio $\frac{R}{A}$ was by no means equal to zero. They found D equal to 0·8, and surmised that R must lie between 0·05 and 0·1. $\frac{R}{A}$ must then be certainly far from zero (case of the absolute black body), and the temperature of the candle-flame would be higher than 1431°. In answer to this Kurlbaum (*loc. cit.*) published measurements of R, showing it to be approximately equal to 0·01. The quotient $\frac{R}{A}$ would then be so small as to be negligible, and 1431° would appear, after all, to be the correct temperature. But now Stewart[2] investigates the measurements of Lummer and Pringsheim, and finds that if Kurlbaum's determination of R is correct, we must take D equal to 0·96. $\frac{R}{A}$ would then again seem to have an appreciable magnitude, and the value 1431° would be too low. Spectrobolometric measurements have yielded results even more variable than these, as can be seen by comparing the results of Lummer and Pringsheim and of Stewart.

The reflecting power of flames, as has already been pointed out, is always small. Any error in Kurlbaum's method must then be due to the transparency of the flames, and must increase with it. The denser and more opaque the flames become, from the suspended particles of carbon, so much more correct will be the results of this method, because the values of $\frac{R}{A}$ become just so much smaller. On the other hand, for any given value of $\frac{R}{A}$ the measurements become relatively more accurate the higher the temperature. Even if we made a mistake of 100 per cent. in measuring the surface brilliancy

[1] *Physik. Zeitschr.*, 3 (1901–1902), 233.
[2] *Ibid.*, 4 (1902), 1.

because we neglected $\frac{R}{A}$, this would only involve an error of 120° degrees in our temperature measurements at 1430°.

Measurement of non-luminous flames.

It is possible to apply this optical method to non-luminous flames when we artificially colour them. Fery[1] used the vapours of salts, preferably sodium chloride, for this purpose. He assumes that the sodium line shines from pure temperature radiation; that all the luminescent radiation of the flame is zero[2] for this wave-length; and, finally, that the transparency of the coloured flame is small for this wave-length, and consequently the quotient $\frac{R}{A}$ is nearly equal to zero. He was able to confirm experimentally that the reflecting power is very small, as would be expected. If all these assumptions are true, then the optical measurement of the temperatures of non-luminous flames by means of the spectrophotometer can be easily carried out, by simply colouring them with the vapour of sodium chloride. In making these measurements Fery observed the spectrum of an incandescent lamp, which he assumed to radiate as a black body, and inserted the flame, coloured with salt vapours, into the path of the rays from it. So long as the brilliancy of the lamp was below a certain point the sodium line appeared bright. As the brilliancy of the lamp was increased, a point was reached at which the bright line changed to a dark one. This reversal of the lines in the spectrum is a well-known phenomenon. Kirchhoff recognized, even before he had deduced his Law of Radiation, that the bright lines which coloured flames show when viewed by themselves, are supplanted by dark lines of identical wave-lengths when more intense light of the same wave-lengths falls upon them. Kirchhoff's explanation of the Frauenhofer lines was based on this very experiment. The only new feature here is the quantitative relation between the temperature of the flame and that of the other source of

[1] *Comp. Rend.*, 137 (1903), 909.

[2] This assumption exactly contradicts Pringsheim's opinion that the vapours of metals in the flames shine only through luminescence. In this connection, see the treatment of the subject in the "Rapports présentés au Congrès international de Physique, Paris, 1900," vol. 2, p. 100, where a bibliography is given; and Kayser, "Handb. der Spektroskopie," vol. 2.

light. It has only been possible to determine what this quantitative relation is, although by no means accurately, now that investigations have been carried out which show us how the radiation from incandescent solids depends upon their temperature. Fery's and Kurlbaum's methods are really identical, for the latter method too depends essentially on a reversal of the lines in the spectrum. The substitution of an incandescent lamp for the absolutely dark body lessens the accuracy. If the radiation of the incandescent lamp is determined as a function of the temperature, using the black body, then Fery's procedure in this regard is naturally justifiable. Fery's results for the Bunsen flame are very good. He found—

Full draught.	Half draught.	Without draught.
1871°	1812°	1710°

The values he found for certain other flames seem less accurate.

The thermo-electric measurement of flames.

The thermo-electric measurement of the temperatures of flames is certainly to be preferred in the present state of the question, whenever the method is applicable.[1] But this method requires certain definite and important experimental precautions, if we would not obtain most erroneous results. Baikoff's experiments upon the temperature of the Bunsen flame illustrate the error possible in such a case.[2] He measured

[1] In regard to the measurement of high temperatures by other methods (air-thermometers, calorimeter, resistance thermometer, contraction pyrometer, Seeger's cones, etc.), see the excellent little book by Le Chatelier and Boudouard entitled "Températures élevées" (Paris, 1900). I take from this source the following table of frequently used fixed points of thermometry :—

	Boiling-point.		Melting-point.
Water	100°	Silver	962°
Naphthaline	218°	Gold	1065°
Sulphur	445°*	Platinum	1780°
Zinc	930°†		

Numerous valuable references to the literature will be found in the monograph, "Chemie der extremen Temperaturen," by Bredig (Leipzig, 1901).

[2] *Chemiker Zeitung*, 28 (1904), 1107.

* At 760 mm. pressure, 1 mm. change of pressure changes the b.p. 0·095°.

† At 760 mm. pressure, 1 mm. change of pressure changes the b.p. 0·15°.

the temperature of the colourless flame from the tip of the inner cone upwards along the vertical axis, inserting the junction of his thermoelement at these points. After that he fitted a cap of quartz over the junction of his thermoelement, and made a second series of observations. Thirdly, he made another series of observations with a cap of bright platinum over his quartz cap. In the first case he found temperatures between 1350° and 1391°; in the second, temperatures between 1110° and 1133°; and in the third, temperatures immediate between the two, though approximating the more closely to those with unprotected junction. The thermoelement therefore showed the lowest temperatures when surrounded by the quartz alone. Baikoff considers these temperatures to be the *true* temperatures of the flame, the higher temperatures registered being due to catalytic action of the platinum. He believes the very natural explanation of unequal radiation to be precluded by the fact that the phenomenon is not altered when the flame is surrounded by a "hot" chimney. But one sees immediately that the intensity of radiation increases so tremendously with the temperature that a chimney, even if heated to incipient redness, would have no effect on the radiation of substances at temperatures of 1500–1800°. Though perhaps superfluous, I have further convinced myself that it is only necessary to cover the quartz cap with finely divided platinum, instead of bright platinum, in order to change the rise of temperature into a fall.[1]

Conditions necessary for the accurate thermo-electric measurements of the temperature of flames.

In order to accurately determine the temperature of a flame thermo-electrically, it is of primary importance to know how much heat is lost through the radiation from the thermoelement itself. Waggener,[2] Nichols,[3] as well as White and Traver,[4] have found that the temperature indicated by the junction is almost exactly a linear function of the thickness of the element. If, then, we measure the temperature of the same

[1] In Baikoff's more extended publication in the Russian, experiments are mentioned with slitted platinum caps, which Baikoff could not bring in conformity with his theory of the catalytic elevation of the temperature (see "Chem. Centralblatt" (1905), i. 1357)

[2] *Wied. Ann*, 58 (1896), 579.

[3] *Journ. Franklin Inst.*, 150 (1900), 374.

[4] *Journ. Soc. Chem. Ind.* (1902). 1012.

flame successively with elements of diminishing thickness, we can extrapolate for the true temperature as measured by an element of zero thickness. If the thinnest element used in this process be not thicker than 0·1 mm., the extrapolation is not a very large one, amounting to between 30° and 100°, depending on the thickness of the wire and the temperature of the flame. If we base our observations simply on the readings of a thick thermoelement, the indicated temperatures will be some hundreds of degrees too low. Waggener found the temperature of the hottest part of the Bunsen flame to be a little above 1785°. Berkebusch[1] found it to be 1830°, using an entirely different, though, to be sure, not very accurate, method. This agrees with the fact first observed by Bunsen, and later confirmed by Waggener, Fery, and others, that a fine platinum wire can be brought to incipient melting in the hottest part of a Bunsen flame. The melting-point of platinum lies at 1780°. Fery, as mentioned above, found 1871°, a temperature lying in this same region. We obtain a similar value if we determine the heating power of illuminating gas with Junker's calorimeter, the amounts of carbon dioxide and of water produced from one volume of illuminating gas by complete combustion, and the amount of oxygen consumed in this combustion. From these data we can again calculate the heat evolved, and can then again calculate from the known specific heats the maximum temperature attainable. Numbers of the same magnitude are obtained as before.

These facts show us that the linear extrapolation to zero thickness of thermoelement leads to correct values for the highest temperature of the Bunsen flame. It leads to still more certain results at low temperatures, for there the loss by radiation is much smaller, and the extrapolated temperature but slightly exceeds that indicated by the thinnest thermoelement.

The effect of too long heating in the flame. The thermoelement should not be left too long in the flame during a measurement. Waggener has shown that prolonged heating in the flame produces certain heterogeneities in the platinum-rhodium wire, which give rise to incorrect readings.

The effect of internal resistance. If we are using thick thermoelements, we may measure the

[1] *Wied. Ann.*, 67 (1899), 649.

electromotive force directly by means of a high-resistance voltmeter. Thin thermoelements have such a high resistance of their own, especially when heated, that with them we must always use the compensation method.

The position of the two wires of the thermoelement in the flames relative to each other is not without effect on the readings. It is not advisable to have the wires leading to the junction parallel to one another. Binding the wires so that the ends in the neighbourhood of the junction are in the same straight line usually suffices to give good results. The effect of the position of the wires.

It is important, in making the measurements, to heat the two wires of the thermoelement symmetrically — that is, keep each at equally hot places in the flame. Any one-sided heating of the wires near the junction introduces an error due to conduction of heat, and this is not eliminated by the extrapolation to zero thickness, but instead is rather accentuated thereby.

I have, in collaboration with Richardt, measured the temperature of the Bunsen flame thermo-electrically in order to determine the water-gas equilibrium existing there. Our results, which have been confirmed by Allner, were presented in the Fourth Lecture.

Let us examine the Bunsen flame a little more carefully, in order to better understand the method of measurement. This flame presents quite a different appearance if the air-supply, which enters at the draught hole of the burner and mixes with the illuminating gas inside the tube of the burner, is altered by even a very few per cents. If we let the flame first burn luminously, and then by slowly opening the draught-hole increase the air-supply, the luminous part of the flame first vanishes, and an indistinct separation occurs into an inner and outer cone. As the air-supply is further increased, the inner cone becomes very much smaller and more sharply defined and brighter. Its colour is now a pronounced green, differentiating it clearly from the violet of the outer flame. Gas-flames of this type are generally obtained in the laboratory by means of the so-called Teklu burner. The modern domestic gas stove always furnishes such a flame. Every Auer burner, after its mantle and mantle-holder have been removed, also gives a flame of this character. The Bunsen flame.

Steadying a Bunsen flame. If we produce a flame of this kind with a Teklu burner, we find that the inner cone burns with a loud noise and a fitful, rapid to-and-fro motion. To convert this into a quiet and perfectly steady flame we need only to lengthen the tube of the burner by a few decimetres. The unequal mixing and cross-currents in the short tube are evidently the cause of the unsteady burning. All the equilibrium measurements were therefore made with the flames from Teklu burners whose tubes had been lengthened about a half a metre by means of a glass tube of the same bore.

The inner parts of the flame. We are accustomed to say that the illuminating gas burns in the flame. The expression is inexact, for nothing is burning *in* the flame. The flame is nothing but a mass of glowing gas surrounded on all sides by an extremely thin zone in which the combustion is taking place.

This boundary is represented diagramatically in Fig. 17: *ab* is the opening of the burner; *abc* represents the "inner cone," *adb* the "outer cone." In the inner cone the mixture of air and illuminating gas burns to a mixture of nitrogen, carbon monoxide, carbon dioxide, water-vapour, and hydrogen.[1] Since the four gases CO, H_2, CO_2, and H_2O are connected by the water-gas reaction, we may say that the inner cone furnishes water-gas diluted with nitrogen. In the whole space between inner and outer cone oxygen can never be detected analytically.[2] This is the basis for the statement that nothing can burn *in* the flame. In the outer cone *adb* everything that can burn to carbon dioxide and water-vapour does so.

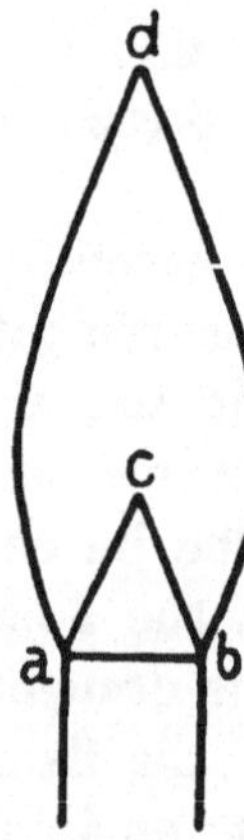

FIG. 17.

The position of the inner cone. The relative position of outer and inner cone is governed by a simple equation. The inner cone represents a stationary explosion. Its position

[1] In addition to these, methane is sometimes found in small quantities. (See Appendix, No. IX.)

[2] Such traces as correspond to the equilibria $2H_2 + O_2 \rightleftarrows 2H_2O$ and $2CO + O_2 \rightleftarrows 2CO_2$. These traces must be extremely small, because the excess of H_2 and CO present in the flame stands in the way of the dissociation, which even in pure H_2O and CO_2 only amounts to a very small value at the temperature of the flame.

conforms to the condition that the rate at which the ignition is propagated downward shall equal the rate at which the entering gas moves upward. Michelson[1] used this relationship to determine the rate of propagation of explosions.

The position of the outer cone. The position of the outer cone is governed by the condition that the amount of oxygen furnished per second by the surrounding air shall just suffice to burn the combustible constituents coming from the interior completely to carbon dioxide and water-vapour in the same interval of time.

The temperature of the inner cone. The zone *abc* is but a fraction of a millimetre thick, and the gas passes through it in about the 0·001 part of a second. The thickness of the zone *adb* is not very easily estimated with the eye, but it is undoubtedly of the same order of magnitude.

The temperature in the zone *acb* is approximately 1550°, under a strong draught. This part of the flame was formerly considered to be much hotter, judging from its brightness. But the thermoelement gives no indication of this. Arguing from theoretical grounds, it has been claimed that the temperature of this zone should be higher than 1550°, because the gas must be heated by the combustion zone *acb* before it reaches it, and should therefore attain a higher temperature when it burns itself. That is, it has been concluded that the gases reach a higher temperature than they would did they depend simply on their own heat of combustion to raise them from the temperature of the room to that of the burning gases.[2] This is a mistaken conclusion. Every burning layer in *abc* must give off *during* its burning just as much heat to following portions of the gas as it received itself from earlier burning portions. It therefore does not attain any higher temperature than it would if it burned without previous heating, and was simply heated by its own heat of radiation. We can calculate what this temperature would be knowing the composition of the gas leaving the zone *acb*. Richardt and I have actually made this calculation, and find the same temperature of 1550° within the limits of experimental error of the thermoelement. The bright green colour of the inner cone is therefore only luminescence.

[1] *Wied. Ann.*, 37 (1893), 1.

[2] Mallard and Le Chatelier, *Ann. des mines* (8), 4 (1883), 344; Gouy, *Ann. Chim. Phys.* (5), 18 (1879), 1; Michelson, *loc. cit.*

Heat losses from the inner cone.

The heat losses from the combustion in the zone *acb* are small. Mache [1] has investigated theoretically how much heat is abstracted by the incoming current of unburnt gas. If c_p be the specific heat of the gas flowing into the combustion zone at constant pressure, c_0 the velocity, and ρ_0 the density of the gas mixture, both referred to 0° and 760 mm., and k the constant of heat conduction, then—

$$\frac{k}{c_p} \times \frac{d^2T}{dx^2} + \rho_0 c_0 \frac{dT}{dx} = 0$$

T here signifies the temperature at a distance x from the flame on the side of the inflowing gas. Integrating this formula and substituting values for the case of hydrogen burning in oxygen, Mache found that room temperature prevailed at a distance not more than a few hundredths of a millimetre from the outer boundary of the flame. No perceptible loss, therefore, takes place in this way. There is, of course, a certain loss due to radiation, but it is so small as to be negligible.

The gas has no opportunity to lose any heat between the inner and outer cone. Indeed, it must become warmer as it approaches the outer cone, for there the gases already heated to 1550° burn completely, the temperature rising to 1800°, as mentioned above.

Distribution of temperature between the cones.

The fact that we find the temperature rises as we move our junction horizontally outward from the vertical axis of the flame toward the mantle *adb* agrees with this conclusion. The reason that we find in fusion experiments the outer parts of the flame not to be as hot as other parts is simply due to the vibration of the flame. Unless the object is sunk deeply into the flame, it comes in contact with the cold outer air. Then, too, heat is conducted away by the supporting device, unless this is also heated by the flame.

The temperature gradient from the outer toward the inner cone is much less abrupt than from the inner cone into the inflowing gas. There are two reasons for this. In the first place, the hot gases in the region *abc* conduct heat very much more readily than do the inflowing gases. The coefficient of heat conduction increases approximately proportionally to the

[1] *Sitzungsber. Vienna Acad.*, 108 II*a*. (1899), 1152.

square root of the temperature (Mache, *loc. cit.*). Then, too, the velocity of the gas is smaller in the ratio of the surfaces of the two mantles *adb* and *acb*, so that the heat conduction can progress farther and more quickly into the inflowing gas.

We cannot directly investigate the composition of the gas mixture in the Bunsen flame. This is due to the fact that the product of combustion in the lower parts of the outer mantle, as they rise, work their way into the interior of the flame, and dilute the gases there present. For this reason the normal temperature conditions are somewhat displaced, particularly in the upper part of the flame.

Splitting the Bunsen flame.

We may, however, arrange conditions which permit an investigation of this point if we dispose of the outer cone. Teklu, as well as Smithells and Ingle, hit upon this device.[1] It was very simply done by fitting a glass tube of suitable dimensions over the end of the burner tube. A stopper is made to fit tightly between the burner tube and the outside glass tube. The relative sizes can be seen from Fig. 18, where a scale divided in centimetres is included in the photograph.

The inner cone *acb* burns quietly at the top of the burner tube. The outer cone burns higher up at the opening of the outer tube. Since the gases now find opportunity to cool off on their way from inner to outer cone, the temperature conditions in this latter are naturally quite different from before.

The outer cone is not nearly so hot. The inner cone, on the other hand, suffers no change. The entrance of air into the space between inner and outer cone is wholly prevented in this arrangement, and samples of gas may be removed from it for analysis. For this purpose we may hang a capillary tube of porcelain down from the edge of the glass tube, or, as in Fig. 19, we may insert a double-walled platinum tube through the side of the glass tube. This tube may be moved in or out, and water kept flowing between its double walls. If this water is kept lukewarm, and the tube is brought close over the tip of the inner cone, the gases experience an extremely sudden drop in temperature, and yet the temperature is not so low that water-vapour is deposited from these gases on the walls of the platinum tube. In this way we get the gases in a suddenly cooled condition without any loss of water-vapour. If at the same time we hang

[1] *Jour. Chem. Soc.*, 61 (1892), 204; *Jour. pract. chemie*, 44 (1891), 246.

a tube down from the edge of the glass cylinder, we get simultaneously a sample of gas which has had opportunity to cool down from the temperature in the flame to that at the point

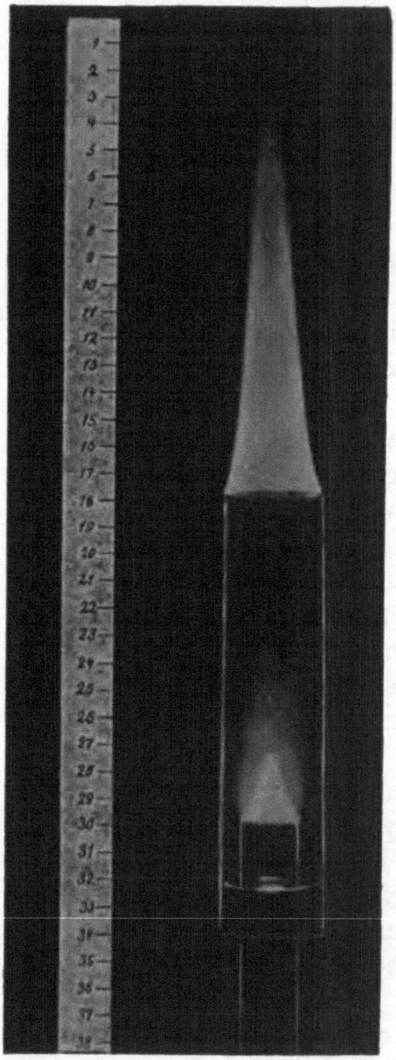

FIG. 18.

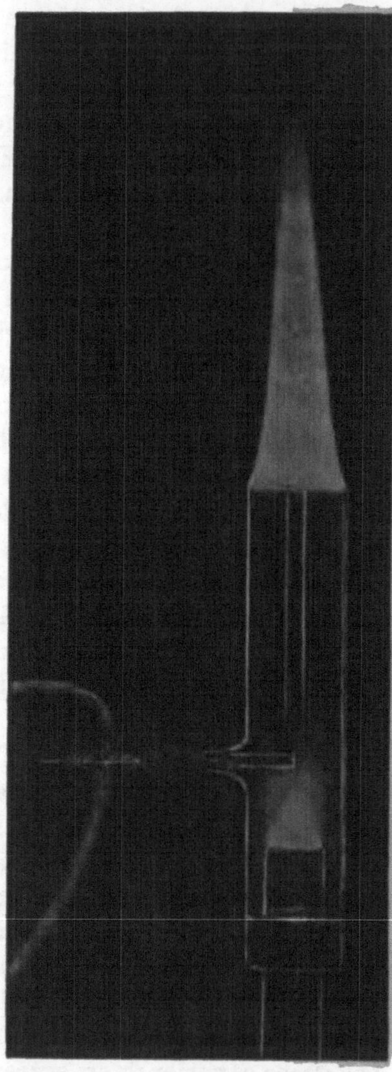

FIG. 19.

where it is drawn off. The arrangement in Fig. 20 is used to measure the temperature of the flame. It consists of two thermoelements of different thicknesses, 0·3 and 0·1 mm., enclosed in quartz capillary tubes, which may be lowered into the outer tube. We can easily bring these capillary tubes to any symmetrical places in the flame with exactness by means of the fine adjustments provided. Extrapolating from the readings of the two thin elements, we can determine the temperature for a thermoelement of zero thickness.

Analysis of the samples of gas.

It is not as easy to determine the composition of the gas samples with the required accuracy as one might at first think. The great quantity of atmospheric nitrogen dilutes the gas and reduces the percentage of the other gases. If we attempt to

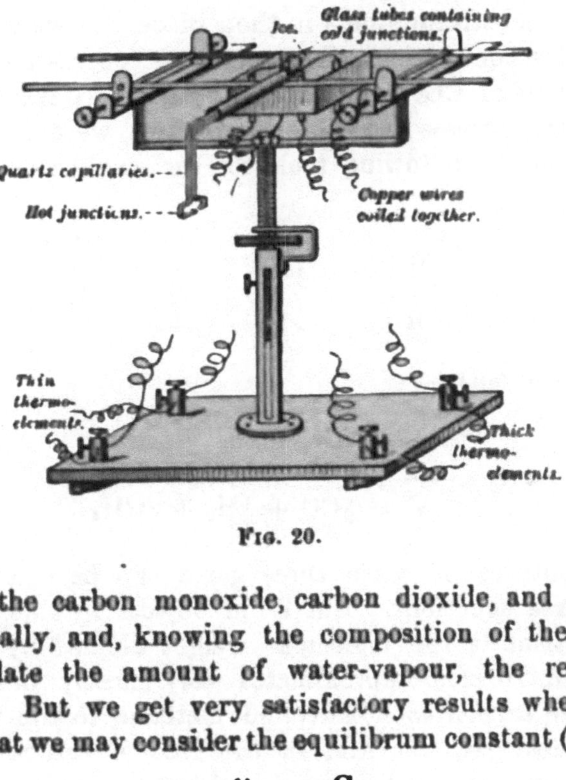

FIG. 20.

determine the carbon monoxide, carbon dioxide, and hydrogen volumetrically, and, knowing the composition of the unburnt gas, calculate the amount of water-vapour, the results are uncertain. But we get very satisfactory results when we remember that we may consider the equilibrum constant (p. 113)—

$$K = \frac{C_{H_2O} \times C_{CO}}{C_{CO_2} \times C_{H_2}}$$

as the product $\kappa \times \kappa'$ of the two quotients—

$$\kappa = \frac{C_{H_2O}}{C_{CO_2}} \text{ and } \kappa' = \frac{C_{CO}}{C_{H_2}}$$

We may determine the value of one of these quotients by simply conducting the gas through the absorption tubes used in an ordinary organic combustion analysis. The other quotient may be determined by a single combustion over mercury of a sample of the gas freed from carbon dioxide. Knowing the contraction accompanying this, the consumption of oxygen and the amount of carbon dioxide formed, we can calculate, not only the amount of carbon monoxide and hydrogen present, but even that of a third component, say methane. The equations here used are those developed by Bunsen[1] for the combustion analysis of gas.

The calculation of the analysis.

Wohl[2] has recently attacked these equations of Bunsen, and sought to substitute others in their place. If we arrange under "Con." the contraction taking place on combustion, under CO_2 the amount of CO_2 produced, and under V_0 the oxygen consumed, all expressed in cubic centimetres, we obtain, according to Bunsen, the following table for the combustion of CO, H_2, and CH_4:—

	CO_2	V_0	Con.
CO	1	$\frac{1}{2}$	$\frac{1}{2}$
H_2	0	$\frac{1}{2}$	$1\frac{1}{2}$
CH_4	1	2	2

and consequently—

$$CO_2 = CO + CH_4$$
$$\text{Con.} = \tfrac{1}{2}CO + 1\tfrac{1}{2}H_2 + 2CH_4$$
$$V_0 = \tfrac{1}{2}CO + \tfrac{1}{2}H_2 + 2CH_4$$

The quantities of these three gases can be calculated from these three equations. But Wohl objects to this, saying that the quotient of the molecular weight divided by the density at 0° and 76 cms. approximates very closely in the case of hydrogen, carbon monoxide, and methane to the value 22·41 for an ideal gas, but that in the case of carbon dioxide we get a very divergent value for this quotient, namely 22·26. He therefore concludes that Avogadro's rule does not apply here, and consequently that Bunsen's equations cannot be accurate.

Discussion of Wohl's formula.

We consider Wohl's objection to be quite invalid. We

[1] "Gasometrische Methoden," 2nd edit., 1877. Second section.

[2] *Ber. d. deutschen Chem. Ges.*, 37 (1904), 429.

obtain from Van der Waal's formula[1] for the relation between the pressure, volume, and temperature of a gas—

$$\left(p + \frac{a}{v^2}\right)(v - b) = RT$$

the following relation between the molecular weight M and the density d_0 of a gas at 0° and 76 cms.—

$$\frac{M}{d_0}(1 + a)(1 - b) = R$$

Van der Waals found a and b to have the values—

$$a = 0{\cdot}00874$$
$$b = 0{\cdot}0023$$

for carbon dioxide, and consequently—

$$\frac{M}{d_0} \times 1{\cdot}00646 = R$$

Wohl's value for $\frac{M}{d_0} = 22{\cdot}26$, therefore, gives an entirely correct value for R if we multiply by the factor 1·00646. The deviation from the value 22·41 is therefore solely due to the fact that carbon dioxide is not a perfect gas at 0° and 76 cms., and has nothing to do with Avogadro's rule.

Justification of Bunsen's formula.

When the pressure of carbon dioxide is reduced to one-third of an atmosphere or less, its deviation from the behaviour of an ideal gas is so small as to be negligible in analytical work. One can easily convince himself that this is true by making use of Van der Waal's equation. Of course there is no question but that we are introducing a perceptible error when we assume that 100 c.c. of carbon dioxide are formed by the union of 100 c.c. of pure carbon monoxide and 50 c.c. of pure oxygen (all the volumes being measured at 0° and 76 cms.). Indeed, we should only obtain 99·4 vol. But it is quite unjustifiable to ascribe any universal significance to this deficit of 0·6 per cent., and to suggest an alteration of the Bunsen formulæ on this ground. Besides, it is very seldom that we have pure carbon dioxide formed in a gas analysis. Instead, we usually obtain by explosions mixtures of which only a fraction is carbon dioxide. The use of Bunsen's equation assumes the gases to obey Dalton's

[1] Van der Waals, "Kontinuität," etc. (Leipzig, 1899–1900), vol. i. p. 85.

law of partial pressures, according to which the total pressure equals the sum of the partial pressures. If the single gases deviate perceptibly from the fundamental gas law—

$$pv = RT$$

as carbon dioxide does at 0° and 76 cms., then Dalton's law is no longer strictly fulfilled. In this case it is absolutely impossible to calculate the results of analysis by using fixed correction constants in Bunsen's equations, as Wohl advocates doing. Instead, it would be necessary to determine fresh constants for each gaseous mixture by a special investigation. This is evidently quite impossible in the ordinary practice of gas analysis. The thing to do, then, is not to attempt to improve Bunsen's *equations*, but rather the *conditions of the analysis*. This can be satisfactorily accomplished by never allowing the carbon dioxide content of the final gas mixture to exceed 35 per cent.

The water-gas equilibrium in the Bunsen flame.

The analysis of the gas taken from the space between the two cones showed that it made no difference where in this space we took the sample from. The ratio (p. 118)—

$$\frac{C_{H_2O} \times C_{CO}}{C_{CO_2} \times C_{H_2}}$$

therefore, does not change as the gases rise from the inner cone and cool several hundreds of degrees in temperature. This applies equally well to a flame from a mixture of air and illuminating gas as to one from a mixture of air, illuminating gas, and carbon dioxide, although the temperature of the latter is lower. Thus, samples taken in one case (*a*) from just at the base of the outer flame, and in another (*b*) from just above the tip of the bright green inner cone, gave the following values for these ratios:—

a	3·54	2·58	2·74	2·89	2·92	2·97	2·54
b	3·58	2·68	2·86	2·77	2·82	3·12	2·87
t° C. . .	1551	1313	1305	1265	1265	1230	1255

The temperatures as thermoelectrically measured are appended.

On closer scrutiny of these numbers, it appears undeniable that those six experiments where the temperature was depressed by the presence of carbon dioxide yielded much smaller values as the first one carried out in absence of CO_2. They correspond in both cases to the values we found before (p. 143) for the water-gas equilibrium. It seems probable from this that in the combustion of the hydrocarbons in the inner cone of the Bunsen flame an equilibrium is attained which readjusts itself so slowly that the gases do not alter their composition as they rush through the cooling space.

The limits within which equilibrium can be attained and maintained in the Bunsen flame.

A further series of experiments was carried out in order to decide whether or not equilibrium was really reached. Illuminating gas, and a mixture of illuminating gas and carbon dioxide were used. They showed that equilibrium *was* reached. Allner supplemented and confirmed them with similar results for mixtures of methane, carbon monoxide, and hydrogen, of methane and hydrogen, and of benzene, carbon dioxide, and hydrogen. He further studied the possible limiting concentrations. The result was a most characteristic one. There was no accurate adjustment of the equilibrium in the relatively cold flames (below 1100°), which he got by burning mixtures of carbon dioxide and hydrogen, and of carbon monoxide and hydrogen. On the other hand, when the flames were very hot, as is the case with a mixture of benzene vapour and air (2000°), the gases changed their composition after they left the inner cone. It follows from this fact that below 1100° the interval of time during which the reaction can take place in the zone of the stationary explosion is insufficient. Above 1200° equilibrium is reached. The equilibrium is not perceptibly displaced during the period of cooling, if the temperature does not exceed 1500°. If the flame is hotter than this, the equilibrium is progressively displaced until the gas cools to 1500°, when the equilibrium "freezes," to use our earlier expression. If, however, we increase the rapidity of the cooling by introducing a cooling tube, we should evidently get different results in the flame of 2000° C. than when we allowed the gas to cool slowly.

Numerical results.

These results may be more readily understood from the accompanying table. The temperatures in brackets were calculated from the heat of combustion and the specific heats. The other temperatures were measured thermoelectrically. The calculated values of the equilibrium constant were taken from the fifth column in the table on p. 143. The values labelled "found" represent mostly the mean values from several experiments, or the values extrapolated from several experiments.

	Temperature, degrees C.	$\frac{c_{CO}\times c_{H_2O}}{c_{CO_2}\times c_{H_2}}$ Found.	Calculated.	
$CO_2 + H_2$	(890–1037)	0·2	1·2–1·8	Equilibrium is not reached.
$CO + H_2$	990–1083	3·2	1·6–1·9	
$C_6H_6 + CO_2 + H_2$ . .	1190	2·4	2·5	Equilibrium is reached in the inner cone and then "freezes."
$C_6H_6 + CO_2 + H_2$ . .	1280	2·3	2·9	
$CH_4 + CO_2 + H_2$ . .	1246	2·6	2·8	
$CH_4 + H_2$	c. 1400	3·1	3·46	
Illuminating gas + CO_2	1255	2·6–2·8	2·8	
" " "	1265	2·9	2·87	
" " "	1305–1324	2·6–2·9	3·1	
" " "	1370	3·2–3·3	3·3	
Illuminating gas . . .	1386	3·6	3·4	
" " . . .	1500–1510	3·6–4·2	3·9	
" " . . .	1525	4	3·97	
Benzene	(c. 2000)	5	5·24	Equilibrium continues to readjust during the cooling, being overtaken between 3·5 and 4·0.

A theoretical discussion of the technical process of water-gas manufacture.

It is not so easy to apply our theoretical reasoning to actual practice, as it was in the case of the nitric oxide equilibrium, because the process is much more complicated. Water-gas is technically prepared by blowing air and water-vapour alternately through a shaft of burning coal.

The air-blast generates either carbon monoxide as the chief product when run according to Humphrey and Glasgow's method, or carbon dioxide when run according to the Dellwick-Fleischer method. In both processes the coal glows intensely. The steam blast generates water-gas, and as it uses up heat in the formation, cools off the coal. It is therefore customary to alternate the two currents at quite frequent intervals. The steam-blast is continued each time until a flame shows the characteristic appearance of a flame fed by a very poor gas.

The usual conception.

According to the usual view, two processes are here taking place—

$$(a)\qquad C + H_2O = CO + H_2 \qquad K_1 = \frac{c_{CO}c_{H_2}}{c_{H_2O}}$$

and

$$(b)\qquad C + 2H_2O = CO_2 + 2H_2 \qquad K_{11} = \frac{c_{CO_2}c^2_{H_2}}{c^2_{H_2O}}$$

$$\frac{K_1}{K_{11}} = K = \frac{c_{H_2O}c_{CO}}{c_{CO_2}c_{H_2}}$$

When both of these reactions run to equilibrium, it depends on the temperature whether we get a good or a bad gas for heating and lighting purposes. Water-gas is poor when it contains a great deal of carbon dioxide—that is, when reaction *b*) takes place to any considerable extent. Here our object is, not simply to get as near as possible to the equilibrium, as it was with the nitric oxide; but instead we wish to reach and fix the equilibrium at a particular temperature where reaction (*b*) does not take place markedly. Now, it is to be borne in mind that these two reactions, (*a*) and (*b*), are connected by the relation—

$$(c)\ C + CO_2 \rightleftarrows 2CO \qquad K_{11} = \frac{c^2_{CO}}{c_{CO_2}}$$

Boudouard[1] has studied this third equation, and finds that the following mixtures were in equilibrium with coal (see Appendix, No. VIII.):— **Boudouard's experiment.**

At degrees C.	650	800	925
Per cent. CO_2	61	7	4
Per cent. CO	39	93	96
$\frac{CO}{CO_2}$	0 64	13	24

These numbers refer to a total pressure of carbon monoxide and

[1] *Compt. Rend.*, 130 (1900), 132; *Bull. Soc. Chim.*, 21 (1899), 712. See, too, Schenck and Zimmermann, *Berichte d. D. Chem. Ges.*, 36 (1903), 1231 and 3663; also *Zeitschr. f Elektrochemie*, 9 (1903), 691; and finally Bodenstein's discussion given in the same place. It is of further importance, in connection with p. 255 of this lecture, that Boudouard found that even at 500° there was still 5 per cent. CO in equilibrium with 95 per cent. CO_2 at one atmosphere pressure. Dixon (*Journ. Chem. Soc.*, 75 (1899), 630) has attacked the views propounded on p. 255, and has sought to discredit the experiment of Lang, which favoured these views. Dixon conducted carbon dioxide mixed with 8 per cent. oxygen over coal at 500°. If carbon monoxide were the primary product of the combustion, there would necessarily be at least as much carbon monoxide in the issuing gas as corresponded to the equilibrium. Dixon found less than one per cent. CO. Experiment, therefore, speaks rather for than against the primary formation of carbon dioxide. Brereton Baker (*Phil. Trans.*, 179 (1889), 571) has observed that coal will form carbon monoxide with extremely dry oxygen under conditions where carbon dioxide similarly dried has no action on coal. It is best, however, not to be influenced by this consideration, for the extreme dryness entirely alters the relative reaction velocities, and introduces passive resistances which destroy the comparative value of the observations.

dioxide equal to one atmosphere. In the actual preparation of water-gas, the hydrogen present lowers the sum of these partial pressures to 0·4 atmosphere. This tends to increase the value of the ratio $\frac{CO}{CO_2}$ over what it would otherwise be, for the percentage constant of CO in equilibrium with coal increases with decreasing pressure. It is essential in actual practice that this ratio should always be greater than 10. Looking at our table, we may therefore conclude that 750° is the lowest practicable temperature at which the gases are in equilibrium with themselves and the coal. And yet no such equilibrium mixture is actually established. A glance at the values which Luggin calculated from Harries' experimental results (p. 138) shows that although the four gases reach an equilibrium among themselves at temperatures between 760° and 1000°, they are not in equilibrium with the coal. For the ratio $\frac{CO}{CO_2}$ is always smaller than it would be were equilibrium with the coal established.

The peculiar action of the coal. Yet we recall how the water-gas equilibrium "froze" at a white-hot temperature in the cooling region of a Bunsen flame. Here we seem to find just the reverse; namely, that it adjusts itself at any temperature down to a moderate red heat (750°). But the gases were in a free gas space there, while here they are in intimate contact with the coal. To be sure, the time given the reaction to reach equilibrium is shorter in the first case than in the second, but the difference is by no means large enough to explain the difference in behaviour. We must conclude that the glowing coal accelerates the adjustment of the water-gas equilibrium without itself reaching equilibrium with the gases. The coal, therefore, acts upon the water gas equilibrium quite as platinum does. It would be of interest and value to determine what intermediate reaction is responsible for this behaviour. Here we should have always to keep in mind the fact that coal is by no means pure carbon, and that the intermediate reaction after all might be due in some way to a constituent of the ash.

Survey of the matter from a technical point of view. Glancing again at Harries' numbers, we see that at about 1000° the content of carbon dioxide and of water-vapour has sunk low enough to meet the demands of practice. If the steam is driven through at a higher velocity, and under the irregular temperature conditions of an actual gas oven, it would be necessary to place this limit somewhat higher. It is plain that above some particular temperature, even in the technical gas, a condition must be reached at which there is equilibrium both between the various gases and between the gases and the coal. Now, the pressure is necessarily always very near the atmospheric. Consequently, according to our above conclusions, the content of carbon dioxide must sink at temperatures higher than 1000° to mere traces which are no longer to be detected by volumetric analysis.

But experience teaches that water-gas technically prepared always contains several percents of carbon dioxide. We may take into account the fact that the temperature in the water-gas sinks during the injection of steam, and that, consequently, the observed carbon dioxide may be found toward the end of this period. For the present we cannot give any more quantitative explanation.

To recapitulate, then, regarding the water-gas equilibrium, we may say that equilibrium is quite readily attained between the gases CO_2, CO, H_2, and H_2O. This equilibrium, however, only has technical importance when the content of CO is high and that of CO_2 simultaneously low. The location of the equilibrium between coal, carbon dioxide, and carbon monoxide shows that this is theoretically possible even below 800°. But, practically, the inertness of the coal prevents us from obtaining satisfactory values for the ratio $\frac{CO}{CO_2}$, and consequently a favourable composition of water-gas at temperatures below 800°. The small content of CO_2 which theory would predict in the case of this equilibrium above 1000° is always exceeded to some extent in the technical preparation of the gas-water.

The principal technical interest in the water-gas is centred upon the question of the heating qualities of the gas mixture. The object striven for is to produce a gas mixture whose fuel value shall represent the largest possible fraction of the fuel value of the coal consumed. The air-blast which heats the coal is here of the most importance. It is particularly important whether we choose to regulate this blast so as to get carbon monoxide, and then make further use of this, or, instead, to obtain dioxide and allow this to escape into the atmosphere. Calculations concerning these points have been made on the basis of the simple laws of thermochemistry, and have been treated so extensively in the technical literature of the subject that they may be omitted here.

APPENDIX TO LECTURE V

Dissociation of CO_2: experiments of Nernst and v. Wartenberg.

RECENTLY the dissociation of carbon dioxide has been determined more exactly by Nernst and von Wartenberg.[1] Their apparatus consisted of a porcelain vessel in the form of a pipette, which was heated electrically in a platinum tube. The bulb of the pipette was some 7·5 cms. long and 2 cms. in diameter. The outlet tube was a capillary of 0·5 mm. inside diameter. The inlet tube was 6 mm. in inside diameter, and through it a thermoelement was inserted.

Pure dry carbon dioxide heated to different temperatures was blown through the apparatus. The issuing gases were collected in a weighed apparatus filled with concentrated KOH and fitted with a narrow tube for collecting the few cubic centimetres of carbon monoxide and oxygen produced by the dissociation. Arrangements were also made so that a spark could be passed through the CO_2 before its entrance into the pipette; by the action of the spark as much as 4 per cent. CO_2 was decomposed. This being far in excess of the amount produced by the dissociation at the temperature to which the pipette was heated, formation of CO_2 instead of decomposition took place in the pipette. Thus the equilibrium could be approached from both sides.

It was difficult to find the interval of temperature where the dissociation was high enough to allow reliable determinations of the percentage of CO and O_2 without recombination taking place in the outlet tube. The shifting of the equilibrium by this recombination was diminished as much as possible by the high speed and rapid cooling of the gas in the narrow capillary tube; but only in the case when the CO_2 was quite dry were the experiments successful, water vapour accelerating the reaction to an extreme degree. Moreover, the porcelain tube, which was glazed on the outside, after some time became porous. This was shown by the fact that the CO and O_2 collected in the absorption apparatus did not stand in the theoretical ratio of 2 to 1.

[1] *Göttinger Nachrichten*, 1905, Heft I., and *Zeitschr f. physikal. Chemie*, 56 (1906), 548.

Notwithstanding these difficulties, experiments of sufficient exactness were made at 1127° C. and 1205° C., and a third value was derived from similar determinations of the water dissociation, using Hahn's (see p. 139) data on the water-gas equilibrium.

These figures are collected in the following table :—

t° C.	T° abs.	x % found.	x % calculated.
1027	1300	0·00414	0·00389
1127	1400	0·01–0·02	0·0138
1205	1478	0·029–0·035	0·0324

The values of x mean the percentage of CO_2 which has dissociated under a pressure of one atmosphere.

The ratio of this value x to the constant of dissociation is easily deduced in the following way :—

In the equation

$$K_p = \frac{p_{CO_2}}{p_{CO}} \times \frac{1}{p_{O_2}^{\frac{1}{2}}}$$

the first term of the quotient, which is independent of the units chosen, may therefore be replaced by the expression—

$$\frac{100 - x}{x}$$

On the other hand, the dissociation of x per cent. of CO_2 increases the volume in the ratio of $100 + 0·5x$ to 100. If the total pressure remains one atmosphere, the partial pressure of the oxygen becomes—

$$\frac{0·5x}{100 + 0·5x}$$

therefore the complete expression for the equilibrium constant takes the following form :—

$$K_p = \frac{100 - x}{x} \times \frac{\sqrt{100 + 0·5x}}{0·707x^{\frac{1}{2}}} = \frac{100 - x}{0·707x^{\frac{3}{2}}}\sqrt{100 + 0·5x}$$

For temperatures below 1600° where x is less than unity this may be simplified to—

$$K_p = \frac{1000}{0·707x^{\frac{3}{2}}}$$

The values for x calculated by Nernst and v. Wartenberg are based upon the determinations of the specific heats by Holborn and

Austin. These heats, reckoned by Nernst and v. Wartenberg for one mol at constant volume between absolute zero and T, are for a permanent gas—

$$c_v = 4{\cdot}68 + 0{\cdot}000268\text{T}$$

and for CO_2—

$$c_v = 5{\cdot}106 + 0{\cdot}00334\text{T} - 7{\cdot}85 \times 10^{-7}\text{T}^2$$

Therefore the heat of formation of one mol of CO_2 at constant volume has the value—

$$\text{Q}_v = 67{,}700 + 1{\cdot}915\text{T} - 0{\cdot}00294\text{T}^2 + 0{\cdot}735 \times 10^{-6}\text{T}^3$$

The heat of formation of the same quantity of CO_2 at constant pressure therefore is—

$$\text{Q}_p = 67{,}700 + 2{\cdot}907\text{T} - 0{\cdot}00294\text{T}^2 + 0{\cdot}735 \times 10^{-6}\text{T}^3$$

Calculations based upon the specific heats as determined by Langen give approximately the same values, as can be seen from the following table :—

° absolute.	*x* found.	*x* calculated.
1300	0·00414	0·00407
1478	0·029–0·035	0·043

The free energy formula in the latter case is—

$$\text{A} = 67440 - 2{\cdot}42\text{T}\,ln\text{T} + 0{\cdot}0017\text{T}^2 - 4{\cdot}56\text{T}\log_{10}\frac{p_{CO_2}}{p_{CO} \times p_{O_2}^{\frac{1}{2}}} - 5{\cdot}95\text{T}$$

This is the same formula as given in the text on page 169, only with a different value for the thermodynamically undetermined constant.

Comparing the results of Nernst and v. Wartenberg with those of Deville, we notice that the temperature which corresponds to Deville's results is not 1300°, but 1400°. This is surprising, because at 1400° C. the equilibrium changes very rapidly in the outlet tube, and Deville did not use any special device to hinder this change ; thus his result is difficult to understand.

The figures of Nernst and v. Wartenberg have been confirmed by Langmuir,[1] who passed CO_2 along a platinum wire heated electrically. The temperature of the wire was determined from its resistance. The escaping gases were collected and analyzed by the **Langmuir's experiments.**

[1] *Jour. Amer. Chem. Soc.*, 28 (1906), 1357.

same method as that employed by Nernst and v. Wartenberg. Experiments were made between 1050° and 1300° C., the values obtained being in close agreement with those of Nernst and v. Wartenberg.

The glass cell.

As these determinations of the equilibrium are restricted to a rather small interval of temperature, it is fortunate that a direct method has been found for determining the free energy of formation of CO_2 and O_2 in a reversible galvanic cell. It had already been shown by Warburg[1] that glass is an electrolyte at high temperatures, even when in the solid state. He found that a current may be produced in a cell with sodium amalgam as anode, mercury as cathode, and glass as electrolyte, which transports the quantity of sodium through the glass corresponding to Faraday's law. The author, in common with Moser,[2] platinized the lower ends of ordinary test-tubes both inside and outside, and heated them to the boiling-point of sulphur or phosphorus pentasulphide, that is to say to 445° C. and 518° C. respectively. The outside surface was surrounded by air, whilst the inside was filled with mixtures of CO_2 and CO, or of O_2 and N_2. The platinum films on both sides of the glass were connected to a compensation apparatus by means of isolated platinum wires, so as to determine the values of the E.M.F. originated between them by the change of the nature or of the composition of the gases inside.

Theory of the glass cell.

The theory of the cell will become clear from the following consideration. According to Warburg, if a current is flowing through the glass, only sodium ions migrate, transporting the electricity from the positive to the negative pole. If there is no oxygen at the negative pole, these ions will be set free at the electrode; in the presence of oxygen, sodium oxide will be formed to an extent of one equivalent for every 96,540 coulombs. At the other electrode the same quantity of electricity sets free one equivalent of SiO_3 from SiO_3'' ions; if no oxidizable substance is present at this pole oxygen will be liberated, while in the presence of CO an equivalent amount of CO_2 will be formed, SiO_2 in both cases remaining in the glass. Hence it follows that the passage of 96,540 coulombs will use up one equivalent of O_2 at one pole, and produce one equivalent of CO_2 from CO at the other pole, the glass becoming at the same time more alkaline on the oxygen side and more acid on the carbon monoxide side.

[1] See Warburg and Tegetmeier, *Wied. Ann.*, 32 (1887), 447; 35 (1888), 455; Schultze, *Wied. Ann.*, 36 (1889), 661; Tegetmeier, *Wied. Ann.*, 41 (1890), 18; and Roberts-Austen, *Engineering*, 59 (1895), 442.

[2] *Zeitschr. f. Electrochemie*, 1905, 594.

This system is of a similar type to a cell composed of an electrode of chlorine and one of silver covered with silver chloride in a solution of HCl as electrolyte. It is well known that in this case the source of the energy may be attributed to the fact that the equilibrium pressure of Cl gas over Ag and AgCl is different from the pressure of the gaseous Cl_2 at the chlorine electrode. So this system may be looked upon as a concentration cell, by the operation of which the pressure of the chlorine falls from its value at the chlorine electrode to the value corresponding to the equilibrium $2Ag + Cl_2 \rightleftarrows 2Ag\,Cl$. In exactly the same way the element of Haber and Moser derives its potential from the difference in the pressure of the oxygen, present in a large amount at one electrode, but at the other only in the imperceptible traces which satisfy its equilibrium with CO and CO_2 according to the equation— Comparison with the chlorine cell.

$$p_{O_2} = \left(\frac{p_{CO_2}}{p_{CO}} \times \frac{1}{K_p}\right)^2$$

The fact that Haber and Moser did not measure the E.M.F. of the cell directly, but instead measured alternately the values for one and for the other electrode against the same air electrode, does not alter this consideration. There is, however, a distinction between the case of the chlorine element which we have instanced and that of the glass element. In the case of the chlorine element the dissociation pressure of chlorine over the two-phased system of Ag and AgCl is invariable at a given temperature, while in the case of the glass element it depends on the composition of the gaseous mixture according to the above equation. It may be further noted that when a current is flowing through the chlorine cell an increase of acidity takes place at one pole and a decrease at the other, just as in the case of the glass cell, diffusion in both cases tending to counteract this effect. If measurements are made by means of the compensation method, this change of concentration becomes zero in both cases, because no current is taken from the cell.

The proof that the glass element really derives its e.m.f. from the reversible reaction—

$$CO + \tfrac{1}{2}O_2 \rightleftarrows CO_2$$

is shown by the fact that on one hand different mixtures of O_2 and N_2, and on the other hand different mixtures of CO and CO_2, give differences in e.m.f. against the standard air electrode, as required by the formulæ— Results with the glass cell.

$$E = \frac{RT}{4F} \times ln\frac{p_{O_2}}{p'_{O_2}} \qquad E = \frac{RT}{2F} \times ln\frac{p_{CO} \times p'_{CO_2}}{p'_{CO} \times p_{CO_2}}$$

The following measurements may be quoted as an example of the results obtained :—

Composition of the gases in the inner tube.	E.m.f. of the inside against the outside air electrode.	Differences in volts.	
		Found.	Calculated.
$O_2 = 94{\cdot}6$ % $N_2 = 5{\cdot}4$ %	+0·088 volt		
		1·011	1·013
$CO_2 = 97{\cdot}9$ % $CO = 1{\cdot}6$ % $N_2 = 0{\cdot}5$ %	−0·923 "		
		1·011	1·013
$O_2 = 94{\cdot}6$ % $N_2 = 5{\cdot}4$ %	+0·090 "		
		1·011	1·013
$CO_2 = 97{\cdot}9$ % $CO = 1{\cdot}6$ % $N_2 = 0{\cdot}5$ %	−0·922 "		

The calculation was carried out according to the formula on p. 317, where—

$$A = 2 \times 96{,}540 \times 0{\cdot}239 \times \text{e.m.f.} = 46{,}200 \times \text{e.m.f.}$$

Another property of the glass may be mentioned in connection with this example, namely, that the outside and inside of the tube often show a small, rather constant discrepancy from the calculated e.m.f. Thus in the case quoted the 94·6 per cent. O_2 mixture gives a potential of 88 millivolts against the air electrode, whilst it should give at a temperature of 717° absolute, at which the experiment was performed, an e.m.f. of—

$$E = 0{\cdot}03535 \log_{10} \frac{96{\cdot}4}{20{\cdot}9} = 0{\cdot}0235 \text{ volt}$$

The discrepancy of 0·0565 volt is due to this property of the glass. In most cases these residual electromotive forces are very much smaller.

The fact that the equilibrium—

$$CO_2 + C \rightleftarrows 2CO$$

requires at temperatures near 500° only small percentages of CO hinders one from using in the element mixtures containing much CO without raising the temperature up to 600° and higher, while on the other hand the potential of the platinum does not easily reach at 600° and higher the theoretical value corresponding to these mixtures of CO and CO_2.

While these measurements of the e.m.f. give us the energy of the formation of CO_2 in the neighbourhood of 500°, we may, by the aid of Lowenstein's[1] determinations of the density of CO_2 and by Emich's[2] experiments on the formation of iridium dust in air and CO_2, gain some knowledge as to the extent of dissociation at very high temperatures.

Löwenstein used an apparatus similar to that employed by Victor Meyer in his density determinations. The bulb was of platinum, and was heated electrically up to some 1550°. The outlet tube was replaced by a horizontal capillary tube containing a short thread of mercury, by the shifting of which the expansion of the gas could be measured. The apparatus was filled with CO_2, and then a small piece of aluminium dropped in. According to the equation— **Lowenstein's density measurements.**

$$2Al + 3CO_2 = Al_2O_3 + 3CO$$

carbon dioxide was reduced to CO. Therefore dissociation existing in pure CO_2 was diminished to an imperceptible extent by the mass action of the CO. Overlooking the minute increase in volume due to the introduction and oxidation of the aluminium thrown into the hot bulb, there should have been no change in volume, supposing that the CO_2 is undissociated and that equivalent weights of CO_2 and CO have exactly the same volume under the same pressure at the temperature of the experiment. Taking into account the influence of a small amount of oxygen present in the CO_2, Löwenstein derives a dissociation of 0·4 per cent. from the observed changes of volume. According to Nernst and Wartenberg the dissociation should amount to 0·56 per cent. at the same temperature.

Emich's experiments were carried out at still higher temperatures. He made use of the results given by Holborn, Henning, and Austin,[3] regarding the formation of metallic dust from strips of different metals when heated electrically in different gases to various temperatures. Iridium was found to produce dust in especially large quantities, varying with the nature of the gas in which it was heated. No gas had so marked an influence on the phenomenon as oxygen, the quantity of dust produced in pure oxygen being eleven times as great as that in air at the same pressure and temperature. Emich points out that at 2150° pure N_2, as well as pure CO, does not produce any appreciable quantity of dust. On the other hand, **Emich's dust experiments.**

[1] *Zeitschr. f. physik. chemie*, 5 (1906), 707.

[2] *Monatshefte für chemie*, 26 (1905), 1011.

[3] *Wissenschaftliche Abhandhungen der physikalisch technischen Reichsanstalt*, Bd. IV. (1907), 87

he finds that the amount of dust produced for the same temperature, in an atmosphere of CO_2 is the same as that produced in an atmosphere consisting of 4·9 per cent. O_2 and 95·1 per cent N_2. From these results he concludes that for this temperature and under a pressure of one atmosphere carbon dioxide is dissociated to such an extent that 4·9 per cent. oxygen is present in the gas. At 1970° he finds that probably 2·2 per cent. oxygen is produced from the CO_2 at one atmosphere pressure.

At 1500° the dissociation as found by the same method is very small, its numerical value being somewhere about 0·1 per cent. Calculation of the percentage of dissociation at the other two temperatures gives for 1970° 4·5 per cent., and for 2150° 10 to 11 per cent.

Emich's determination of the duration of outflow.

The temperatures in these experiments were measured photometrically, according to Holborn and Kurlbaum's method described in the last lecture of this book.

It may be mentioned that these results which Emich obtained are not in agreement with those he found by another method.[1] He tried to find the decrease in apparent density produced by the dissociation of CO_2 at high temperature, according to the well-known method of Bunsen. The time which same volumes of different gases take to pass through a narrow opening is proportional to the square root of the densities of the gases. Emich determined the ratio of the duration of flow through an iridium orifice for CO_2 and N_2 at ordinary temperatures and at 2000°. He did not find any, or, at the most, no appreciable change in the ratio. Discussing former papers on this subject, he emphasizes the fact that a dissociation amounting to 6 per cent. would be entirely inconsistent with his results. It is undoubtedly difficult to make this result accord with his latter work on formation of dust, but it seems probable that the determination of the duration of outflow is less able to give certain values at so high a temperature, because of the rapid deterioration of the orifice.

Summary of the results on CO_2 dissociation.

Owing to the uncertainty which still exists in the specific heats of the gases, it is difficult to give an exact expression for the reaction energy of CO and O_2 forming CO_2. However, marked progress has been made, and the following figures, which Nernst and Wartenberg propose for the percentage of dissociation at different temperatures and pressures, may be probably regarded as a sufficient basis for further theoretical reasoning, though we cannot overlook the fact that the expression given by Holborn and Austin for the specific heat of CO_2 is used in this calculation for temperatures up

[1] *Monatshefte für Chemie*, 26 (1905), 505.

to 2500° abs. This expression, derived from experiments at much lower temperatures, contains a negative term with T^2, which makes the value for the specific heat of CO_2 undoubtedly too small at high temperatures.

T	P = 10 atm.	P = 1 atm.	P = 0·1 atm.	P = 0·01 atm.
1000	$7·31 \times 10^{-6}$	$1·58 \times 10^{-5}$	$3·40 \times 10^{-5}$	$7·31 \times 10^{-5}$
1500	$1·88 \times 10^{-2}$	$4·06 \times 10^{-2}$	$8·72 \times 10^{-2}$	0·188
2000	0·818	1·77	3·73	7·88
2500	7·08	15·8	30·7	53·0

Surely it is of interest to see how nearly these figures are in agreement with those given in the Fifth Lecture derived by help of Le Chatelier's assumptions, before any of the experiments which are recorded in this appendix were published. Comparison of new and old figures.

According to the table on p. 171, the percentage of dissociation at one atmosphere amounts at—

(a) 1300° C. to 0·1 per cent.
(b) 1500° C. to 0·4 „
(c) 2000° C. to 5·5 „

Nernst and Wartenberg calculate from their results the following values :—

(a) 1323° C. to 0·104 per cent.
(b) { 1423° C. to 0·242 „
 1523° C. to 0·507 „
(c) { 1923° C. to 4·88 „
 2023° C. to 7·55 „

The fact that Le Chatelier's own calculations gave the different results which are mentioned on p. 172, is explained by the accidental error that the partial pressures in his calculations referred to the formation of $2CO_2$ from $2CO$, whilst the heat of formation of CO_2 was used in the numerical treatment.

Reviewing our knowledge of the formation of CO_2 from CO and O_2, the uncertainty seems limited to the highest temperatures attainable by combination of CO and O_2. Indeed, Le Chatelier's assumption of 3000° C. for Deville's flame is a very uncertain one, though, as we have seen above, it has proved a useful starting-point to derive correct dissociation values for temperatures below 2000° C.

Dissociation of H_2O.

Still greater progress has been made in the case of the following reaction :—

$$H_2 + O \rightleftarrows H_2O$$

Beginning with determinations at comparatively low temperatures, oxygen-hydrogen cells may first be mentioned, which the author has studied in common with Fleischman [1] and Foster,[2] using hot glass and porcelain as electrolytes.

Glass and porcelain cells.

Most of the glass cells were made in the following way: Two ordinary glass tubes 10 mm. in diameter and closed at one end were fused together by help of the blowpipe by their closed ends to form a straight tube of double the length. The bottom of the tubes formed a glass partition in the middle. This was platinized or gilded on each side, and brushes of either platinum or gold wires were inserted from the open ends so that the points of the brushes touched the partition, the other ends of the brushes being connected to a compensation apparatus with which the e.m.f., set up between the two sides of the partition by the introduction of different gas mixtures from each end, could be measured. Porcelain cells were made by the help of a straight porcelain tube, the middle part of which was platinized or gilded inside and outside for a length of 10 cms. These films were connected to a compensation apparatus by wires of the same metal, different gas mixtures acting on the inside and outside setting up an e.m.f. Another form of the porcelain cell was made by the help of a porcelain tube closed at one end, and platinized or gilded inside and outside at the bottom. In all cases the heating was done electrically, and the temperature was measured by the platinum platinum-rhodium thermoelement. For details as to arrangement the reader is referred to the original papers. For temperatures between 330° and 580° glass cells were used, and between 800° and 1100° porcelain ones.

Influence of water-vapour discussed theoretically.

The theory of the cell is the same as that given for the CO-CO_2 cell, except in one particular. Water-vapour acts upon the glass and porcelain glaze, while the CO_2 was inert. The electrochemical changes at the electrodes would correspond in the simplest case to the equations—

$$\text{(A)} \quad \tfrac{1}{2}O_2 + SiO_2 + 2\theta \rightleftarrows SiO_3''$$
$$\text{(B)} \quad H_2 + SiO_3'' \rightleftarrows SiO_2 + H_2O + 2\theta$$

From these equations it follows that a change in the pressure of

[1] *Zeitschr. f. anorg. Chem.*, 51 (1906), 245.
[2] *Ibid.*, 51 (1906), 289.

the oxygen would influence the oxygen electrode according to the formula—

$$E_A \times 2F = RT ln \frac{P_{O_2}^{\frac{1}{2}}}{p'_{O_2}{}^{\frac{1}{2}}}$$

On the other hand, the ratio of H_2 to H_2O at the second electrode, changing from $\frac{p_{H_2}}{p_{H_2O}}$ to $\frac{p'_{H_2}}{p'_{H_2O}}$, would produce a change in the e.m.f. equal to—

$$E_B \times 2F = RT ln \frac{p_{H_2}}{p_{H_2O}} \times \frac{P'_{H_2O}}{p'_{H_2}}$$

However, experiments show that these simple assumptions are incorrect, and must be replaced by

$$\text{(A)} \quad \tfrac{1}{2}O_2 + SiO_2 xH_2O + 2\theta \rightleftarrows SiO_3'' + xH_2O$$
$$\text{(B)} \quad H_2 + SiO_3'' \rightleftarrows SiO_2 xH_2O + (1 - x)H_2O + 2\theta$$

The corresponding formulæ for these changes then assume the form—

$$E'_A \times 2F = RT ln \frac{p_{O_2}^{\frac{1}{2}}}{p'_{O_2}{}^{\frac{1}{2}}} - RT ln \frac{p_{H_2O}^{x}}{p'_{H_2O}{}^{x}}$$

$$E'_B \times 2F = RT ln \frac{p_{H_2} \times p'_{H_2O}}{p'_{H_2} \times p_{H_2O}} + RT ln \frac{p_{H_2O}^{x}}{p'_{H_2O}{}^{x}}$$

Looking at these expressions, we see that the second terms on the right-hand side of both equations vanish when the water-vapour does not change in partial pressure; that is, when p'_{H_2O} is the same as p_{H_2O}. Using on both sides of the cell gas mixtures which have the same partial pressure of water-vapour, no influence of these terms is to be expected theoretically, or is it found experimentally. The figures given later on will show that the agreement between theory and experiment with equal pressure of water-vapour on both electrodes is a close one. If the pressure of the water-vapour was different at both electrodes, qualitative agreement between theory and experiment was still found, experimental difficulties being in the way of quantitative proof of the theory, that is to say of an exact determination of the exponent x.

Before giving the detailed figures, the theoretical deductions of the e.m.f. of the cell from the energy of reaction may be discussed, in order to make clear to what extent uncertainty still exists about the theoretical values, and how far the electrical measurements agree with conclusions drawn from other sources.

The free energy of the formation of H_2O from the elements.

Using the same water-gas equation as on p. 175, and the equation for the formation of CO_2 given on p. 169, but with the value $-5{\cdot}95$ for the thermo-dynamically undetermined constant justified by the above-quoted experiments of Nernst and v. Wartenberg, the energy of the reaction—

$$H_2 + O \rightleftarrows H_2O$$

becomes—

$$A = 57{,}790 - 0{\cdot}87 T \ln T - 0{\cdot}00025 T^2 - 4{\cdot}56 T \log \frac{p_{H_2O}}{p_{H_2} \times p_{O_2}^{\frac{1}{2}}} - 5{\cdot}95 T$$

The values of the e.m.f. of the oxygen-hydrogen cell derived from this formula are given as E_1 in the following table. The values E_2 are founded on the determinations of the equilibrium in the formation of water-vapour from the elements as described later on, which, by the help of the specific heats of Langen, allow the formulation of the reaction energy as—

$$A = 57{,}066 - 2{\cdot}974 T \ln T + 0{\cdot}00125 T^2 + 7{\cdot}6 T - 4{\cdot}56 T \log_{10} \frac{p_{H_2O}}{p_{H_2} \times p_{O_2}^{\frac{1}{2}}}$$

The values E_3 are calculated from Nernt's new theory as developed in the Appendix to Lecture III., the expression for the reaction energy according to this theory being—

$$A''' = 57{,}300 - 1{\cdot}75 T \ln T - 0{\cdot}0003 T^2 + 0{\cdot}457 T - 4{\cdot}56 T \log_{10} \frac{p_{H_2O}}{p_{H_2} \times p_{O_2}^{\frac{1}{2}}}$$

In the numerical calculations of E_1, E_2, E_3, the term—

$$\frac{p_{H_2O}}{p_{H_2} \times p_{O_2}^{\frac{1}{2}}}$$

is taken as unity, and therefore the term $4{\cdot}56 T \log_{10} \frac{p_{H_2O}}{p_{H_2} \times p_{O_2}^{\frac{1}{2}}}$ equal to zero. In ordinary experimental cases this term will have a considerable influence, which may be easily found out by help of the figures in the last column, giving the equivalent of $\frac{4{\cdot}56T}{2F}$ in volts. Multiplying this equivalent by the Briggs' logarithm of the ratio $\frac{p_{H_2O}}{p_{H_2} \times p_{O_2}^{\frac{1}{2}}}$ calculated according to the given experimental data, we find the figure which, subtracted from the values of E_1, E_2, or E_3, produces the theoretical e.m.f. of the cell under the required conditions.

Temperature.		E.m.f. of the cell in volts.			$\frac{4{\cdot}56T}{2F}$ in volts.
° C.	° absolute.	E_1	E_2	E_3	
25	298	1·180	1·177	1·178	0·029
327	600	1·099	1·096	1·098	0·059
427	700	1·073	1·068	1·070	0·069
527	800	1·043	1·039	1·040	0·079
627	900	1·015	1·010	1·012	0·089
727	1000	0·986	0·981	0·981	0·099
827	1100	0·957	0·953	0·951	0·108
927	1200	0·928	0·924	0·920	0·118
1027	1300	0·898	0·895	0·889	0·128
1127	1400	0·869	0·865	0·857	0·128

It may be pointed out that, as the ratio $\frac{p_{H_2O}}{p_{H_2} \times p_{O_2}^{\frac{1}{2}}}$ is in most cases a fraction, its Briggerian logarithm is therefore negative, and hence, as this value must be subtracted, it means an increase in the values of E_1, E_2, or E_3. Thus, in the case of the ordinary oxygen-hydrogen cell at 25° C.—

$$p_{O_2} = 0{\cdot}969 \text{ atmosphere}$$
$$p_{H_2} = 0{\cdot}969 \quad ,,$$
$$p_{H_2O} = 0{\cdot}031 \quad ,,$$

therefore—

$$\frac{p_{H_2O}}{p_{H_2} \times p_{O_2}^{\frac{1}{2}}} = 0{\cdot}0325$$

further—

$$0{\cdot}029 \times \log 0{\cdot}0325 = -0{\cdot}0437$$

subtraction of this from E_1, E_2, or E_3 gives—

E_1	E_2	E_3	E found
1·224	1·221	1·222	1·14

The result shows the same difference of 80 milli volts as pointed out in the text of Lecture V., on the basis of Le Chatelier's old assumptions. This agreement proves that the determinations made since the publication of the German edition of this book, though they have greatly increased our knowledge on the subject, did not change the general aspect of the case derived from the few earlier figures. In connection with this it may be mentioned that Nernst and v. Wartenberg[1] deduced, in a paper published at the same time

E.m.f. of Grove's cell.

[1] *Göttinger Nachrichten* (1905), Heft 1.

as the German edition of this book, almost the same theoretical value of 1·230 volts for the oxygen-hydrogen cell at 25° C. The agreement becomes still better with a second calculation of Nernst and v. Wartenberg,[1] which gives 1·225 volts for this e.m.f.

Returning now to the practical work on the glass and porcelain cells, the following experimental figures may be quoted:—

(A) Hydrogen Concentration Cells.

Temp. ° C.	$\frac{p_{H_2}}{p'_{H_2}}$	E.m.f. (volt).		Electrodes.	Electrolyte.
		Found.	Calc.		
470	22·85	0·093	0·099	Pt	Glass
472	24·12	0·099	0·101	Au	,,
560	47·65	0·138	0·138	Pt	,,
572	24·12	0·116	0·115	Au	,,
860	11·04	0·111	0·116	Pt	Porcelain
860	11·77	0·123	0 119	Au	,,
1000	11·04	0·134	0·130	Pt	,,
1105	8·77	0·122	0·127	Pt	,,

(B) Oxygen Concentration Cells.

Temp. ° C.	$\frac{p_{O_2}}{p'_{O_2}}$	E.m.f. (volt).		Electrodes.	Electrolyte.
		Found.	Calc.		
460	51·9	0·059	0·062	Pt	Glass
475	51·9	0·058	0·063	Au	,,
560	49·5	0·069	0·069	Pt	,,
572	51·9	0·075	0·072	Au	,,
860	71·5	0·101	0·103	Pt	Porcelain
860	51·0	0·090	0·094	Au	,,
1000	38·7	0·100	0·099	Pt	,,

[1] *Zeitschr. f. physikal. Chemie,* 56 (1906), 545.

• (C) Oxygen-Hydrogen Cells.

(The figures given under e.m.f. calc. are the limits derived from the three different expressions for the reaction energy given before.)

Temp. ° C.	$\frac{p_{H_2O}}{p_{H_2} \times p_{O_2}^{\frac{1}{2}}}$	E.m.f. (volt).		Electrodes.	Electrolyte.
		Found.	Calc.		
460–470	0·0282	1·164	{1·168, 1·174}	Pt	Glass
473–480	0·0282	1·165	{1·167, 1·173}	Au	"
560	0·0274	1·143	{1·162, 1·157}	Pt	"
570–580	0·0321	1·151	{1·151, 1·156}	Au	"
860	0·0387	1·087	{1·098, 1·104}	Pt	Porcelain
860	0·0448	1·097	{1·105, 1·111}	Au	"
1000	0·0345	1·052	{1·080, 1·089}	Pt	"

The oxygen-hydrogen cell has been studied further by W. H. Patterson.[1] He continued the work done by Haber and Brunner (see p. 178), and found that iron inserted in molten alkali acted for some time as a hydrogen electrode, while platinum, like other metals, acted as an oxygen electrode. He found the following values for the e.m.f. of such cells at different temperatures, with sodium hydroxide as electrolyte :— Patterson's work on the oxygen-hydrogen cell.

Temp. ° C. ...	348°	382°	420°	458°	510°	575°
E.m.f. found ...	1·20	1·19	1·17	1·16	1·13	1·10
E.m.f. calc. ...	1·16	1·15	1·14	1·14	1·13	1·12

The theory of this cell and the basis of the calculation may be seen in the text of Lecture V. (p. 175). Following a later publica- of Brunner and the author,[2] Patterson uses for numerical calculation the equation given in this Appendix on p. 326.

Whilst this work of Patterson is perhaps more supported by the theory than *vice versâ*, Lewis[3] has proved conclusively, in an independent way, that the theoretical evaluation of the e.m.f. of the oxygen-hydrogen cell for ordinary temperature is correct, and that Lewis's work on the oxygen-hydrogen cell.

[1] *Phil. Mag.* (1907), January.

[2] Haber and Brunner, *Zeitschr. f. Electrochem.* (1906), 79.

[3] Lewis, *Zeitschr. f. physik. Chemie*, 55 (1906), 465.

Grove's cell does not possess the e.m.f. corresponding to reversible formation of oxygen from its elements. Lewis succe in proving that the equilibrium pressure of O_2, corresponding to reaction—

$$2Ag_2O \rightleftarrows 4Ag + O_2$$

is equal to $5{\cdot}0 \times 10^{-4}$ atmospheres at 25°. Now, accordin Böttcher,[1] the concentration of $Ag^{\cdot}$ and OH' ions in a satur solution of Ag_2O at 25° C. is $1{\cdot}4 \times 10^{-4}$ normal. From the of Lewis upon the dissociation pressure of Ag_2O, it immedia follows that the e.m.f. of a cell—

$$AgAg_2O \mid \text{Saturated solution of } Ag_2O \mid O_2 \text{ (1 atmosphere)}$$

must be equal to—

$$E = \frac{0{\cdot}059}{4} \log_{10} \frac{1}{5 \times 10^{-4}} = +0{\cdot}049 \text{ volt}$$

On the other hand, from Böttcher's determination it follows th combination, $Ag \mid AgH^{\cdot}$ ($^1/_1$ norm.)$-OH'$ ($^1/_1$ norm.) $\mid O_2$ (1 at would have an e.m.f. of $-0{\cdot}405$ volt if such a system could made. Now, Lewis, determining carefully the value of Ag ⁝. ($^1/_1$ norm.), finds it equal to $-0{\cdot}515$ volt, the so-called nor electrode ($Hg \mid HgClKCl$ $^1/_1$ norm.) being taken as zero.

Thus O_2 (1 atm.) $\mid OH'$ ($^1/_1$ norm.) $= -0{\cdot}110$ volt, the nor electrode being taken as zero as before. By the help of the w known value of Kohlrausch and Heydweiller for the dissociat of water at 25° ($1{\cdot}07 \times 10^{-7}$ equivalents of $H^{\cdot}$ and OH' being pure water at this temperature), we find the e.m.f. of O_2 (1 at $\mid OH'$ ($^1/_1$ norm.) $- H^{\cdot}$ ($^1/_1$ norm.) $\mid O_2$ (1 atm.) equal to $+ 0{\cdot}$ volt, and therefore—

$$O_2 \mid H^{\cdot}\ ^1/_1 \text{ norm.} = 0{\cdot}934 \text{ volt.}$$

Taking into account that the potential $H_2 \mid H^{\cdot}$ ($^1/_1$ norm.) known to be $+ 0{\cdot}283$ volt, the normal electrode being taken zero, we finally find $1{\cdot}217$ volts for the e.m.f. of the combinat O_2 (1 atm) $\mid H_2O - H_2O \mid H_2$ (1 atm.), which value, according Lewis, can hardly be wrong by more than $0{\cdot}01$ volt. It diff indeed, from the values for E_1, E_2, E_3, given on p. 327, by not m than $0{\cdot}004$ to $0{\cdot}007$ volt.

Nernst and v. Wartenberg's experiments

Cells of glass and porcelain tell us the reaction energy oxygen and hydrogen up to 1000°, while dissociation measureme

[1] Böttcher, *Zeitschr. f. physik. Chemie*, 46 (1903), 521.

made by Nernst and v. Wartenberg[1] are at our disposal for temperatures between 1124° and 1984°. Nernst and v. Wartenberg used the same apparatus for the determination of the decomposition of water-vapour as in the parallel investigations on the dissociation of CO_2. The water-vapour introduced into the heated pipette came from a boiling flask of 250 c.c. capacity, containing water with a little alkali. The boiling flask was fitted with two platinum wires dipping into the liquid. A current of electricity of known strength passing through the water developed a definite quantity of oxy-hydrogen gases, which was passed along with the steam through the pipette, and there either increased or decreased their percentage according to the temperature. Equilibrium was thus approached from both sides. The steam coming from the apparatus was condensed over mercury, and the volume of the hydrogen-oxygen mixture, as well as that of the water, measured. Some corrections were found necessary, since the gases collected did not show the ratio 2 : 1 for $2H_2 : O_2$, but contained an excess of H_2, probably due to some action of the steam on the walls of the pipette. The equilibrium was reached from both sides; at temperatures below 1207° no shifting took place in the outlet tube, whilst at 1288° such a shifting could not be avoided.

on water-vapour dissociation.

The amount of dissociation of water-vapour has been determined for higher temperatures first by Löwenstein[2] and later on by Wartenberg (*l.c.*) according to an ingenious scheme devised by Nernst. Löwenstein used a cylindrical platinum bulb, 8 cms. in length and 1·2 cms. in diameter, fitted with a capillary tube of the same metal 12 cms. long and of 0·5 mm. bore. This apparatus was connected through a drying tube with a manometer, arrangements being made such that after complete evacuation of the whole apparatus one limb of the manometer remained in connection with the bulb, the other with the pump, thus maintaining a vacuum over this limb. The bulb was enclosed in a tube heated electrically. Steam was passed through this tube, and the temperature measured with a thermo-couple in contact with the wall of the bulb. Hydrogen produced by the dissociation of the water-vapour diffused rapidly through the platinum walls and produced a difference in pressure between the two limbs of the manometer. The pressure of the hydrogen so measured is equal to its dissociation pressure in the steam. A slight complication arose from the fact that the hydrogen which diffused through the walls of the heating tube left an excess of oxygen in the steam, and thereby

Experiments of Löwenstein on water-vapour dissociation.

[1] *Göttinger Nachrichten*, 1905, Heft 1; *Zeitschr. f. physik. Chemie*, 56 (1906), 513 and 534.

[2] *Zeitschr. f. physik. Chemie*, 54 (1906), 715.

diminished the dissociation. Making allowances for this, the following figures were found :—

° C.	1432°	1510°	1590°	1695°
Dissociation % ...	0·102	0·182	0·354	0·518

v. Wartenberg applied the same method at higher temperatures, using an iridium bulb and an iridium furnace. He first tested the permeability of the iridium bulb at 2000° against O_2 and N_2, and found it almost zero. The experiments were carried out in the following manner: the iridium bulb was evacuated by help of the mercury pump; the manometer, arranged differently than in the experiments of Löwenstein, was read, and the temperature of the bottom of the bulb determined photometrically. Then steam was blown through the furnace for five minutes, the pressure in the manometer becoming steady within this period. Air instead of steam was now blown through, and the pressure to which the manometer returned determined. The following figures were derived from these experiments :—

t° C.	x % of dissociation.
1882	1·18
1984	1·77

Experiments of Langmuir and Holt.

It still remains to mention experiments of Langmuir carried out in the same manner with water-vapour as with carbon dioxide (see p. 317). The results of these experiments are in agreement with those of Nernst and v. Wartenberg.

Different results for high temperatures have been found by Holt,[1] who passed a current of steam, like Langmuir, along a platinum wire heated to temperatures from 950° up to 1760°. His figures do not differ much from those of Nernst and v. Wartenberg and from those of Langmuir for low temperatures, but fall much below them at higher temperatures.

Summary of data on water-vapour dissociation.

In conclusion of this part we quote the following table, calculated by Nernst and v. Wartenberg from the experimental work of Nernst and his co-workers.

T.	P = 10 atms.	P = 1 atm.	P = 0·1 atm.	P = 0·01 atm.
1000	$1{\cdot}39 \times 10^{-5}$	$3{\cdot}00 \times 10^{-5}$	$6{\cdot}46 \times 10^{-5}$	$1{\cdot}39 \times 10^{-4}$
1500	$1{\cdot}03 \times 10^{-2}$	$2{\cdot}21 \times 10^{-2}$	$4{\cdot}76 \times 10^{-2}$	0·103
2000	0·273	0·588	1·26	2·70
2500	1·98	3·98	8·16	16·6

[1] *Phil. Mag.*, 1907, 630.

Referring to pages 176 and 177 of this book, we find that there a dissociation of 0·01 per cent. to 0·02 per cent is calculated for 1473° abs., and of 2·5 per cent. for 2246° abs. The first of these figures, given in the German edition before any of the experimental work mentioned in this appendix had been done, agrees closely with the calculations lately given by Nernst and v. Wartenberg, namely—

° abs.		Per cent. of dissociation.
1400	... ...	0·0086
1500	... ...	0·0221

while the second one corresponds, according to Nernst and v. Wartenberg, to a temperature only 100° higher. Thus the great progress of our knowledge has confirmed rather than changed our ideas of the dissociation water at different temperatures. Looking back at the results reported here, we see, as in the case of CO_2, that only the dissociation at the highest temperatures attainable by combination of O_2 and H_2 still lacks investigation.

In the text of Lecture V., following the discussion of the energy of the formation of CO_2 and H_2O, the equilibrium in the Deacon process is discussed. Considering the small amount of experimental work which formerly underlay the theoretical treatment of both these fundamental reactions, the fact that the expression for the water-vapour equilibrium derived from that of carbon dioxide by help of the water-gas equilibrium could be combined with the expression for the HCl equilibrium to a formula which agreed so closely with the experimental results of Lunge and Marmier was valuable confirmation. This long, indirect method gave us 1·73 at 480°, and for the 2·42 at 430° constants of the Deacon process. The direct measurements of Lunge and Marmier gave us 2·00 and 2·51 respectively for these two temperatures. **The Deacon process.**

It may perhaps be mentioned that Bodländer, whose work on gas equilibria was quoted on p. 69, did not overlook the possibility of deriving the Deacon process equilibria from the other gas equilibria. He failed in his attempt to do this, because his assumptions for the other gas equilibria were not near enough to the truth. During the translation of this book the Deacon process has been studied by Gilbert N. Lewis,[1] and by Vogel v. Falkenstein.[2]

Lewis worked with HCl and O_2 as original substances, reaching the equilibrium from one side only. He used a glass cylinder of **Lewis' experiments.**

[1] *Amer. Chem. Journal*, 28 (1906), 1380.

[2] *Zeitschr. f. physik. Chemie*, 59 (1907), 313; and *Zeitschr. f. Electrochem.*, 1906, 41.

200 c.c. capacity, half filled with pieces of pumice-stone about the size of a pea. This pumice-stone had been saturated with copper chloride and dried in a current of hydrochloric acid gas. The cylinder was heated in a fused mixture of sodium and potassium nitrates. The HCl and O_2 mixture, kept over concentrated sulphuric acid, was passed into the cylinder, and after remaining there a suitable time, was withdrawn through a drying tube containing glass beads wet with H_2SO_4 into a pipette, from which samples of the gases could be taken out for analysis. The process was an intermittent one, samples of the mixture always remaining half an hour in the reaction cylinder before being withdrawn. The cylinder was repeatedly filled and exhausted before the real experiments were performed, so as to establish constant conditions in the cylinder and drying tube. The composition of the final gas mixture was analyzed by absorbing the HCl and Cl_2 with an aqueous solution of potassium iodide and measuring directly the volume of the oxygen. The solution was then titrated first with thiosulphate and afterwards with potassium hydrate, using phenolphthalein as indicator. The procedure was later on simplified, as it was found that the amount of O_2 could be calculated with sufficient accuracy from the composition of the original gas mixture; thus only the ratio of HCl and Cl_2 in the final mixture had to be determined. The following values for

$$K = \frac{p_{Cl_2}^{\frac{1}{2}} \times p_{H_2O}^{\frac{1}{2}}}{p_{HCl} \times p_{O_2}^{\frac{1}{4}}}$$

were found :—

°C.	352	352	386	386	419
K found ...	4·15	3·95	2·94	3·01	2·40
K calculated ...	4·02	4·02	3·02	3·02	2·35

Lewis obtained these calculated values on the assumption that the change in heat capacity at constant pressure during the reaction is zero; the heat of reaction being therefore independent of the temperature under the experimental conditions. Taking this heat of reaction as

$$HCl + \tfrac{1}{4}O_2 = \tfrac{1}{2}H_2O + \tfrac{1}{2}Cl_2 + 6900 \text{ calories}$$

Lewis deduces the expression [1]—

$$\log_{10} K = \frac{1509}{T} - 1{\cdot}811$$

[1] This formula is of the same type as the formula of Bodländer—

$$\log_{10} K = \frac{1484{\cdot}7}{T} - 2{\cdot}434$$

The difference in the results, however is a marked one, the value 2·35

The values for K calc. given above are derived from this formula. Lewis quotes experiments of Löwenstein in favour of his formula made according to the method as before described (see p. 331). They indicated that at 1537° HCl is 0·274 per cent. dissociated. Taking into account the values for the dissociation of water found by the same author, Lewis derives the constant K = 0·133 for the temperature 1537°, while his equation given above gives K = 0·106 for the same temperature. Surely Lewis is right in saying that this is a very surprising agreement considering that this value is obtained by exterpolating through more than 1000°. On the other hand, calculating from his expression for the equilibrium in the Deacon process the value of K for 25°, and combining this with Dolezaleek's results (see Lecture IV.), he finds the value 1·207 volts for the e.m.f. of the oxygen-hydrogen cell at 25° C. That is to say, only one centivolt less than the value obtained by his other method described before.

Vogel v. Falkenstein has determined the equilibrium in the Deacon process from both sides at temperatures of 450°, 600°, and 650°. His catalyst at the lowest temperature was cupric chloride, and platinum chloride at the higher ones, both catalysts being finely divided on asbestos. The oxygen-hydrochloric acid gas mixture was made by passing electrolytic oxygen through hydrochloric acid of known strength and at a known temperature, and drying the resulting gas mixture with concentrated sulphuric acid. The chlorine-water vapour mixture was prepared by passing electrolytic chlorine through water at a known temperature. In both cases the gases passed without interruption through the reaction chamber, which was heated electrically to the desired temperature. **Vogel v. Falkenstein's experiments.**

In these experiments the time of reaction was shorter than in those of Lewis's, and in consequence equilibrium was only reached at temperatures higher than 430°. The following table of values is computed for—

$$K = \frac{p_{Cl_2}^{\frac{1}{2}} \times p_{H_2O}^{\frac{1}{2}}}{p_{O_2}^{\frac{1}{4}} \times p_{HCl}}$$

from the results of v. Falkenstein, the experiments starting with O_2 and HCl being given under "direct process," and those from Cl_2 and H_2O being under "reverse process:"—

corresponding, according to Lewis's calculation, to 419°, while according to Bodländer's formula it should correspond to 256°. The source of the difference lies in the fact that Bodländer, who was convinced of the agreement of the e.m.f. of Grove's cell with the theoretical value for the oxy-hydrogen cell, formed his expression to coincide with this assumption.

Degrees C.	K.		
	Direct process.	Reverse process.	Mean value.
450°	2·31	2·22	2.26
[600	0·98	1·04	1·02]
650	0·804	0·789	0·794

The values for 600° are in brackets, as only a few experiments were made at this temperature.

The agreement of Vogel v. Falkenstein's results with Lewis's is by no means good, the constant 2·40 corresponding, according to the former, to 442°, while Lewis found it at 419°.

Results and calculations. For the theoretical discussion Vogel v. Falkenstein uses first the expression—

$$\log K = \frac{1447{\cdot}5}{T} - 0{\cdot}4375 \log_{10} T - 0{\cdot}0000325T - 0{\cdot}35$$

which corresponds to Nernst's theory as developed in the Appendix to Lecture III. As this equation does not express Lewis's results, he derives the following formula:—

$$\log K = \frac{1437{\cdot}5}{T} - 0{\cdot}534 \log T - 2{\cdot}1425 \times 10^{-4}T + 1{\cdot}7075 \times 10^{-8}T + 0{\cdot}074$$

on the basis of the specific heats of Holborn and Henning. The calculations made according to the above formula are compared with the experimental facts in the following table. In the last column are added values derived from the formula given by the author in the text of the Fifth Lecture. It is interesting to note how nearly the old formula agrees with the new experiments.

Degrees C.	K found.	Experimenter.	K calc. (Falkenstein).	K calc. according to p. 185.
352	4·02	Lewis	4·70	4·57
386	3·02	„	3·53	3·40
419	2·35	„	2·76	2·62
430	2·5	Lunge and Marmier	2·53	2·42
450	2·26	v. Falkenstein	2·22	2·10
480	2·0	Lunge and Marmier	1·35	1·73
600	1·02	v. Falkenstein	0·99	0·90
650	0·794	„	0·806	0·728

Furthermore, we can derive from our old formula the equilibrium constant for 1537°, and we find—

$$K_{1537} = 0{\cdot}123$$

while, as Lewis pointed out, the experiments of Löwenstein show that the value at this temperature is—

$$K_{1537} = 0{\cdot}133$$

Thus the agreement for both high and low temperatures is astonishing, strongly supporting our statement that the great increase in experimental data rather confirms than changes our views on the location of these gaseous equilibria.

Experiments of Levi and Battoni.

It may be mentioned, in addition to this discussion of the Deacon process, that Levi and Battoni[1] investigated the alleged chemical changes of the cupric chloride acting as catalyst in the reaction, but were unable to find either cuprous chloride or cupric oxychloride. They state that the affinity for water is the reason for the catalyst's activity.

[1] *Gaz. Chim. Ital.*, 35, i. 320.

APPENDICES TO LECTURE VII

I. THE theory of heterogeneous reactions has been further developed. Bodenstein and Stock[1] have shown that, in the decomposition of stibine at ordinary temperatures, which is greatly accelerated by finely divided antimony, the velocity of decomposition is determined by the velocity in the layer of absorbed gas covering the particles of antimony. The mass M of this absorbed gas is connected with the concentration C of the gas in the free gas space by the relation—

$$M = aC^p$$

What we know regarding the effect of changing temperature on p may be summarized by saying that at very low temperatures p is small, at high temperatures nearly equal to unity, and at ordinary temperatures about equal to 0·5. The factor a is different for different substances.

The mass decomposing in unit time is proportional to the absorbed mass; that is—

$$-\frac{dx}{dt} = kM = kaC^p$$

Referring the amount of decomposition to the total quantity present, which is equal to the concentration C when its volume is one in arbitrary units, we obtain for the reaction velocity V the expression—

$$V = k\frac{aC^p}{C} = \frac{k'}{C^{1-p}}$$

Putting $p = 0·6$, we have—

$$V = \frac{k'}{C^{-0·4}}$$

This expression agrees excellently with the observations at 25°. It appears that the catalysis of oxyhydrogen mixtures by porcelain belongs to the same class as the catalysis of stibine, whilst the catalysis of oxyhydrogen mixtures and of mixtures of SO_2 and O_2 by platinum appears to be governed by diffusion velocities. The

[1] *Berl. Ber.*, 40 (1907), p. 570.

difference in the catalytic effect of porcelain and platinum discussed in the Seventh Lecture, and its reference to differences in the velocity of the chemical change, are thus fully confirmed.

II. When the location of an equilibrium is known, measurements of the rate of formation or of decomposition may serve to determine the order of the reaction involved. H. v. Wartenberg [1] succeeded in demonstrating in this way that the formations of water from the elements at 1100° and of carbon dioxide from carbon monoxide and O_2 at 1200° to 1300° are dimolecular reactions. But it should here be noted that kinetic equations in their simple form only apply to reactions at *constant volume*. If the number of molecules changes during the reaction at constant pressure, as it does in the formation of water or of carbon dioxide in a stream of gas, then, as Wegscheider [2] has shown, we must calculate somewhat differently.

In a given quantity of an oxyhydrogen mixture the amount changed $d\mu$ in the time dt is given by the equation—

$$d\mu = \frac{dx}{dt}\mathrm{V}dt$$

where $\frac{dx}{dt}$ signifies the amount changed per unit of volume, and V the volume of the oxyhydrogen mixture. We also may write—

$$\frac{d\mu}{dt} = \mathrm{V}\frac{dx}{dt}$$

If now $(\mathrm{M}_1 - 2\mu)$ is the amount of unchanged hydrogen still present at the time t, and $(\mathrm{M}_2 - \mu)$ the corresponding amount of oxygen, then—

$$\frac{dx}{dt} = k\left(\frac{\mathrm{M}_1 - 2\mu}{\mathrm{V}_t}\right)^2\left(\frac{\mathrm{M}_2 - \mu}{\mathrm{V}_t}\right)$$

$$\text{and } \frac{d\mu}{dt} = \frac{k(\mathrm{M}_1 - 2\mu)^2(\mathrm{M}_2 - \mu)}{\mathrm{V}_t^2}$$

But the volume V_t is known from the known initial volume and the change of volume during the reaction. The former is, where M_1 and M_2 are reckoned in gram-molecules—

$$\frac{\mathrm{RT}}{p}(\mathrm{M}_1 + \mathrm{M}_2)\ .$$

[1] *Zeitschr. f. Physik. Chem.*, 56 (1906), 513.

[2] *Ibid.*, 35 (1900), 578.

The latter equals $-\frac{RT}{p}\mu$, since when a gram-molecule of oxygen ($n = 1$) is used up, the volume of the gas mixture decreases by one molecular volume. We obtain then, for constant pressure, including $\frac{R^2T^2}{p^2}$ in the constant—

$$\frac{d\mu}{dt} = \frac{k'(M_1 - 2\mu)^2(M_2 - \mu)}{(M_1 + M_2 - \mu)^2}$$

If M_1 and M_2 are equivalent, $M_1 = 2M_2$, and if we express the amount changed in fractions of the initial amount present $\left(x = \frac{\mu}{M_2}\right)$, we obtain—

$$\left(\frac{dx}{dt}\right)_p = \frac{k''(1 - x)^3}{(3 - x)^2}$$

while the corresponding kinetic equation for constant volume was—

$$\left(\frac{dx}{dt}\right)_v = k'''(1 - x)^3$$

It is important, however, to keep in mind that in the derivation of both formulæ it has been assumed that the reaction takes place in a homogeneous system. Consequently, as soon as a gas reaction takes place on a catalytic surface neither formula is entirely rigorous.

III. Recent investigations have shown that among those gases which are stable at high temperatures, but which exhibit a tendency to decompose as they cool off, nitric oxide is unique in the slowness with which it decomposes. Thus Clement[1] found that at 1000° a gaseous mixtures containing 1 per cent. of ozone decomposes spontaneously in 7×10^{-4} seconds to a content of a thousandth part of a per cent. It is, consequently, difficult to prepare ozone thermically. Nevertheless Fischer,[2] in conjunction with Braehmer and Marx, has succeeded in obtaining ozone by means of heat in a number of different ways. They burned hydrogen and other combustible gases in liquid air or liquid oxygen, and found ozone in the residual air or oxygen. They obtained the same result by heating a platinum wire or a Nernst filament to a white heat in liquid oxygen. Indeed, in this last case the ozone content corresponded very nearly to the amount required by theory. That is, one can calculate from the e.m.f. of an ozone-oxygen cell approximately what the ozone

[1] *Annalen der Physik.*, (IV.) 14 (1904), 334. Compare also Jahn, *Zeit. f. anorg. Chem.*, 48 (1906), 260.

[2] *Berl. Ber.*, 39 (1906), 940 and 2557.

concentration would be when in equilibrium at different temperatures with oxygen at atmospheric pressure.[1] It is found that at 1296° a content of 0·1 per cent., and at 2048° a content of 1 per cent. (by volume), should be stable. This agrees approximately with the results of Fischer and Braehmer. It is possible to prove that ozone is formed in the combustion of hydrogen and around a Nernst filament glowing in oxygen, even without the use of liquefied gases, by simply forcing air or oxygen at a high-enough speed over the hydrogen flame or filament. If the cooling gas is not forced by rapidly enough, nitric oxide is formed. The ozone has time to decompose, but the nitric oxide does not.

Hydrogen peroxide was also formed in the combustion experiments when the velocity of the cooling gas was very great, but, strange to say, none was detected in the experiments with liquid air or oxygen. Its rate of decomposition is also very great, approaching that of ozone.[2] Its formation by the sudden chilling of a hydrogen flame with ice has been well known since the experiments of Traube. Its concentration at equilibrium can be calculated from the difference between the potentials of oxygen and hydrogen peroxide. Nernst[3] made the calculation on the assumption that the difference of potential amounted to 0·374 volt. Meanwhile it has been found that the potential then assumed for oxygen was some 0·12 or 0·13 volt too low. The author now finds, from his measurements of the hydrogen peroxide and the oxygen potential, that the difference is 0·42 or 0·43 volt. Calculating from this value, the stable content of hydrogen peroxide at these high temperatures is found to be somewhat less than Nernst's values.

Nitrous oxide is the only other endothermic compound of this class where rate of decomposition has been measured. Hunter[4] finds that its rate of decomposition is considerably less than that of ozone or hydrogen peroxide, but still at about 800° is a thousand times greater than that of nitric oxide. This shows clearly why, in preparing nitric oxide at high temperatures, the formation of nitrous oxide has never been observed.

IV. Warburg and Leithäuser[5] have shown, in contradiction to the older observations of Berthelot, that when a direct current

[1] Nernst, *Zeit. für Electrochem*, 9 (1903), 891.

[2] *Ibid.*, 11 (1905), 713. Compare Finckh, *Zeit. f. anorg. Chem.*, 45 (1905), 116.

[3] *Zeit. phys. chem.*, 46 (1903), 720.

[4] *Ibid.*, 53 (1905), 441.

[5] *Annalen der Physik.*, [4] 20 (1906), 743.

is sent either through dry or moist air at atmospheric pressure the nitrogen is partially oxidized, even at temperatures between 19° and 200°. It therefore becomes doubtful whether the formation of NO from N_2 and O_2 in the electric arc is purely a thermic process. Warburg and Leithäuser used potentials of 5500 to 10,000 volts, and current strengths of 0·14 to 0·4 milliampere. One electrode consisted of a pair of platinum points, the other of a piece of sheet platinum connected with the earth. The quantity of oxidation products formed was, to be sure, very small, about a mol of NO being formed in the most favourable cases per ampere hour, or, on the basis of a potential of 10,000 volts, about 0·1 mol per kilowatt hour. Usually the yield was not so good (about 0·05 mol per kilowatt hour, or 3 grams of HNO_3 per kilowatt year). The reason of this formation of nitrous vapours by still electrical discharges is not yet clear. One might think that photochemical action was perhaps responsible, because we know that this is usually the cause of the formation of ozone. But if it is, then only wave-lengths less than $0{\cdot}2\mu$ can be operative, for Kreussler[1] has shown that only such are absorbed to any extent by air. On the other hand, Warburg's experiments furnish ground for the assumption that ozone is produced in still electrical discharges by electronic impacts. Warburg came to the conclusion that the negative glow and the positive column in point discharges indicate a high velocity of the gaseous ions, and that the collisions of these rapidly moving ions with the gas molecules cause chemical reactions to take place (ozonization), which would not occur were the motions less rapid. A similar action of rapidly moving ions might contribute to the formation of the oxides of nitrogen, both in the still electrical discharge and in the high-tension arc. Nevertheless, since there are no facts known which disagree with the thermic explanation of the action of the arc, we had best adhere to that explanation for the present.

V. New data are now available bearing on the velocity of formation and decomposition of the oxides of nitrogen. These data, which are derived from experiments performed in tubes heated from the outside, do not, however, provide us with any simple picture of what takes place when these reactions are brought about by the electric arc.

A series of experiments by Finckh[2] first deserves notice. He repeated Bunsen's experiments upon the oxidation of nitrogen by the explosion of an oxyhydrogen mixture in the presence of an

[1] *Wied. Annalen*, 6 (1901), 412.

[2] *Zeit. f. anorg. Chem.*, 45 (1905), 116.

excess of air. He covered the mercury used to seal off the exploding gas mixture with a thin layer of 10 per cent. potassium hydroxide in order to prevent the nitrous or nitric acid formed from acting on the mercury. Equilibrium was not reached because of the relatively low temperatures attained, these being, in the two series of experiments according to calculations of Nernst,[1] 2307° and 2402°. Nernst assume, as a first approximation, that the formation of NO takes place isothermally at the maximum temperature, and puts the time of reaction proportional to the square root of the initial pressure of the gas mixture. In this way he calculates values for the yield which are very close to those actually obtained by Finckh. Finckh found the concentration of NO at equilibrium at 2307° to be 2·05 per cent.; at 2402°, 2·23 per cent. These numbers are very similar to those found by other investigators in entirely different ways.

Certain experiments by Jellinek[2] afford a surer basis for calculation. He studied the rate of formation and decomposition of nitric oxide in porcelain, platinum, and iridium vessels at temperatures varying from 689° to 1750°. He found that the velocity constants throughout the whole temperature interval studied were doubled for every 50° rise in temperature. To be sure, the walls of his vessels exerted a somewhat disturbing catalytic effect.

The equilibrium was reached even at 2800° in the thousandth part of a second. We can predict that at a temperature of 3300° the same result would be attained in 10^{-9} second. If, then, the arc light consists of a heated thread of gas only 0·01 mm. in thickness, equilibrium would be very nearly attained even if the air were forced through at a rate of 1000 m. per second, provided no other factor enters in. Similar considerations hold for the rate of decomposition, so that it is surprising that the air can be withdrawn from the arc without much trouble, and its NO content still kept up to 7 per cent. On the other hand, it is known that if a stream of air is blown through an alternating current arc with the greatest velocity, only a very small amount of NO is formed.

VI. Grau and Russ[3] have undertaken a thorough investigation of the formation of nitric oxide in stationary high-tension arcs. They used cooled capillaries in sucking their samples of gases out of the arcs. With arcs 8 cm. long they succeeded in obtaining a gas mixture containing as much as 5 per cent. NO, corresponding, according to the equilibrium measurements of Nernst, to a temperature of

[1] *Zeit. f. anorg. Chem.*, 45 (1905), 126. [2] *Ibid.*, 49 (1906), 229.

[3] *Sitzungsber. d. kais. Acad. Wien*, 115 IIa. (1906), 157.

3100°. They concluded that this content represented the true equilibrium concentration, using air and an arc of the given length. Using arcs 5 cm. long, they improved the yield to as much as 5·6 per cent. NO. The yield in their experiments corresponds to between 500 and 600 kg. HNO_3 per kilowatt year, and they promise still better yields using long arcs. This proposal is, indeed, contained in the French patent of the Badische Anilin und Sodafabrik. We cannot deal here with the already extensive literature, or the rapidly increasing number of patents dealing with the technical preparation of nitric oxide. They can be found in the recent volumes of the *Zeitschrift für Elektrochemie* either as original articles or as abstracts.

VII. Baur has attempted, in his book on "Spectroscopy and Colorimetry,"[1] to explain the radiation phenomena of flames without the assumption that luminescence takes place in them. He explains, for instance, the intense green light which the inner cone of the bunsen burner sends out when the primary air-supply is large, by the hypothesis that under these conditions methane is present in the brightly radiating layer. Its absorptive power, according to all the measurements we possess, is large, so that we can assume a high coefficient of emission. Since, now, there is no methane in the layer of gas in immediate contact with the burning zone, a difference in the brilliancy could be brought about in this way.

Dr. Lacy, however, has tried in vain, in the author's laboratory, to produce a lighting effect by blowing methane into a flame of benzene at a point above the inner combustion zone, even when the methane had been previously heated by passing it through hollow Nernst rods heated for several centimetres to a very high temperature.

Further, it should be observed that the difference between the absorptive power of carbon dioxide and methane (1 : 4·5) is not sufficient to explain the difference between the radiation from the inner cone and the adjacent layers of the bunsen flame. The unburnt gas-air mixture does indeed contain 7 per cent. methane which is not present in the burnt mixture, but, on the other hand, the latter contains a considerably greater amount of water-vapour and carbon dioxide.

VII. Mayer and Altmayer[2] have investigated the stability of

[1] Vol. v. of Bredig's "Handbuch der angewandten physikalischen Chemie." Leipzig, 1907.

[2] *Berl. Ber.*, 40 (1907), 2134.

methane, and find that at one atmosphere pressure the following percentages are stable in the presence of hydrogen :—

Temp. C. ...	250°	450°	550°	750°	850°
Per cent. ...	98·79	76·80	46·69	6·08	1·59

The thermodynamic expression for the equilibrium—

$$C + 2H_2 \rightleftarrows CH_4$$

is—

$$21{\cdot}1 + \frac{18{,}507}{T} - 5{\cdot}9934 \ln T - 0{\cdot}002936T = R \ln \frac{p_{CH_4}}{p^2_{H_2}}$$

It is evident from this expression that with decreasing hydrogen pressure the equilibrium pressure of methane rapidly falls. Thus at 850°, where at atmospheric pressure 1·59 per cent. methane is an equilibrium with hydrogen whose pressure is then 0·9841 atmospheres, a decrease of the pressure of the hydrogen to 0·1 atmosphere lowers the methane content to 0·016 per cent. This is the very limit at which methane can be analytically detected. By combining the methane equilibrium with the carbon dioxide equilibrium—

$$C + CO_2 \rightleftarrows 2CO$$

we readily obtain the equilibrium—

$$CO_2 + CH_4 \rightleftarrows 2CO + 2H_2$$

between carbon dioxide, methane, carbon monoxide, and hydrogen. This can be done by simply subtracting the expression for the carbon dioxide equilibrium from the expression we have just given for the methane equilibrium.

The carbon dioxide equilibrium has recently been more accurately investigated by Mayer and Jacoby (private communication). The measurements show that above 750° a complication sets in, but, using the values obtained below this temperature, we derive the expression—

$$15{\cdot}8 - \frac{37{,}936}{T} + 3{\cdot}54 \ln T - 0{\cdot}003136T = R \ln \frac{p^2_{CO}}{p_{CO_2}}$$

Subtracting as indicated above, we obtain—

$$-5{\cdot}3 - \frac{19{,}429}{T} + 9{\cdot}5334 \ln T - 0{\cdot}0002T = R \ln \frac{p^2_{CO} \times p^2_{H_2}}{p_{CO_2} \times p_{CH_4}}$$

We see immediately, from this expression, that in hot water-gas containing a few percents of CO_2, CO, and H_2 no perceptible amount of CH_4 can exist *in equilibrium* even at a red heat. Any methane found in the flame must, therefore, represent simply traces of the original substance which have escaped oxidation and decomposition in the inner cone.

We can consider the reactions—

$$2C + H_2 \rightleftarrows C_2H_2$$

$$\text{and} \quad 2C + 4H_2 \rightleftarrows 2CH_4$$

in an entirely analogous way. Their difference is—

$$2CH_4 \rightleftarrows C_2H_2 + 3H_2$$

The acetylene equilibrium as represented by the first of these reactions has recently been studied by v. Wartenberg.[1] It appears from his results that the expression—

$$Rln\frac{p_{C_2H_2}}{p_{H_2}} = -\frac{50,270}{T} + 0{\cdot}00457T + 4{\cdot}57$$

is not far wrong. We then obtain—

$$Rln\frac{p_{C_2H_2} \times p^3_{H_2}}{p^2_{CH_4}} = -\frac{80,284}{T} + 12lnT4{\cdot}57 + 0{\cdot}01T + 37{\cdot}63$$

From this expression we see that any analytically detectable trace of methane would have a very strong tendency to dissociate into acetylene and hydrogen. The fact that traces of methane do nevertheless persist is evidence of the extraordinary slowness with which methane reacts at high temperatures. The well-known difficulty of igniting a mixture of methane and air in the absence of hydrogen is closely connected with this same phenomenon.

[1] *Zeit. f. anorg. Chemie*, 52 (1907), 299. Data are here given regarding the equilibrium—

$$2CN + H_2 \rightleftarrows 2HCN$$

and related reactions.

INDEX

SUBJECTS

INDEX

NAMES

THE END